Titanic Ships, Titanic Disasters

Titanic Ships, Titanic Disasters

An Analysis of Early Cunard and White Star Superliners

William H. Garzke, Jr.
John B. Woodward

The Society of Naval Architects and Marine Engineers
Jersey City, New Jersey

ISBN 0-939773-35-X

Contents

Preface

This book concerns five famous transatlantic liners. Two were near-sisters, *Lusitania* and *Mauretania* of the Cunard Line, ships that were renowned for their speed and pioneering use of turbine propulsion by first-tier liners of the North Atlantic. The others were the near-sister ships of the White Star Line, *Olympic*, *Titanic*, and *Britannic*, renowned for their size and their luxurious accommodations. Three of these five ships were made legends by their spectacular demises. The *Titanic* was the most famous of all because of her unique encounter with an iceberg in April 1912. The *Lusitania* became infamous from the controversies accompanying her sinking by a single torpedo delivered by a German submarine in 1915. HMHS *Britannic* went to the bottom from a fatal brush against a German-laid mine in 1916.

In the early years of the 21st century, the wreckages of these three ships lost to unplanned encounters still lay upon the seabed, accessible to some degree for examination and forensic analysis. There have been uncertainties that have intrigued ship people for most of the 20th century, but these are now potentially resolvable. Did the *Titanic* break in two as she sank? Did her boilers explode? What was the source of the second explosion that wracked the *Lusitania*? Was that ship carrying munitions, a cargo that might have justified its sinking and thereby providing an obvious solution to the issue of that second explosion? Several of these questions have been answered, although several have not, but exploration has narrowed the list of possible solutions. In part, the answers and potential answers constitute the incentive for this book. One of the authors, William H. Garzke, Jr., is the chairman of the Marine Forensics Panel of the Society of Naval Architects and Marine Engineers, and has had a unique vantage for discussing what examination of these wrecks have revealed. In 1996 and again in 1998 he was part of expeditions to the wreck site of the RMS *Titanic* to conduct a marine forensic analysis on her demise. The other author, John B. Woodward, Professor Emeritus (naval architecture and marine engineering) at the University of Michigan, provides a theoretical background, particularly in marine machinery and in the history of ship design and operation.

As any reader who is not a recent immigrant via *Starship Enterprise* must know, the *Titanic* is honored by an extensive literature, essentially all of it focused on the tragic events of 14–15 April 1912. Thus, in spite of our attempt to present a technical treatment that will be equal among the three ships that sank, *Titanic* usurps a major fraction of this book. Indeed, such unbalance is unavoidable, but desirable. Her story helps to explain the loss of her sister, *Britannic*, some four years later. Marine forensics analysis was born in attempts to analyze the *Titanic* experience, and hence she is the one in which most of the interest still lies. She is also where most of the errors in earlier literature on her can be found. Hence, our concentration on that most famous of ships.

Are there errors in the earlier literature? Yes, some errors that are almost inevitable because the early authors did not have the benefit of Robert Ballard's discovery of the *Titanic* wreck in 1985, and of the evolving technology of deep ocean exploration during the latter part of the 20th century. There are some errors because those authors overstepped their knowledge of ship design and technology or overlooked technical aspects of some survivor testimonies. Beyond those rationales, most of these errors are those of the omission variety, prevalent in the works of authors who do not have a technical background to delve into that technology, inadequate even for recognizing the questions that might be explored. For example, before the advent of marine forensics and the deep-diving submersibles, no other book had explored the questions, inter alia, of how the *Titanic*'s lights were kept burning almost to the end, nor what part her expansion joints played in the sinking.

Messrs. Garzke and Woodward both earned degrees in naval architecture and marine engineering at the University of Michigan. Each has put his degree to work for at least 40 years, Mr. Garzke at the noted marine design firms, Gibbs & Cox, Inc. and George G. Sharp, Inc. Mr. Garzke also experienced a maiden voyage, being aboard the new passenger line *France* (now the *Norway*) when she made her maiden voyage in February 1962. Dr. Woodward made his mark at Newport News Shipbuilding and on the University of Michigan faculty. They are not only equipped with the requisite intellectual tools to take the *Titanic* beyond the reach of earlier authors, but are thoroughly "ship people." No, neither ever commanded a ship at sea through the cold and dark in the presence of deadly hazards (neither is old enough to have done North Atlantic convoy duty in the '40s). However, they are nonetheless fit to comment on such things as a lookout's doubtful usefulness when staring with unprotected eyes into a 22-knot wind of freezing temperature. They know about a ship's floodable length, knowledgeable on the properties of riveted joints, on the role of a bilge and ballast system as a ship is sinking, and in other ways on topics beyond the reach of the typical ship-enthusiast author.

Thomas Andrews, an engineer intimately familiar with the design and construction of the *Titanic*, was aboard the ship and gave counsel to Captain Edward Smith after the fateful collision. Being naval architects and marine engineers, the authors can describe the actions of Thomas Andrews from a professional viewpoint: how he estimated rate of flooding, the rate of sinkage, and how he was a vital factor in a fateful conference in the engine room with Captain Smith and Chief Engineer Bell.

In spite of the authors' implied criticism of other authors who did not have a technical background, there are those whose works we admire. Walter Lord (*A Night to Remember*), Charles Eaton and John Haas (*Titanic: Triumph and Tragedy*), and Geoffrey Marcus (*The Maiden Voyage*) are prominent among them. We base our description of the personal nature of the *Titanic*'s brief voyage on works such as theirs. In fact, the narrative description that we present is superficially much like theirs. On the other hand, it is given with a definite and simple purpose in mind: to provide a matrix in which to imbed the technical factors that bear heavily upon the *Titanic* event, but which only the naval architect/engineer can evaluate. There are many such factors, and we believe that they enrich the oft-told tale, even for the nonprofessional.

The same statements can be offered with respect to the *Lusitania* and *Britannic* stories. Their popular literature does not overwhelm the public to the extent that *Titanic* literature has. We are indebted to Simon Mills, who has written much on the *Britannic*. Nonetheless, this literature also lacks the imprimatur of the professional ship person. As with the story of the *Titanic*, we repeat the stories as vehicles for presentation of the technical factors that can only be evaluated by naval architects and marine engineers.

In the case of the *Lusitania*, the lingering controversy of British-guilt versus German-guilt has not been considered, since its intricacies lie in fields of expertise far beyond those of the naval architect/marine engineer.

PART I

The Beginnings

The year is 1900. Wooden sailing ships still survive, and a few are still being built. More iron or steel sailing ships are being built. However, our interest is in the ships that travel the crowded ocean tracks between Europe and North America, built for the carriage of passengers. These ships were, fifty or sixty years before, themselves wooden, and although within our scope of interest they were all steamships, struggling across the ocean with primitive engines turning paddlewheels. Now they have passed through a phase in which their hulls were of wrought iron to become vessels of steel. Their propulsors are screw propellers. Although the engines are still reciprocating steam engines, they (and the boilers that supply them) are at a much higher level of sophistication than at that point a half-century earlier. But the ships are not yet the great ocean liners of the 20th century. Those ships are soon to come, however. By the end of an elongated decade that will end in 1914, a fleet of liners—large, fast, luxurious, and flying several European flags—will form the first generation of the "North Atlantic Ferry." This institution that began to carry emigrants to North America will survive two World Wars, and thrive until new societies, new tastes, and new means of crossing oceans, end its brief life.

A particular subset of those great liners-to-come is our topic—the Cunarders *Lusitania* and *Mauretania,* and the White Star ships *Olympic*, *Titanic*, and *Britannic*. Interest is focused on the sinkings of three of them. That interest leads through the histories of the ships, commentary on their design features (especially those called into play by their sinkings), and the marine forensic analysis that became feasible in the late 20th century with the emergence of the manned and unmanned submersibles capable of diving to the deep depths of the oceans.

Meanwhile, we begin the story by looking into the background that is evident in the years leading up to 1900.

In 1900 ship design as a field of engineering endeavor was quite new. The designers and builders of ships in the British Isles, home of the pioneers in ship design and construction, had come together into their first technical society (the Institution of Naval Architects) only in 1860. Many of these pioneers were still active in 1900. What they knew, and what they were able to do, form an important part of the background for our writings, and are a topic for Part I.

Although ship design had reached a state that might be called "maturity" by 1900, many advances were awaiting the field in the new century. For instance, electric arc welding would displace riveting as the near-universal means of joining steel. But that revolution was not to begin for another 30 years, and hence is not part of the story here. On the other hand, in the realm of propulsion machinery, some of the revolutionary advances of the 20th century were already on hand, challenging the ship designers to apply them in their next projects. One of them—the steam turbine—was to be used by both sets of ships that are the subject here. Such challengers are thus pertinent to discussions of ship design/building in the early century, and hence are also a topic for Part I. Other fields of engineering associated with this study also began a renaissance. Metallurgy of materials, for example, began to evolve with a better understanding of the elements that make up steel, such as carbon, nickel, manganese, and molybdenum, as well as those elements that weaken it such as sulfur and phosphorous.

CHAPTER 1

Introduction

Three hundred years of development lay behind transatlantic trade when a new century opened in 1800. The ships that carried it were built of wood and propelled by sail—a statement as true in that year as it had been 300 years before. But the Industrial Revolution was stirring on both sides of the ocean, and by 1900 the great majority of ships were built of steel and propelled by steam engines that fed on coal. Standing out from an armada of mundane cargo carriers was a fleet of ships that was fast, comfortable, and safe by the ancient standards: the ships that catered almost exclusively to passenger traffic. Passenger ships . . . passenger "liners," even,[1] shuttled between Europe and North America. They were a class of ship unimagined 100 years earlier.

However, by the standards of just 50 years later, the passenger ships of the North Atlantic in 1900 were small and slow. As a randomly chosen example, we can note the typical Cunard liner *Ivernia* of 1900. Its gross register tonnage was 14,066; its speed was about 16 knots, obtained by twin engines totaling 10,000 indicated horsepower.[2] Sizes, powers, and speed had been increasing steadily since the beginnings of steam-powered traffic across the ocean in 1840,[3] but incrementally. Terms that became common in the 20th century—Luxury Liner, Ocean Greyhound, Floating Palaces, and so on—were not common coin.

Then came that stretched decade, 1900 to the beginning of World War I, in which great ocean liners were built that were comparable in size, speed, and luxury to the North Atlantic ships that were to appear in the last great burst of liner-building, 1950 through 1960. The *Lusitania* of 1907 was measured at 31,540 gross tons, and was driven by unprecedented turbine engines of 68,000 hp at speeds up to 25 knots. Four years later, the *Olympic* was measured at about 45,000 gross tons. To be sure, these early liners were obviously lacking such later-day fixtures as watertube boilers (or even diesel propulsion), welded structure, and air-conditioned passenger spaces, but they were nonetheless great engineering achievements. In size, they did not approach the 80,000 ton range gross tons of the 1930s' great liners such as the *Queen Mary*. Nonetheless, the *Lusitania*, *Olympic*, and their sisters are much more of a piece with the later-day superships than with *Ivernia* and the like. And no book that purports to present the history of "great ocean liners" omits them.

Now, this book is not another history of great ocean liners. It concerns itself with only five ships that were the prides of only two transatlantic lines, Cunard and White Star. Both of these were British. The ships of their continental European competitors, such as the *Deutschland* and *Kaiser Wilhelm II*, may be outstanding as well, their designers and builders equally deserving of recognition that may be delivered a century after. But these German ships do not offer what two White Star and one Cunarder can—wrecks that are accessible on the seabed.

The two Cunard liners, *Lusitania* and *Mauretania*, came first; both crossed to New York for the first time in 1907. They were far bigger and faster than any ship built before them in the United Kingdom. They recaptured the North Atlantic speed prize that had belonged to the German *Kaiser Wilhelm II*. They were innovative in their propulsion plant, the first steam turbines in a transatlantic liner. The *Mauretania* survived service in World War I to serve again as a well-favored liner on the North Atlantic until the Great Depression and obsolescence drove her to a peaceful end at a ship breaker's yard. The *Lusitania* was not so fortunate, being the victim of a German submarine's torpedo in 1915 while still serving the transatlantic passenger trade. Her wreck lies on the seabed near the site of this tragedy off the southeast corner of Ireland.

The *Lusitania* sinking was an instant source of controversy because the ship was, from one viewpoint, an innocent passenger carrier torpedoed without warning by an enemy that deliberately ignored the long-established customs of maritime warfare. From another, the enemy was justified in its actions because the ship was an auxiliary cruiser of the Royal Navy, and moreover, was secretly carrying a cargo of munitions. From another, the *Lusitania* was led to her destruction by the machinations of her own government. The Admiralty (specifically in the person of Winston Churchill), knowing of the presence of a large number of Americans on board, did not provide the master, Captain William Turner, with the latest information on the activities of German submarines off

Fastnet. Perhaps his purpose was to provoke a *causus belli* between the United States and Germany.[4]

These points—all intertwined in conspiracy theories—are of no concern here. Although for background we furnish brief reviews in chapters that treat the *Lusitania*, our interest is in engineering problems and revelations. All parties to the controversy accept the fact that a second explosion followed the torpedo hit. Was it a second torpedo or boilers, or, perhaps, coal dust or the alleged cargo of munitions? Those who would skewer the alleged conspirators in London maintained it was munitions. Those who declare for deliberate slaughter-of-innocents by the Germans favored the second torpedo, fired in an effort to kill as many passengers as possible.[5] An answer seems impossible to reach until the wreck can be examined internally, and we will make no effort to do so. Until then only the facts as we know them will be presented.

The three *Olympic*-class ships of the White Star Line put to sea several years after the two Cunarders. The maiden voyage of *Olympic* took place in June 1911, that of *Titanic* in April 1912. *Britannic* was not ready for sea until after World War I had begun, and spent her short service life as a hospital ship. The *Titanic*—a familiar story if ever there was one—sank in the North Atlantic, April 1912, after striking an iceberg. The *Britannic* sank in the Aegean, 1916, after striking a mine that had been planted several weeks before by a German submarine.

The *Titanic* sank peacefully and slowly after a brush against the flank of an iceberg opened six compartments to the sea. It is exceedingly difficult to find conspiracy here, but the event holds fascination a century later; books of both fiction and nonfiction, movies, web pages, television documentaries, a continuing stream from every kind of "media" attests to that assertion. Conspiracy, no, but there is a broad area that those who exploit the interest of the reader/listener/viewer cannot exploit: the ship itself. Was there some great flaw in its design? Some weakness in construction that exacerbated the predicaments of the ship and its people after the collision? Engineering analysis cannot compete against our fascination with humans in extremis, but a large part of the *Titanic* story lies in what the engineers—the naval architects among them, in particular—can deduce from their understanding of ship design and building as practiced in the 21st century, and in the early 1900s as well. Added to that is the input that is available from examination of the wreck. We have looked for flaws, but have found very few. On the other hand, features that were superceded by later advances in ship construction were significant in the event. The response of riveted seams to impact—quite different from that of welded seams—is an example. We perforce examine such matters.

The *Titanic*'s younger sister, *Britannic*, also met a tragic end, a victim, 1916, of a mine in the Aegean. The sinking was an "enemy action" event like that of the *Lusitania*, but it lacked the inflammatory elements of the *Lusitania* story. To be sure, the ship was prominently marked as a hospital ship, was traveling in daylight, and some of the survivors claimed to have seen torpedo tracks before the explosion. However, the loss of life was small, and much more attributable to mismanagement of the ship's evacuation than to deliberately inhumane actions of the enemy. A rather impersonal sinking by a mine, laid in a war zone, was (and is) generally accepted as the cause of the sinking.

The *Britannic* was not "unsinkable," obviously, but she had better protection against sinking than the *Titanic*, all as a result of lessons demonstrated by that ship. *Britannic* had been built with an inner skin, and with watertight bulkheads that extended a deck higher than those of her sister. If the question, "will the explosion of a single mine or torpedo sink this ship?" had been asked before the event, the professional answer should have been "very probably not." Yet she did sink. Like the *Lusitania*, however, the *Britannic* may have experienced a severe second explosion whose origin remains a subject of speculation. On the other hand, this putative explosion obviously (from the appearance of the wreck) occurred well forward of the forward-most boiler room. The naval architect's judgment would be that no amount of damage forward of the boiler rooms should have sunk the ship. There thus remain questions to challenge the naval architect and marine engineer.

One Cunard liner, the *Mauretania*, and one White Star liner, the *Olympic*, survived war and the hazards of peacetime service until old age and changing economic conditions ended their careers. Mystery and controversy are absent from their histories. They might be regarded as bearing the stigma of sisters that failed, but the opposite is preferred: their success into the days of their obsolescence demonstrates the merits of those sunken sisters.

So we have five ships here, products of the best in British ship design and building of the pre-1914 decade. Two served their intended purpose until the 1930s; three sank early in life but live on in the form of wrecks to which forensic analysis may be applied. A unique opportunity to evaluate the field of ship engineering as it was approximately a century before our time thereby offers itself. Evaluate we can because we are naval architects and marine engineers, and have available a large share of the technical literature from the early 1900s. Moreover, we can apply the information flowing from the wrecks to fold long hidden insights into the stories of these ships.

Notes for Chapter 1

1. A "liner" is a ship that runs on a schedule between fixed points. "Cargo liner" is thus a suitable term for a ship that serves that way, but carries cargo as its principal function. The term "passenger liner" is sometimes used, but more typically "liner" without a modifier implies a passenger ship.
2. Throughout this book the tonnage of a ship refers to its gross register tonnage, a measure of internal volume in units of 100 cubic feet (until the 1960s, when an equivalent metric measure was introduced). *Indicated* horsepower is the standard way of designating the power of a reciprocating steam engine. It is the horsepower as measured within the engine cylinder, and hence 10 to 15 percent higher than the power reaching the propeller shaft. See also a glossary of pertinent terminology offered in Appendix 1.
3. The date 1840 might be disputed, but it is nonetheless a suitable time to use as a starting point because it is the year in which Cunard ships began regular transatlantic service under a mail contract from the British government.
4. For conflicting views of whodunit aspects of the *Lusitania* sinking, see Colin Simpson, 1972, *Lusitania*, Longman Group, London, and Thomas A. Bailey and Paul B. Ryan, 1975, *The Lusitania Disaster*, The Free Press, New York.
5. The official British investigation concluded that a second torpedo struck the ship, and that it was fired for that very purpose: to kill as many innocents as possible. ("Report of a Formal Investigation into the circumstances attending the foundering on 7 May, 1915, of the British Steamship *Lusitania*, of Liverpool, after being torpedoed off the Old Head of Kinsale, Ireland," Lord Mersey, Wreck Commissioner.)

CHAPTER 2

The Origins of Ship Design

The ships with which this book is concerned were built in the United Kingdom. Even a century later we look back upon them as artifacts of great sophistication, veritable Parthenons and Luxors along the developing path of naval architecture and marine engineering. Nonetheless, they were primitive in the sense that suitable architecture/engineering had only recently been organized by the pioneers who had struggled to make use of the Industrial Revolution's new materials and techniques. The substance of these ships' building was steel, common enough throughout the 20th century, but a material that in 1900 had only recently assumed its dominance over wrought iron, a material that had in turn become dominant over wood at only a slightly less recent time. Insofar as the hulls were concerned, the mature technology of wooden construction had offered a firm base for the use of metals, but the complex thermodynamic appliances that drove the ships had grown from only the tiniest of seeds, and in less than a century. The several applications of electricity—lighting, heating, mechanical power, communication—had reached a near maturity in an even much shorter time. The engineering—the methods of analysis that allowed such enormous structures to be built and driven against the seas—had grown in the struggle with these new things. The engineering was "modern" by 1900 in the sense that its terminology and techniques are still the stuff of ship design. But it, too, was of *very* recent maturity.

Our purpose in these early chapters is to sketch the development of naval architecture/engineering in its century of growth from essentially nothing early in that century to the point, arbitrarily set at 1900, where it could undertake the building of great transatlantic liners that were to be almost as large, and almost as sophisticated, as their descendants were to be at the end of the next century. Of course, the wrecks of three of the five ships we honor here lie on the seabed, subject to forensic investigation, and our overriding purpose is to speak of what the investigations tell us about possible inadequacies in their building. It will, perhaps, answer a near obvious question: were they too big and too powerful for the technology and knowledge base from which they sprang?

The pre-1900 story is spread over five chapters. This, the general one, treats the origin and growth of iron (i.e., wrought iron) building in its evolution from wooden construction, and honors the pioneering engineers of this immense step forward. These men were, for the most part, untrammeled by formal education.

Let us start at the beginning. The temptation is to start with a modestly biblical declaration; an irresistible one—in the beginning there was wood.

Wood was not the only shipbuilding material of the pre-industrial ages, but it was the dominant one, and by the year 1800 was probably the only material used for serious sea-going. However, the methods of its formation into viable hulls differed among the many seaside cultures of the planet.[1] If we limit our back-looking glance solely to North America, the bark canoes[2] of the ancient natives of the wooded regions, and the log-built bugeyes[3] of later-day natives of the Chesapeake watershed, come to mind. An interesting speculation: if the Industrial Revolution had flowered first among either of these peoples, how would they have adapted a ready supply of industrial iron to the building of their watercraft? Of course, that most consequence-laden of all revolutions began instead in the United Kingdom, and hence the parallel speculation about the building of watercraft is easily answered, at least in a general way: what the shipwrights had been building in wood they now duplicated, almost part-for-part, in wrought iron.

The British builders in wood (and most of their cousins in America) began with a stout keel, often a number of short pieces artfully scarped into one of sufficient length. At the ends of the keel were mounted stem and stern posts, perhaps shaped with only minimal working from naturally curved pieces, perhaps built up short, straight pieces.[4] Between these, the shape of the hull was fleshed out by rib-like frames that were usually formed to their curves from short pieces (in smaller craft, they could be steam-bent from a single piece). Planking was fitted to the frames to form the outer envelope. Beams running transversely supported deck planking. Knees, often formed from naturally grown crooks, joined deck beams to frames, and clamps and shelves (both "stringers") joined frames longi-

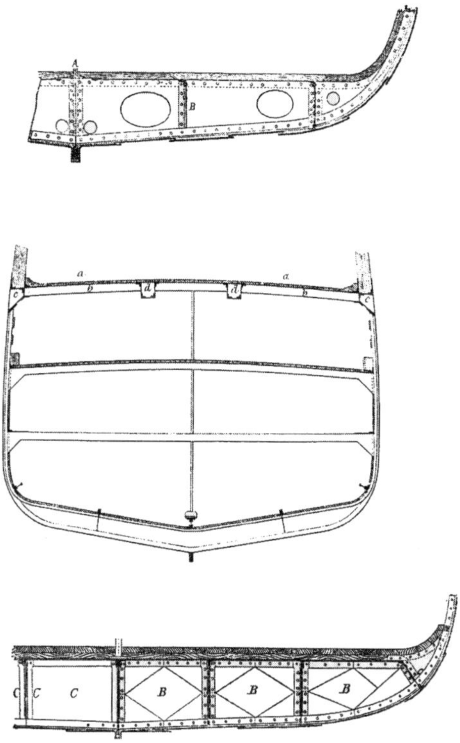

FIGURE 2-1 Elements of structural drawings from the nineteenth century, showing examples of composite iron-wood construction. (*From several nineteenth-century volumes of* Transactions, *Institution of Naval Architects*)

tudinally, and also supported deck beams. In many craft, the outer planking was mirrored by an inner planking (the ceiling) that protected cargo from bilge water, added strength to the hull, and protected the hull envelope from the impacts of shifting cargo.

In the decade 1820–1830 (approximately) wrought iron, a nearly pure (e.g., much less carbon than steel) iron, whose processing left it laced with fibers of slag that enhanced its strength and corrosion resistance, became available in the United Kingdom in such quantity, and at such price, that it could be used for engineering structures. The rails that soon were to make roads that were infinitely superior to muddy tracks, and the boilers and engines that would move trains of wagons and carriages at phenomenal speeds on those roads, would not have had their beginnings without this wonderful material. What about using it in ships? There was some resistance to that idea, for iron has a density of about ten times that of good English oak! Fortunately, venturesome men recognized that high density of material was of trifling consequence in building a viable hull, especially when relative thinness of iron members offsets the penalty of higher density.

There was little question of how the iron would be applied: what was fabricated in wood became iron. Iron keels, iron stem and stern posts, and iron frames took the places of their wooden predecessors. The introduction of iron allowed ships made of wood to increase in length to about 300 feet. Iron brackets replaced the knees that joined deck beams to vertical frames. On the other hand, long, narrow strips of iron did not take the place of wooden planking; rather, rectangular plates of thin iron were easy to manufacture, and much more suitable because all pieces had to be joined to its neighbors by rivets, regardless of size or shape.

Structural members such as frames and deck beams perforce presented their thin edges to the plating they were to support, making riveting impossible unless these structural members were fitted with flanges to receive the rivets. No matter, since varied shapes never seen in wood were easily achieved in the rolling that converted an amorphous blob of soft iron into channels, tees, or whatever cross section was appropriate to the application.

Wood, as a ship structural material, did not disappear instantly by any means. For example, wood planking might be used as the hull's ceiling and for the decks, or for sheathing the decks. (See Figure 2-1,[5] from a treatise published 50-some years after the first iron ships, showing wooden sheathing in an iron hull.) In addition, deckhouses might be built entirely of wood. Regardless of such qualifying statements, the age of iron had arrived, and during its time of dominance iron was to drive wood to the margins.

The keel-ribs-plates model was to dominate iron shipbuilding, and steel shipbuilding as well, once the newer material began to supplant iron. However, there were aberrations that are worthy of note. One was the cellular hull construction employed by Isambard Kingdom Brunel in his most famous ship, the *Great Eastern*. Her hull was constructed of an array of cells, best and simply explained by Figure 2-2.[6] There were no "ribs," though the membranes that divided the cells formed a grid-like frame for the hull. Of course, the inner surfaces of the cells formed an uninterrupted inner hull surface that *might* have made the ship unsinkable. Certainly, no iceberg, struck *Titanic*-wise, would have sunk the *Great Eastern*. A rock that the great ship did strike in Long Island Sound, puncturing the outer hull, caused no internal flooding. The ship survived with only the inconvenience of a docking to repair the outer-hull puncture. Given the low speed, and comparatively small size, of contemporary ships, a ship-to-ship collision might have done no more damage than the

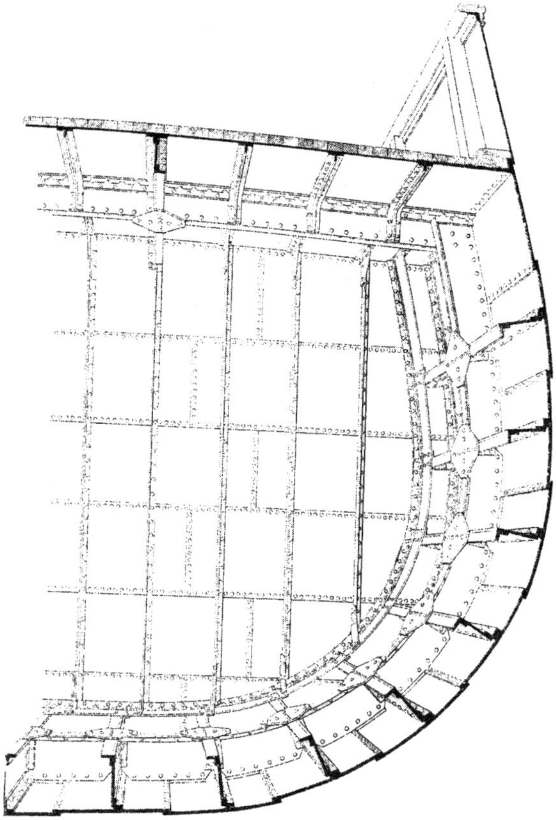

FIGURE 2-2 Half-section of the *Great Eastern* hull, chosen to illustrate the cellular concept of ship hull structure.

rock. And the self-propelled torpedo, the armor-piercing shell, and the aerial bomb that were to make a joke of later-day claims to unsinkability were many years in the future when the *Great Eastern* came to be in the 1850s. Regardless of such favoring evidence, the cellular scheme was rarely used after the *Great Eastern*'s day. Its originator, I. K. Brunel, with his success in many engineering ventures, might have been a forceful advocate for his ideas in hull construction. However, he died in 1859. That event, at age 53, was also unfortunate for the marine historian because it occurred before the first *Transactions* of the Institution of Naval Architects (later the Royal Institution of Naval Architects) was published in 1860. Other prominent engineers, principally William Fairbairn, Edward Reed and John Scott Russell, whom we shall soon honor here, dominated the realm of publication in the early days of the Institution, and likewise the practices of the shipbuilders. Some interest did remain: an 1880 paper[7] shows a latent interest in Brunel's cellular method, though the author did not attempt a strong advocacy for its continued use.

Another method of hull construction that was popular early in the iron age, but that was also an aberration, employed diagonal framing. Here was an obvious example of copying wooden construction into iron, but unlike the passage of wooden "ribs" into iron, the diagonal technique made no sense at all because the weakness—hogging of the hull—that it overcame was unique to the wooden ship. This unfortunate phenomenon is easily demonstrated by handling a deck of cards. In spite of a low length/depth ratio of about 5.9, the assembly of cards is easily bent in the vertical/longitudinal plane because its 52 members leave 51 planes that offer zero resistance to shear as the hands attempt to bend the deck. The wooden hull, conventionally planked fore-and-aft, similarly has joints between planks that cannot offer an adequate resistance to shear, and hence the ends of the hull gradually droop as it ages. This inadequacy plagued even small wooden yachts, where remedies, at first based on the edge-gluing of planks by waterproof glues, were developed in the 20th century. For wooden ships, the only feasible remedy was a sufficiency of diagonal framing members.[8]

Iron hulls required no diagonal members because their plating was thoroughly riveted along its horizontal seams. Had the early iron plating been as thick as the wood it was superceding, the builders might have left these seams unriveted, though caulked for watertightness, all in imitation of wooden-ship practice. But this traditional technique was infeasible with thin iron plates, and hence they were riveted to their neighbors along all edges. That this necessity precluded the need for additional anti-hogging measures was apparently not universally recognized. An author of a paper in the first volume of the Institution of Naval Architects *Transactions* felt it necessary to condemn diagonal framing in iron shipbuilding on the grounds that it was unnecessary for the reason just outlined, and undesirable because of its added expense.[9] This paper of 1860 is evidence for development of building techniques based on simple substitution—but also for evidence that the thoughtful pioneers of iron construction were identifying fundamental differences.

Other ramifications of wooden construction that were perhaps less obvious were nonetheless being recognized by 1860, and were also the subject of papers in the INA *Transactions*. We can illustrate by pointing out the differing circumstances in which wooden hulls and their iron (and steel) counterparts are likely to fail.

One might think of fire as being a fundamentally different threat, but so much combustible material was part of an iron ship that that problem was only slightly different. On the other hand, collisions and groundings were events that were to be regarded quite differently when iron gained dominance. Take collisions. Now, steamers may be built of wood and sailers may be built of iron, but at the time in question "wooden ship" meant—almost universally—a sailing ship. Collisions between such vessels, particularly at speeds that resulted in the bow of one ship slicing into the hull of the other, were extremely rare when compared to the many other hazards of potentially fatal magnitude. The watertight transverse bulkheads that have saved many iron/steel ships from sinking as the consequence of major hull intrusions were, therefore, *not* a feature of wooden shipbuilding, circa 1860. Similarly, the watertight inner bottom that saved many later-day ships after bottom damage was not a feature of wooden shipbuilding because (for one important reason) its preclusion of ventilation would seriously exacerbate the rot problem that plagued wooden hulls.

The application of watertight bulkheads and inner bottoms was not common among iron ships in 1860, but their utility as safety features might be regarded as obvious improvements, even to builders with no understanding beyond the problems of fitting pieces of iron together. But then there was that hogging problem. We've noted that the absence of deck-of-cards sliding planes in an iron hull obviated the resultant distortion of wooden hulls. However, the absence of hogging distortion severe enough to be noticed by a casual observer did not signal that related *stresses* were absent in the iron hull. Recognition of their presence and the extension of that recognition into design practice required a higher degree of sophistication than did appreciation of bulkheads and inner bottoms. How and when that recognition first came about is uncertain, but we shall date its effective appearance as 1860 when an appropriate paper appeared in the first volume of the INA *Transactions*.

Indeed, because of the founding of the Institution of Naval Architects in 1860, and its early publication of several significant papers, the decade 1860–1870 is reasonably named as the time of beginning for "modern" ship design. To be sure, there are arguments against it. For one, the new-age ships of that decade were built of wrought iron rather than the steel that was to be applied almost universally by the end of the century. For another, a few outstanding ships were afloat already. The *Great Eastern* was one, and the *Great Britain* another. Nonetheless, the knowledge that was disseminated in the decade named easily justifies the honor we award it.

Several notable pioneers were active in that time. They published not only papers as members of the INA, but also published significant books as well. The pioneer who wrote the 1860 paper on certain stresses in an iron hull, and in so doing opened the door to a concept of hull design that we still call "modern," was William Fairbairn. His paper was "The Strength of Iron Ships," from which we borrow two illustrations (see Figure 2-3) that show an iron ship aground, supported precariously by a rock amidships, then show it in similar distress from a rock at each end of the hull. Fairbairn offered calculations of bending stresses in both bottom and in deck and thereby demonstrated the need for the appropriate hull section modulus. He went on to show how the deck structure of contemporary iron ships was usually inadequate, and how this deficiency might be rectified. He claimed in his paper that such notions had not been publicly stated before, and indeed, a later (1867) INA author and fellow pioneer in iron shipbuilding, Edward J. Reed, credits Fairbairn with being the originator of the concept of the hull as a beam.[10] Of course, the "classical" notion of a hull supported by a wave of equal length—at its ends, then amidships—replaced the ship-on-rocks artifice of Fairbairn. The wave concept made its first published appearance (1874) in a paper by yet another INA pioneer, William John[11] of Lloyds.

An instant impression on seeing Fairbairn's two illustrations might be that he was picturing situations in which a wooden ship would be irretrievably wrecked by those rocks crunching through its hull. There should be no stresses in the deck. The pinnacles of rock would pierce the wooden bottom with fatal consequences. Of course, many an iron/steel ship has been lost through groundings, and Fairbairn's rocks may be a bit extreme in their sharpness, but the unfeasibility of inner bottoms and watertight bulkheads in wooden construction meant that design to preclude the loss of a ship on the rocks was simply beyond consideration by the designer/builder.

Now, we have introduced names of engineers who were prominent through their publications in the 1860–1870 decade. There were indeed giants in those days: the names we have just mentioned, surely, as well as others who will receive their due in this chapter. Incontestable judgments are difficult to make 140 years after the events in question, but we choose William Fairbairn as the first among them, the founding father of shipbuilding in the Industrial Age. Major points in his favor are his publication of a transactions paper that was not only a first, but that lay within the mainstream of ship hull design, and his pioneering work in the practice of iron shipbuilding that he brought with him into that decade. Pioneering, it was, in ways beyond mere prescient writing. The *Encyclopedia Britannica*,[12] the source for our biographical information, credits him with being the first to build an iron ship hull. The same source cred-

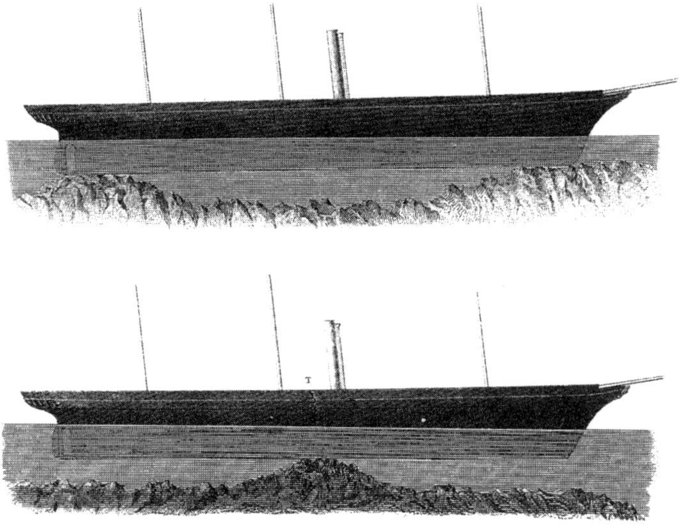

FIGURE 2-3 William Fairbairn's illustration of a ship hull supported at the ends and at its mid-length point. (*From his paper, The Strength of Iron Ships,* Transactions, *Institution of Naval Architects, vol. 1*)

its him with being first to use wrought iron for bridges, structural beams, and industrial shafting. He established his own shipyard on the Thames in 1835, and there built several hundred iron vessels. He was author of several more papers, and books[13] as well, and near the end of a long career (he was 71 when his 1860 paper was read) was recognized by Queen Victoria for technical excellence; he then became Sir William Fairbairn.

Isambard Kingdom Brunel, he of flamboyant name and astounding ship (*Great Eastern*), also built the *Great Western* (of wood, however) that pioneered transatlantic steam traffic, and the iron *Great Britain* that is preserved as a monument to the man and to his era. He might contest with Fairbairn for the title of The Pioneer, but he died in 1859, and hence no paper of his was published by the Institution of Naval Architects, and the cellular hull construction that so marked the *Great Eastern* did not enter the mainstream of ship hull design. He built bridges, railroads, tunnels, as well as ships, and in all was a remarkable man. However, he did not contribute to the development of ship design in the lasting way that Fairbairn and certain other contemporaries did.

Fairbairn and Brunel each was the son of a father who was a successful builder of engineering structures, a circumstance that surely contributed to their own careers. Beyond that, they were, by the standards of 20th century engineering, uneducated men. Of course, there were no schools of engineering. However, one recognized pioneer in iron ship design was a university graduate and himself a professor and author, all before the magic date of 1860. He was John Scott Russell (1808–1882), of whom the encyclopedia[14] says "a civil engineer best known for his research in ship design." He is credited with designing the Royal Navy's *Warrior* (1860), the world's first all-iron warship, and with participating in the design of the *Great Eastern*. But for all of that, he is remembered more as a scholar than a doer, having been a graduate of Glasgow University (at age 16), and later a professor of natural history at Edinburgh. At the latter school his major interest was in the surface waves of water, and their effects on ship hulls. A product of that interest was his "wave line theory" for the shaping of ship hulls, and it is for this that he was remembered a century later. Remembered, but not honored, since the theory was soon discarded as engineering science advanced. Indeed, his name bears a hint of taint because of its association with the wave line theory. But it should not, because he was a founder of theory in a field that needs a theoretical foundation as much as any other.

Russell was author (1841) of *On the Nature, Properties, and Applications of Steam, and on Steam Navigation*, and (1864) *The Modern System of Naval Architecture*, books that had as little relevance in the 20th century as his wave line theory. After 1860 he also contributed papers to the Institution of Naval Architects and frequently discussed the works of other authors. His major interest in those days appears to have been advocacy of the longitudinal method of ship hull framing. In an 1862 paper[15] he criticizes transverse framing ("ribs") as being a blind adaptation of the framing method common in wooden shipbuilding. His preferred alternative is shown by Figure 2-4. Frames of both deck and hull shell run fore-and-aft, with a bulkhead and a web frame (marked by the diamond-shaped connectors) providing transverse support. He demonstrated that this alternative provided a lighter hull than one built on transverse framing. Later experience showed him to be correct on that point: vessels where light weight of structure is important, and where waste of space enforced by the depth of web frames can be tolerated, were and are built on the longitudinal system. However, transverse framing has been the norm in the great majority of ships, placing Professor Russell somewhat out of the mainstream of ship design progress.

Every naval architect will demand that the name of William Froude (1810–1879) stand high on the list of pioneers. In the 1870s he built (under Admiralty auspices) a model towing basin near his home at Torquay and demonstrated there how data from the towing of hull models could be scaled up to an estimate of the power that would drive a full-size ship.

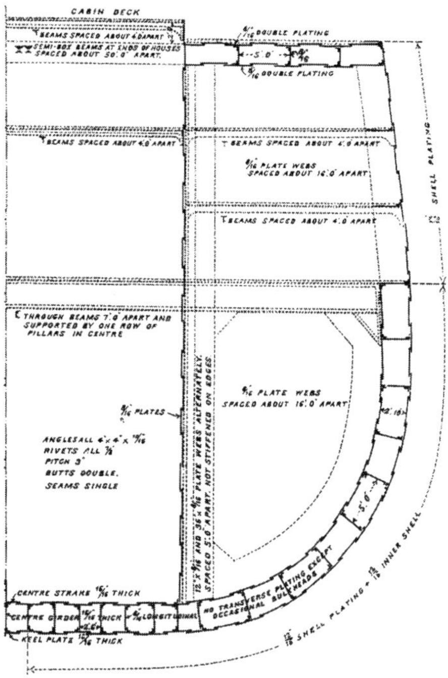

FIGURE 2-4 Longitudinal framing as conceived by John Scott Russell in 1862. (*From a paper by Jenkins, Note 16*)

Therein lay the key to predicting the engine power required to attain a specified speed, and to investigating the effect of hull shapes on that power. With that done, Russell's work on ship/sea interaction was reduced to the status of historical curiosity. However, as a pioneer in the *building* of ships, Froude does not stand with William Fairbairn, say, nor even with John Scott Russell.

There is some justification for honoring as a pioneer every author of an INA paper before 1900, but many of these men were advocating ideas that were not generally accepted, or were simply amplifying ideas that had been published earlier. These we do not dishonor, but merely omit them as not contributing sufficiently to the purposes of this chapter. On the other hand, several developments of great significance had at least faint beginnings even before 1860, and were near universal practice by 1900. One was the fitting of watertight bulkheads and inner bottoms as safety features. Another was the transition from wrought iron to steel. Both of these are of such importance that they are reserved for chapters of their own.

Then there is the question of pioneers who were not British. Surely there must be some in America or in France...? So far as the United States is concerned, we note that its builders continued far too long to exploit the forests of Maine and the shipbuilding skills in wood of the New Englanders. To learn the engineering of iron and steel, the late converts (e.g., Annapolis graduates) to those materials were sent to Britain for their education. For instance, Glasgow University became the place for Americans (and many other foreigners) to learn about the profession of shipbuilding after a professorship in naval architecture was established there in 1883. The American counterpart to the Institution of Naval Architects, the Society of Naval Architects and Marine Engineers, was not established until 1893. France had its pioneers, but their work seems to have been mostly in naval design. The most widely used French innovations seem to have been in marine watertube boiler technology. The names of Belleville and Niclausse are the eponymous labels for boilers that were developed in France late in the 19th century. These boilers were applied in both Britain and America, but mainly in naval vessels. However, their deficiencies sent these boilers to the scrapyards early in the 20th century. Certainly Germany should have had its pioneers. In the very first years of that century their passenger ships were dominating the North Atlantic trade, and those ships obviously sprang from some sort of fertile soil. However, the German interest in the sea arose only after the aspirations of Otto von Bismarck and Kaiser Wilhelm II turned seaward in the late 19th century, and like the Americans, their technology of those days was a borrowing from Great Britain.

As 1900 approached, we can claim that steel shipbuilding had matured sufficiently to look familiar to the 20th century engineer: hulls were built of steel, fitted with inner bottoms and transverse watertight bulkheads, framed and plated in ways that were to endure. The imitations of wooden structure had been weeded out, and the wrought iron that served the transition from wood to steel was largely abandoned. Some of the story of how these things came to be is to be told later, and some stories are not appropriate because they were still among future events when the last of our five ships were built (the transition from rivets to welds is an obvious example). Machinery for propulsion and auxiliary uses had also reached a maturity of sorts, but that maturity was being challenged by several sea-change developments that will demand some explanation a few chapters hence.

This chapter closes with a note on what the ship designers of 1900 were taught. As observed earlier, that teaching came largely from Glasgow University. The appropriate professorship had been established at that university in 1883, and soon became a magnet for the naval architects, foreign and domestic, who were motivated to learn all that could be taught about ship engineering. In 1889, the first holder of the professorship, Professor P. Jenkins, described his work in an INA paper.[16] To a nonspecialist, a listing of details may appear overly tedious, but on the other hand, it may in every detail fascinate a naval architect; some of Professor Jenkins's descriptions, and a complete listing of curricular matters, therefore, appears in Appendix 2. Here, we'll simply say that in the 1890s the program consisted of three sessions, each beginning in early November, ending in late April, and that most of the topics sound familiar to naval architects practicing a century later.

Notes for Chapter 2

1. Basil Greenhill, 1995, *Archaeology of Boats and Ships,* U.S. Naval Institute Press, Annapolis, Maryland.
2. Edwin T. Adney, 1964, *The Bark Canoes and Skin Boats of North America*, The Smithsonian Institution, Washington, DC.
3. Alexander Crosby Brown, 1940, *Chesapeake Bay Bugeyes*, The Mariners Museum, Newport News, VA.
4. The age of iron ships arrived none too soon for those who struggled to build ships for a worldwide empire from a resource base provided by the oak forests of a small island. This story—fascinating to those who care for shipbuilding history—is related in Robert G. Albion's classic work, 1926, *Forests and Seapower*, Harvard University Press, Cambridge, MA.
5. William John, 1880, "On Cellular Construction of Merchant Ships," *Transactions*, Institution of Naval Architects, Volume 21.
6. Figure 2-1 is from William John's paper (see Note 5). Figure 2-2 is the *Great Eastern* midship section, original source unknown.
7. The author and paper are listed in Note 5.
8. Visitors to the USS *Constitution* might note the diagonal strapping that has been added to its hull to preclude its gradual distortion by hogging.
9. John Grantham, 1860, "The Strength of Iron Ships," *Transactions*, Institution of Naval Architects, Volume 1.
10. Edward J. Reed, 1867, "On Certain Weaknesses in Iron Ships," *Transactions*, Institution of Naval Architects, Volume 8.
11. William John, 1874, "On the Strength of Iron Ships," *Transactions*, Institution of Naval Architects, Volume 15.
12. *The New Encyclopedia Britannica*, 1995, Volume 10.
13. Fairbairn's books were *Iron Shipbuilding,* 1865, and *The Life of Sir William Fairbairn*, 1877.
14. *The Encyclopedia Britannica.* See Note 12.
15. John Scott Russell, 1862, "The Longitudinal Method of Constructing Iron Ships," *Transactions*, Institution of Naval Architects, Volume 3.
16. P. Jenkins, 1889, "On the Course of Instruction in Naval Architecture at Glasgow University," *Transactions*, Institution of Naval Architects, Volume 30.

CHAPTER 3

Wrought Iron Becomes Steel

Iron ships—*wrought* iron ships, as accuracy of expression demands—had but a brief command of the seas, roughly speaking only about a quarter of the 19th century. The first such vessel was built in 1818,[1] and by the 1850s iron was becoming the dominant hull material for transoceanic (in particular) ships. Steel as a practical hull material began to displace iron in the 1880s. By the time of the Cunard and White Star ships that are our principal interest, wrought iron was largely out of fashion among ship designers. Within the context of this book we remember it principally as rivet material (in part) for the *Olympic*-class ships. The reign of wrought iron, roughly 1850–1880, was, therefore, rather brief indeed. It nonetheless has an importance far beyond its role as building material: it was the dominant hull material during the days when ship design advanced from the rule-of-thumb practices of wooden construction to the engineered construction of the 20th century's great steel hulls.

Iron first replaced wood simply as a substitute for that material, but during its time certain novel hull features, such as inner bottoms and watertight transverse bulkheads were introduced, and moreover, the all-important knowledge base of ship design was established, evidenced by the founding of the Institution of Naval Architects in 1860.

Steel as an available structural material came in the late 1850s with the introduction of the Bessemer process, and by the 1880s was becoming the material of choice for shipbuilding. Wrought iron was far, far older than this upstart competitor, with a history of use several millennia long. As a widely used industrial material—the stuff of bridges and ships, say—its history was much shorter; for that period it appears to have begun with the development of the puddling process for its large-scale production. The date was 1784,[2] the place Great Britain, and the wearer of the credit for this development was Henry Cort.[3]

In puddling, iron at a temperature near its melting point is in contact with iron oxide that reacts with silicon, phosphorus, and sulfur to form a slag, and with air that oxidizes carbon. Hand stirring (called "rabbling") mixes the slag uniformly throughout the iron mass. The process ends with a near-molten blob that must be squeezed by rollers to remove most of the slag. The product is a near-pure iron, but one still laced with a fibrous remnant of the slag. Ductility, and resistance to shock, fatigue, and corrosion are its notable properties. In corrosion resistance, at least, it is distinctly superior to mild steel.[4]

Iron becomes steel when it is alloyed with carbon to a low percentage, usually not greater than 2%. It is also often alloyed with other metals, the nickel and chromium that produce stainless steels (typically 18% chromium, 8% nickel) being a common case. Mechanical and chemical properties can be varied in many ways, and over wide ranges, by alloying, but the near-unique feature of steel is the several iron-carbon phases it exhibits as a function of temperature, and that can be maintained at ordinary temperatures by the techniques of heat treating. That brief statement skims lightly over a complex and lengthy subject—the metallurgy of steel—that was understood quite incompletely in the 1850s when steel was first becoming available. However, in the early years of its competition with wrought iron, several fundamental qualities were recognized as keys to its adoption: ultimate strength, ductility, and price. The first of these clearly favored steel, its ultimate strength being almost twice that of wrought iron. With that in mind, ship designers rightfully concluded that scantlings could be lighter in steel, with consequent increase in deadweight for any given displacement. The price (£/ton, say) was to fall below that of iron by the 1880s, but was in the early years high enough to make a steel hull the more expensive alternative. High deadweight might have been a trade-off against high price, but uncertainty over ductility was a troubling technical factor that was difficult to quantify.

The first paper in the Institution of Naval Architects concerning steel in shipbuilding was published in 1866.[5] Its principal topic was brittleness versus ductility, exacerbated by shipbuilders' lack of understanding of the significance of these properties. It also mentioned that the first vessel of steel construction was a river steamer built for the Livingstone expedition to explore the Zambesi region of East Africa, beginning in 1858. Two papers[6,7] appeared in the 1868 *Transactions*, again discussing the unwanted brittleness that might be present in shipbuilding steel. The apparent source of trouble was the circumstance that the properties of steel are much more at hazard than wrought iron from manufacturing processes applied to it between the mill and the finished ship. For example, the second of the 1868 authors mentioned punching of rivet holes as a cause of local embrittlement in steel. (He suggested that post-punching annealing was a suitable remedy, and cited his own experiments with punching and annealing.)

The ductility problem remained a major concern through the 1870s. In 1871 an INA author[8] brought small steel test pieces to his Institution presentation. Their significant properties he stated to be 0.2% carbon, elastic limit 23 tons/square inch, ultimate strength 38 tons/square inch—and the pieces were severely bent into a U shape. That last, of course, was his demonstration of ductility, but then he cited recent tests by the Royal Navy in which mild steel plates of 2½- and 3-inch thickness had shattered on impact from solid shot. His point was that steel *could* be ductile, but that ductility in samples of large thickness might still be an elusive property. This author also touched on the embrittlement caused by punching of rivet holes. His suggested remedy was drilling instead of punching.

John Scott Russell, the most prominent theoretician in the early days of iron/steel shipbuilding (see Chapter 2), presented a discussion of this paper. In it he claimed to be "one of the oldest believers in steel," and declared that steel was now (1871) "trustworthy." Other authorities were to reserve their enthusiasm for a few more years.

Nathanial Barnaby (a name that many naval architects of a century later may recognize) had his work on steel shipbuilding published in 1875.[9] He expressed a desire to see steel supplant iron, but wanted to do so "without care and anxiety." As a cautionary note he cited the recent practice of the French Navy in building warships, largely of steel (steel except for outer bottom and rivets). The particular point was that great care was taken to avoid compromising ductility in the building process. For example, hammering during the process of forming plates and frames was not allowed by the French constructors.

Now, all of the problems brought to shipbuilding by the use of steel were not to be eliminated until the middle of the 20th century, but by 1880 this material was clearly shouldering aside wrought iron. For instance, in 1878, tons built in steel were 4,470, whereas in iron the total was 517,692. In 1885, there were 165,437 in steel and 290,429 tons of iron. Iron was still the preferred material in 1885, but a strong trend toward steel is evident in these figures.[10]

Based on the number of papers published in the INA *Transactions*, the '80s decade was the one of peak interest in the technology of building in steel, and also the one in which the problems of its use seemed to have been solved. Steel was then about equal to wrought iron in cost-per-ton, and could be used in scantlings of about 20% lighter.[11] Testing of steel was highly developed, and its results disseminated widely.[12] The local embrittlement, caused by punching of rivet holes, was recognized and countered by appropriate measures: annealing having proven too cumbersome for large pieces, rivet holes were typically punched undersized, then reamed ("rimered" as it was commonly expressed) to full size. Narrow pieces, such as butt straps, might be punched and annealed, or might be drilled.[13]

Hints of problems that would surface in the future could be found in the literature, however. An 1882 paper, "On Cracks and Annealing of Steel,"[14] asserts that the current view that failures through cracking were a consequence of uneven heating was false, that the actual cause remained a mystery, and that prior annealing was not a preventative measure.

As we shall describe in later chapters, brittle failure of steel is a factor in the marine forensics story, in particular with regard to the *Titanic*. (As this was written, the failure of rivets was finally judged to be the prime factor.) Now we realize that this unfortunate phenomenon usually stems from the presence of minor constituents in the steel, constituents that were not (but might have been) removed in the steel-producing processes.

For some years into the age of steel shipbuilding, the steel was a product of the Bessemer process. The principal feature of the process is a "blow" of air through molten iron to remove, via oxidation, the impurities that have remained from the blast furnace reduction of the original iron ore. At the end, sufficient carbon is added to produce its desired level in the product, and manganese is added to reduce any oxidized iron through the reaction $FeO + Mn \rightarrow MnO + Fe$, and to remove sulfur by $Mn + FeS \rightarrow MnS + Fe$. The removal of the small amounts of sulfur that may be present in the form FeS (ferrous sulfide) is important because this compound appears preferentially at grain boundaries, and thereby forms paths of weakness. (On the other hand, the manganese sulfide distributes itself throughout the steel in the form of lenticular nodules.)

Phosphorus, occurring to a significant degree in some ores (particularly in Great Britain), is removed in the Bessemer process only in the presence of basic lining of the Bessemer converter. Such a lining contains metal oxides as FeO, MnO, MgO, and CaO (lime). These materials, forming a slag, can react as $2P + 5FeO + 3CaO \rightarrow 3CaO \cdot P_2O_5 + 5Fe$, thus absorbing the phosphorus into the slag.

By 1900 the competing open hearth process (introduced in the late

1860s) had largely supplanted the Bessemer process in Great Britain's steel industry, so that it is almost certain that the ships we are to speak of were built with open hearth steel. Now, the Bessemer process begins with a charge of molten iron, and further heat is generated by the impurities in the charge as they are oxidized by the air blown through the charge. In contrast, the open-hearth process begins with a charge in the solid state, with subsequent heating being supplied by an external source. Time to process is thus much longer; a consequent benefit—which may be quite important—is that the time allows careful adjustments of chemical composition. Provided that the elements present are the same, the chemical reactions are the same in both processes. However, we should note that both may be either "basic" or "acid," as determined by the compounds that are provided, either in the converter or hearth linings, or by additions. The reactions symbolized earlier are for the basic version, and hence it follows that when phosphorus removal (in particular) is important, the steel should come from a basic process. Note, however, that sulfur removal cannot come from reactions with slag constituents, but must be achieved by addition of manganese.

Joining of small individual pieces into the whole is a matter of concern whether wood, wrought iron, or steel is the material of construction. Late in the 20th century welding had become the universal method of joining the material that had also become universal in steel shipbuilding. Throughout the era of iron, and during the time that the White Star and Cunard ships were built, production welding was unheard of, leaving only one choice—rivets. Riveting was a familiar method of joining metals long before the first iron ship (witness ancient body armor), but it did undergo development and rationalization as iron construction moved toward its maturity. For instance, William Fairbairn,[15] writing in the first volume of the Institution of Naval Architects, discusses the general topic of hull design for adequate strength, including a table that he labeled "Table exhibiting the best proportions for riveted joints." The table gives rivet diameter, length, and center-to-center spacing of rivets, as functions of plate thickness. Figure 3-1 shows his suggested geometry for transverse plate joints (joints between adjacent plates in the same hull strake), a joint type that he believed to be inadequately fastened by the practices of the time.

The topic of rivets was not featured in the INA *Transactions* to the extent that the advent of steel was, but several papers did appear. For example, J. G. Wildish[16] in 1885 reported his own research on the strength of riveted joints, and included, inter alia, extensive tables giving the shearing stress per rivet and per square inch of joint for a number of independent variables.

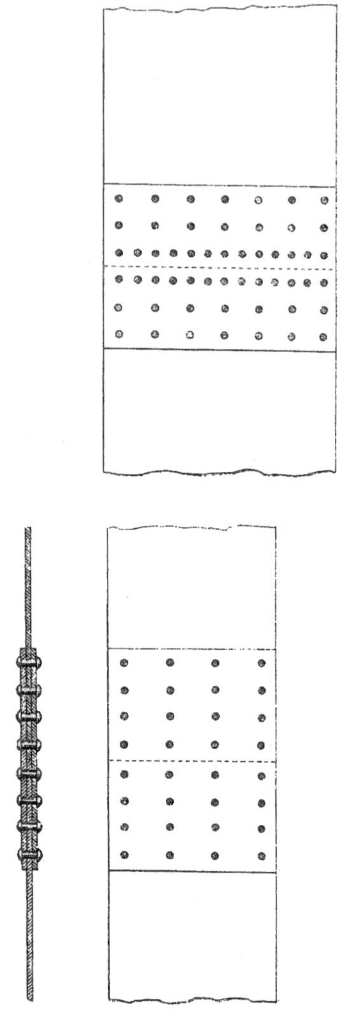

FIGURE 3-1 Rivet configurations from the 1860 paper by William Fairbairn. (*The Strength of Iron Ships,* Transactions, *Institution of Naval Architects, vol. 1*)

Riveting was one of the earliest items of standardization by navies and by regulatory bodies such as Lloyds. Indeed, by 1900 the ship designer had no need to do first-principles engineering in joint design, since handbooks supplied multiple pages of tables that seem to have covered every detail of "standard practice" riveted joints.[17] However, the method of finishing a rivet might be a matter of choice. Of course, the rivet was softened by heating before being inserted into the hole it was to occupy, but the forming of the head could be done in several ways. The traditional way was hand hammering, and was the usual method in 1900. The pneumatic hammer—the "riveting gun" that was to provide the violent sound of steel construction for at least half a century—had its advent just about 1900. The same statement can be made with respect to the hydraulic riveter that was to be used on all of our ships. William Hovgaard,[18] an early 20th century authority in ship design and construction, stated the case of wrought iron versus steel for rivets with the following points:

- Iron rivets are often used to join steel because steel rivets require a higher level of workmanship, and that requisite is not always available.
- Iron rivets are not so liable to suffer from overheating, are easier to drive, and are less subject to corrosion.
- Shearing strength of iron rivets of good quality is more uniform and reliable than with steel.
- When used in steel plating, strength of iron rivets is less than that of steel rivets. (As we shall see in a later chapter, this circumstance may have been an important factor in the failure of the *Titanic*'s hull structure under impact from the iceberg.)

To summarize, we can say that by 1900 construction of hulls in steel, with riveted joining, had matured rapidly, but quite successfully, from wrought iron technology (and from wooden construction not long before) to the point where it could be called "modern" a century later. A tradition-seeking designer of cruise ships, working in the 21st century, would do little wrong in copying their hulls (welding advised, however). If there were defects, then only the thorough analysis of the later-day forensic experts can show them.

Notes for Chapter 3

1. This is the date given by *The New Encyclopedia Britannica*, 1995, Volume 12. Recall from Chapter 2 that this source credits the first building in iron to William Fairbairn.
2. Wrought iron was not plentiful before this time, but one still wonders why no known ship (or even small boat) was ever built of it before the 19th century, even into "ancient" times. Surely, many persons over many years must have observed that metal vessels (e.g., cooking pots) would float. And Archimedes . . . where was he, the Father of Naval Architecture, and an inventive person credited (though with much exaggeration) with building ships enormous in size and imagination? Wherever his imagination may have been applied, it did not come to fruition in iron ships.
3. Source for Henry Cort is also the *Encyclopedia Britannica*. See Note 1.
4. Metallurgy is, of course, the foundation of any structure-related engineering, and

in consequence innumerable texts treat the properties of metals, and the processes by which they are produced. The one used for these brief remarks on wrought iron are based on W. T. Frier, 1952, *Elementary Metallurgy*, McGraw-Hill Book Company, New York.

5. T. A. Rochussen, 1866, "Application of Steel to the Construction of Ships," *Transactions*, Institution of Naval Architects, Volume 7.
6. T. A. Rochussen, 1868, "On the Treatment of Steel Plates," *Transactions*, Institution of Naval Architects, Volume 9.
7. Henry Sharp, 1868, "On the Treatment of Steel Plates," *Transactions*, Institution of Naval Architects, Volume 9.
8. J. B. Howell, 1871, "Steel Applied to Shipbuilding," *Transactions*, Institution of Naval Architects, Volume 12.
9. Nathaniel Barnaby, 1875, "On Iron and Steel for Shipbuilding," *Transactions*, Institution of Naval Architects, Volume 16.
10. The tonnage are for ships classed by Lloyds, but are taken here from B. Martell, 1886, "A Brief Review of the Progress of Mild Steel," *Transactions*, Institution of Naval Architects, Volume 27. According to Martell, "These numbers include ships built on the Continent, in America and the Colonies."
11. B. Martell, 1886. See Note 10.
12. J. Ward, 1886, "The Present Aspects of Mild Steel for Shipbuilding," *Transactions*, Institution of Naval Architects, Volume 27.
13. J. G. Wildish, 1885, "On the Strength of Mild Steel Plates and Rivets of Various Kinds Used in Shipbuilding," *Transactions*, Institution of Naval Architects, Volume 26.
14. A. C. Kirk, 1882, "On Cracks and Annealing of Steel," *Transactions*, Institution of Naval Architects, Volume 23.
15. William Fairbairn, 1860, "The Strength of Iron Ships," *Transactions*, Institution of Naval Architects, Volume 1.
16. J. G. Wildish. See Note 13.
17. An example of handbooks that furnished the details of riveted joint design is George Simpson, 1904, *The Naval Constructor*, D. Van Nostrand, New York. Although the title suggests the book's emphasis on naval shipbuilding, it also includes tables from Lloyds rules for commercial ships.
18. William Hovgaard, 1940, *Structural Design of Warships*, U.S. Naval Institute Press, Annapolis, Maryland.

CHAPTER 4

Hulls

From the earliest days of waterborne craft, unknowable millennia ago, the hulls of ships were built of wood, and in a small way the practice continues. In the United States, essentially the only application of large wooden hulls is in the reproduction of historical ships. With the coming of the Industrial Revolution, superior materials became practicable and available. Wrought iron was the first of those, soon to be followed by steel. By the latter part of the 19th century, wrought iron had supplanted wood as the main hull material. This process was aided by the fact that the supply of wood was diminishing. By 1900 with the development in steel making, steel had become dominant. Compared to wrought iron, it could sustain higher design stresses, its cost to the builder was less, it was more ductile at low temperatures, and its earlier problems of suspect ductility had seemingly been overcome. The use of any material other than steel for a major transatlantic liner was beyond consideration.

For a structure such as a ship hull the material must be ductile. It is true that by the year 1900, earlier problems in this regard had been overcome. However, the possibility of a nominally ductile steel being embrittled in service through a phase change caused by low temperature was not recognized, and indeed would not be until the failure of war-built hulls due to brittle fracture in the 1940s publicized the problem. In our context, the apparent brittle condition of the *Titanic* plating and its rivets as the hull suffered impact from the ice is a matter for forensic consideration. Indeed, once examination of the *Titanic* hull revealed its mode of failure under the ice impact, analysis of the ship's structural steel and

This is a view looking inboard of the RMS *Olympic* in dry dock, showing a typical riveted structure on her hull bottom plates.

its rivets became the forensic naval architects' main concern in evaluating the wreck. Their findings are discussed in several later chapters.

When the ships discussed in this book were built, the method of joining the many and diverse steel shapes and plates into a single entity, the hull, could only be accomplished by riveting.[1] This was the only acceptable way of metal assembly for large and complicated structures. It is true that welding was an ancient process—every blacksmith could "forge weld" small pieces of metal together as they lay upon the anvil. It is also true that the structural plans for the *Olympic* ships specify welding at a few (very few) places. However, all of the affected areas on those ships are nonload bearing.[2] The electric arc process that was to revolutionize the building of large load-bearing assemblies remained, until the 1920s, inconceivable to the shipbuilding industry and to the classification societies.

Further, any critical statements concerning the relative merits of welding and riveting are fueled by hindsight, of course. In 1900, comparisons with arc welding were not possible. Unlike promises of water-tube boilers, turbine engines, petroleum fuel, and similar, all prospects that roiled the decision processes in boiler and engine rooms, welding was not a challenge to hull designers.

At the time of our interest, riveting was a mature technology. It had been discussed in the earliest Institution of Naval Architects papers, and appears as a topic in several subsequent 19th century volumes of the INA. The mid-1880s curriculum outline for Glasgow University (see Appendix 2) shows it to be a significant topic. It did have its weaknesses; the joint efficiency (fraction of joint strength actually achieved) could never approach 100% in the way that welding does. True, as the number of rivet rows increase, the joint efficiency will increase, but 100% is far from practicable. For example, the following joint efficiencies can be obtained with two 1-inch plates joined together in a lapped connection, 1.25-inch rivets spaced 3 inches apart:

- Double rows = 27%
- Triple rows = 53%
- Quadruple rows = 72%

It is also true (perhaps obviously so) that achieving higher efficiency through increasing the number of rows or through closer spacing increases the cost of building. Further, the punching holes and subsequent "rimming" or "rimering" (reaming to finished size) were expensive processes that has no analogue in welding. When the failure was traumatic—as from iceberg or a contact explosion—joint failure was likely to be much more extensive than with welding.

There are two features of weakness in a riveted joint. One is the tendency of the rivet shank to shear and the second is for the plates to fail across a line of rivet holes. The shearing failure can be combated by simply increasing the number of rivet rows. The second cannot be eliminated altogether because the presence of rivet holes will always constitute a line of weakness. Another type of failure in rivet connections is the bearing failure where a portion of the plate in line with the rivet hole and plate end falls out.

Hydraulic riveting was introduced in the early years of the 20th century since it resulted in better driven rivets than by hand and the plates would be forced into absolute contact increasing the frictional resistance. Hand riveting might result in shanks that do not completely fill the rivet hole. A rivet that is driven with hydraulic riveting is subjected to such compression that the flanks of its shank flow like wax. Every part of the rivet hole is thus filled. The process is done by a machine that holds the rivet head in the die while the red-hot shank end is crushed by a cup-shaped recess that forms the head. Excess material is squeezed out as a thin collar. The pressure exerted could be as much as 140 tons per square inch. However, the use of hydraulic riveting depended on a shipyard, as it was not always found to be advantageous in terms of expediency and economy.

Riveting was the only option in the early 20th century. Use of high-strength steels was innovative, though not challenging to a designer or builder. But other aspects of hull design/building should be mentioned, and against a background of later-day practice. Figure 4-1 shows the mid-ship sections of the *Mauretania* and of a containership from the 1980s. Given that the two ships were built some 70 years apart, and served two distinct commercial functions, their structural cross sections are remarkably alike. Of course, even a casual examination of the illustrations reveals the difference that came with the advent of welding: plates of the earlier ship overlap because they must be riveted along their seams (as well as along their butts, and to the frames, of course); the plates of the latter simply fit edge-to-edge because they are fastened in all respects by welding. The two distinct methods of fastening are also the source of other details of structure that are not so evident in the illustration. But details only . . . except for the evidence that welds superceded rivets, the illustration demonstrates that no radical change in structural concept lay before the designers of the early 1900s. And in those days the rivet/weld exception lay far in the future.

Casual glances at a profile picture of *Mauretania* and a late 20th century cruise ship reveal obvious differences in their hulls. By far the most obvious are in the bow profiles, the vertical and straight stem of the old ship with a raised forefoot, and the radically raked bow of the new one with a bulbous bow. These extremes are typical of the two eras of passenger ships. In part, the "modern" bow evolved over a century; in part it represents the greatly different environments for which the ships were designed.

The profiles of passenger ships built in the 1920s and the 1930s show that rake and flare were then being adopted for the bows of North Atlantic ships. The divergence from earlier practice is rather small, however. This would be the beginning of a trend toward the extremes that characterized the late century. Perhaps, but the mid-century bows were apparently less a product of stylistics than a rational attempt to improve seakeeping. A ship that must maintain a schedule across the Atlantic perforce confronts any sea state (barring extremes, of course) that lies along her route. Moderate flare forward is essential in high sea states since it allows the bow to throw potentially boarding seas aside, and lifts the bow through added buoyancy. But note the word "moderate." Extreme flare leads to extreme impact loads as a rising wave meets a descending bow. The extreme flare of the later-day cruise

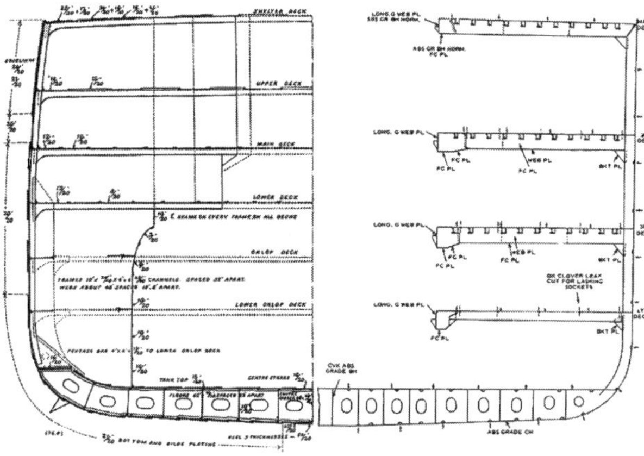

FIGURE 4-1 On the left, a half-section of the *Mauretania* hull (*from* Engineering, *vol. 84, 1907*) and on the right that of a roll-on roll-off cargo ship of the 1970s (*from* Ship Design and Construction, *Society of Naval Architects and Marine Engineers, 1980*).

ship would, therefore, be unsuitable to a ship required to maintain a high-speed schedule across the North Atlantic during winter and early spring storms. However, the typical cruise ship does not maintain a close schedule in rough waters.

In the more benign waters where the cruise ships operate, the heavily flared bow presents no hazard vis-à-vis dangerous sea states. Styling can dominate, but the flare may have a rational element as well. The cruise ship does not have the length of machinery spaces, something on the order of 50% of their length, that distinguished the earlier Cunard and White Star ships, but they do need a maximum of deck space high in the superstructure. In simple terms, the cruise ship of the 21st century has a long and high superstructure, coupled with a short waterline; the flared bow is a major part of the transition between the two.

The bows differ radically below the waterline as well. The cruise ship is likely to have the rather extreme bulb protruding forward, all quite typical of late 20th century technology. Early 20th century passenger ships show no trace of such a structure. The bulb's purpose is reduction in the resistance to forward motion that originates from the energy that goes into wavemaking. The hull produces waves; the bulb also produces waves. With proper geometry, the two sets of waves tend to cancel each other, and to the extent that they do, hull resistance is reduced. This useful concept was first applied in the 1920s. The German liners *Europa* and *Bremen* were among the first to use this new technology and with convincing results. After 20 years of the speed dominance of the *Mauretania*, the *Europa* wrested the "Blue Riband" away from the Cunard liner in 1927. Photographs of the underwater hulls of ships built in those years often show a bulb on the bow, though a rather small one compared to the post-World War II versions. The difference in the later 20th century hulls comes from research done in model basins during the 1960s that (rather improbably, it might seem) showed the benefits of the extreme bulbs that soon began to appear as part of new construction. Part of that technical advancement came in Japan with the construction of the battleships *Yamato* and *Musashi*. Their oversized bulbs showed a very promising reduction in wave-making resistance. In consequence, these super battleships achieved a speed of 27 knots with a 15% reduction in hull length. Research in postwar Japan advanced the concept of large bulbous bows. (As a matter of passing interest, this work was based largely on experimental work done by Professors Takao Inui and Tetsuo Takahei while visiting scholars at the University of Michigan in the early 1960s.)

The testing of models under the principles enunciated by William Froude in the 1870s (Chapter 2) led to greatly enhanced knowledge of the hydrodynamics of hulls moving through water. However, beyond the so obvious presence of a bow bulb on later-day ships, changes in hull form over a century are more subtle—and not necessarily the consequence of increased hydrodynamic sophistication. A comparison of transverse sections of passenger ships built 100 years apart show hulls of essentially the same shape. Hull shapes are dictated not so much by hydrodynamics as by drafts that harbors allow, by beams that stability require, by sea-keeping considerations, and by depths of hull that its bending stresses demand. For an example of how these factors might operate, we note that the beam of the White Star liner *Britannic* was 18 inches greater than that of *Olympic* and *Titanic*. The reason: the possible need of a stability compensation for added weight high in the ship. (A more detailed explanation appears in Chapters 18 and 24.)

Many changes in hull design since the early 20th century are probably of interest only to the professionals. For example, the beam to depth (depth to strength deck) ratios of both the Cunard and the White Star ships are approximately 1.45; liners built in the '20s and '30s typically show these ratios around 1.30. Other hull parameters could be cited, but are not because they reveal no deficiency of design knowledge, nor do they have any forensic significance.

Returning to visual comparisons of the old and later-day hulls, we note that above-water portions of the hulls show that they are radically different in contour at the stern as well as the bow: the earlier counter (or "fantail") type of stern has typically been superceded by the flat-across transom stern. This change is in part just a matter of taste. As a more practical matter, the transom stern allows slightly more deck area and internal volume on a given hull length.

The underwater outfit astern differs in significant ways as well. In the days of the Cunard and White Star ships, steering was usually accomplished by a single plate-type rudder, mounted by gudgeons and pintles on a centerline stern post. In fact, this design was a carryover from wooden ship construction. Beneath the waterline, the criterion was free flow of water to propellers and rudder. The early ships show fine lines aft, but the keels typically carry all the way aft to form a skeg to support a rudder on the centerline, and shafting for outboard propellers are invariably encased in long bossings, as in the *Olympic* class. Later ships usually feature an underhung rudder (or rudders), and hence do not require a long skeg to support rudder gudgeons. Propeller shafting is usually exposed, though is supported by a strut bearing just forward of the propeller. By and large, these changes arose from growth during the 20th century of the ship designers' understanding of hydrodynamics.

In summary, we can distill several points from discussions here of difference in the hulls of early and late 20th century:

- Many of the differences are obvious from casual observation, but have no significance in any critical analysis of the earlier ships.
- Some differences are not so easily seen, but also have no significance within the scope of this book.
- Riveted seams and metallurgical properties of hull steel are differences—and indeed weaknesses—that will have to receive more extensive treatment in later chapters.

Unlike the machinery designers of 1900, hull designers of that year were not faced with several highly significant innovations that they must either accept or reject.

Notes for Chapter 4

1. The dimensions of the plate and all rivet spacing and sizes in this book pertaining to the RMS *Titanic* come from the official document, "Form of Comparison of Scantlings of Unclassed Iron and Steel Ships with the Rules of Lloyds Register for 1885." The form was signed by surveyor F. Carruthers.
2. This statement was extracted from the Shell Plating Plan of the SS *Olympic*.

CHAPTER 5

Propulsion Machinery in 1900

The 18th century belonged to wooden hulls and canvas sails. In that century's ultimate year, thoughts of the iron and steam to come lurked only in the corners of a few imaginative minds. By 1900 wood and canvas still held a few niches, but iron had arrived and had already yielded most of the shipbuilding domain to steel; an even more radical innovation—the reciprocating steam engine—had usurped almost all of the marine propulsive tasks from sail.

Now, 1900 was no year of stasis in the pace of steady change in the engine rooms and boiler rooms of oceangoing ships, but on the other hand progress had reached a point where a certain standard outfit of propulsion machinery could be recognized. In spite of a looming challenge from the turbine, the engine was the steam reciprocating machine that would still be built into a few ships as late as World War II. Some Liberty ships, such as the *Jeremiah O'Brien* and the *John Brown* were so equipped due to the shortage of reduction gears. In spite of a similar challenge from the watertube concept, the boiler was the "Scotch" (or "cylindrical") fire-tube boiler, burning coal. That device would yield its place of dominance a few years sooner than the reciprocating engine, but would nonetheless be the near-standard for perhaps another two decades. Both of these machines had evolved from more primitive beginnings early in the century, but by 1900 were almost beyond further perfection.

The reciprocating steam engine had begun as a crude piston-in-cylinder machine that was fed by low-pressure, saturated, steam.[1] The piston moved slowly—ponderously, even—though at a speed quite adequate for turning the typical slow-turning paddle wheel. It was nonetheless a miracle machine for its time, since it drove vessels at decent speeds with complete independence from the fickle sea winds, and from daunting currents as well. Far from the domain where wind had ruled, the early engine moved vessels up the Mississippi at usable speeds relative to the shore in spite of an ever hostile current. Of course, viewed from a later vantage this engine is seen to have several inadequacies: it was very heavy and bulky for the amount of power it produced, and its efficiency was so low that it could never drive a ship very far from a source of firewood.

The early steam engine was nonetheless a manifestation of the Industrial Revolution that not only spawned machines, but spawned machines by which more and better machines could be built, and attracted the attention of scholars who were eager to study the principles by which the machines operated. When one thinks of the theoretical bases underlying the steam engine (and other engine types as well), the name Sadi Carnot emerges from the mist of two centuries gone. The great principle of this French scientist states that the fraction of the energy possessed by a thermal working fluid (i.e., the steam) that can be converted to mechanical work is a function of the difference between the throttle temperature and the exhaust temperature. The simple formula, to be found in an early chapter of every thermodynamics book, is $\eta = (T_H - T_C)/T_H$ where η is the fraction of the energy extractable from steam at its tem-

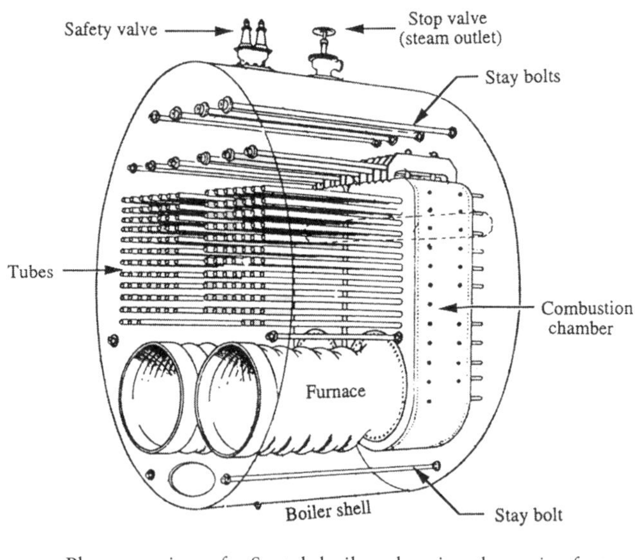

FIGURE 5-1 Phantom view of a Scotch boiler, showing the major features of its internal construction. (*Original source unknown*)

perature T_H as it cools to a temperature T_C (temperatures on an absolute scale are implied). In practice, this meant supplying steam at the highest possible pressure, and exhausting it from the engine at the lowest possible pressure. In the case of a marine engine, that pressure is the steam saturation pressure at seawater temperature.

Supplying steam at high pressure was more easily demanded than done, since that steam had to be produced by a boiler that had to go through its own struggles of evolution toward being safe and efficient. By 1900 typical pressures found aboard first-line ships were in the 200 to 250 pounds per square inch range, and temperature was that of saturation (i.e., not superheated beyond its evaporating temperature). For the 200 psi (gage pressure), the temperature was 388°F. The low T_C was pursued by exhausting the steam into a condenser operating under a vacuum, and cooled by seawater. A condenser of reasonable size could cool the exhausting steam to within 25°F of seawater temperature, which means to a pressure roughly 1.0 pound per square inch absolute. However, the very high specific volume of steam at such a low pressure usu-

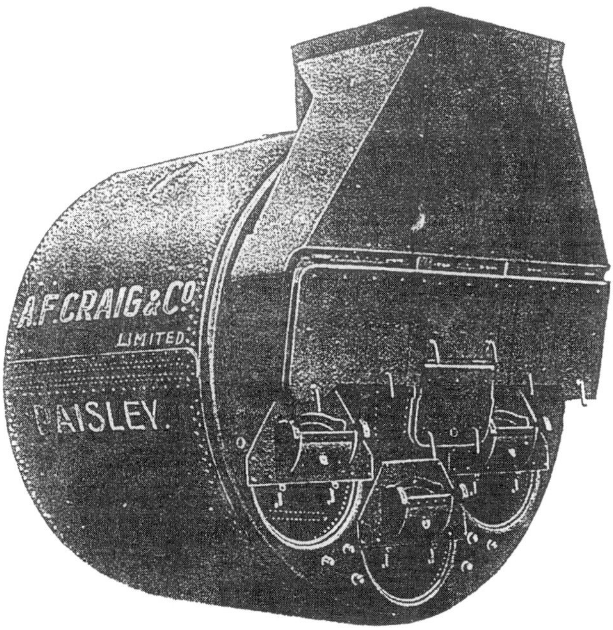

FIGURE 5-2 A marine Scotch boiler, from a 1905 advertisement by its builder. A particular feature of this picture is the smokebox, the chamber just above the furnace, that is not shown in the other boiler illustration.

One of the 29 cylindrical boilers is shown being lowered into the *Britannic* while she was moored to the outfitting quay at Harland & Wolff. This is the area of the shipyard where the superstructure, funnels, and much of the internal equipment were placed. The large floating crane was one of the items purchased by the shipyard for the outfitting of these ships. (*Courtesy of the Mariners Museum*)

ally demanded that the cylinder be of extreme dimension; actual exhaust pressures were, therefore, usually several psi higher than that desirable minimum.

The gradual increase in throttle pressures became noticeable when the screw propeller began to replace the paddlewheel on the oceans in about 1840. Now, the efficient speed of a paddle wheel is quite low, while the efficient speed of a propeller to do the same job is roughly an order of magnitude higher. The early paddle wheel engines, slow-turning, were thus poorly suited to turn a propeller. Gearing to step up to shaft speed was used in an attempt to match the engine to the new propulsor, but a more reasonable expedient was to build engines for higher speeds, engines that were inevitably smaller, and hence better suited for higher pressure. Better design and operational understanding had meanwhile allowed higher boiler pressures, and the lessons of Carnot had begun to reach the engineering level. These factors combined to send pressures upward.

The combined advent of propellers and higher pressure thus had the great benefit of making sea-going machinery far less bulky, and thereby bringing the benefit of a greater volume and weight that could be assigned to cargo.[2]

The higher the initial pressure in a steam cylinder, the more the fluid cools as it expands toward exhaust pressure. The exhausting steam in turn cools the cylinder walls, and the walls in turn cool the next charge of steam even before it begins its expansion. Energy lost to cooling is energy no longer available to do mechanical work, and as marine steam pressures increased, this phenomenon was recognized as an ever more significant detriment to good efficiency. The remedy that began to appear in the 1850s, and that became common practice aboard seagoing ships in the 1870s, is compounding: dividing the expansion between two cylinders through which the steam flows in series. Since each cylinder experiences only half of the total expansion, its walls remain much closer in temperature to that of the incoming steam. Efficiency of the machinery plant is thus increased, thereby reducing the consumption of coal for a given distance traveled. Some students of maritime history aver that it was the widespread application of compound engines—not merely the use of steam propulsion—that doomed the sailing ship as a transoceanic cargo carrier. With a compound engine the space and weight saved from coal and given to cargo carriage was enough to overcome the sailers' advantage of the wind's "free" propulsion.

If two cylinders in series are good, three in series are better. Thus it was that triple-expansion engines soon displaced the compound engine, and became the standard for the seagoing world by 1900. Indeed, the engines that were mass-produced in the 1940s for that decade's many war-built Liberty ships were triple-expansion reciprocating steam engines. They were not significantly different from the typical engines of 1900.

Now, if three cylinders in series are better than two, then four should be better than three. They are, and where efficiency of machinery was of high importance, quadruple-expansion engines were fitted to ships. They constituted a distinct minority, but we'll include them by saying that "multiple-expansion" reciprocating engines, matured to a point near their ultimate form, were the near-standard for oceangoing ship propulsion at the end of steam's first century. Have there ever been five or more expansions? Not ever, since progression beyond four cylinders required just too much mechanical complexity.

Meanwhile, the boiler is the essential partner of any steam engine. Of course, the boiling of water was a commonplace task for several millennia before the first practical engine was built. The capturing of the steam in a closed vessel, and leading it through a pipe to an engine, was no great challenge, but when the pressure began to be raised above atmospheric, severe problems arose. One of crucial significance: a vessel containing steam under high pressure, and with its surfaces being buffeted by hot combustion gases, can explode with devastating consequences. By 1900, the application of safety valves, a good understanding of what circumstances led to explosion, and safety inspections required by law, had reduced the problem to insignificant magnitude. However, the unusual circumstances of a sinking ship might negate all care in design and operation. The possibility of a boiler explosion as a ship sank was, therefore, ever a worry in a marine disaster, and we shall have to speak of it when we describe the violent ends of *Titanic* and *Lusitania,* the two ships of our interest in which boiler explosions were suspected.

Like the engine, the boiler is characterized by efficiency. In this instance, that property is the fraction of energy released by the burning fuel that is transferred into the water that is to become steam. The essence of design for efficiency is to maximize the heat transfer surface through which heat passes from combustion gases to the water, and to minimize the surfaces through which heat can pass into the surroundings. Many competing schemes to these ends were applied during the 19th century. However, by century's end the cylindrical boiler,[3] so-called because its shell was a cylinder and its furnaces were smaller cylinders within the larger (see Figure 5-1 on page 40 and Figure 5-2 on page 41), was well understood by both designers and by those who tended it in the stokeholds of ships. Its efficiency was about as high as can be expected of a boiler.[4] The watertube boiler loomed large on the horizon, and although its advantages were eventually to prove overwhelming, that event was still far away in 1900. All of the ships being treated in this volume were fitted with the cylindrical boiler.

They also were fitted with surface-type exhaust steam condensers. Condensing exhaust steam in a closed vessel—an alternative to allowing its exhaust to atmosphere—has the great benefit to propulsion plant efficiency of allowing exhaust pressure to be set below atmospheric. Carnot's T_C can thus be roughly 100°F instead of 212°F (for formula use we would express these on the absolute temperature scale), with a major increase in the amount of work done by a given amount of steam. If the condensing vessel keeps the steam isolated from the cooling water (as implied by the word "surface" used earlier), then sea-salt contamination can be kept out of boilers, pipes, and engines, all to the great benefit of their internal condition. Throughout their era, the steamers of the Mississippi continued to feed their boilers with river water, and to discharge exhaust steam to atmosphere. Even into the 20th century, steamers on the

Great Lakes of North America condensed their exhaust steam by mixing it with lake water. On the broad oceans of the world, however, surface condensers operating under vacuum were nearly universal by the late 19th century.

The propulsor—the device that converts the torque of a rotating shaft into propulsive thrust—for oceangoing vessels was almost universally the screw propeller in the late 19th century, but compared to engines and boilers, it had undergone comparatively little development in the 50 to 60 years since it became commonplace. The relationships of diameter and pitch to power and speed of the ship were understood, but the refinements that could increase its efficiency, and mitigate the phenomena of cavitation and excitation of vibration, were to come only in the 20th century.[5]

The near universal fuel was coal, so much so that the words "coal" and "fuel" were usually treated as synonyms. A century later, coal has become extremely rare as a marine fuel, and we view it with distaste as dirty, dangerous to mine, and a polluter of the atmosphere. In 1900, however, its soon-to-be competitor—petroleum and its derivatives—was still a novelty. Petroleum was expensive, lacking the infrastructure of supply, and with the technology of its application undeveloped.[6] Petroleum fuel would ultimately sweep the seas just as turbines and watertube boilers would. However, the transatlantic liners were to remain coal-burners until after World War I. By 1923 the *Mauretania* and *Olympic* were converted to oil-burning boilers with an exceptional savings in manpower.

Whatever the fuel, the amount burned per unit of time for a given power is a significant factor in the design of a ship. Even if the fuel is of insignificant cost, its weight and volume burden the vessel in several ways. At a time (1840, say) when steam propulsion was well established, but had not begun the technical advances of higher pressure, condensing, and compounding, coal consumption was in the range of 7 to 8 pounds per hour per indicated horsepower;[7] by 1900 the value of this important parameter was near 1.0. (Advances in steam propulsion during the next century were to cut this figure by a factor of two; diesel propulsion would cut it by a factor of almost three.)

To summarize at mid-chapter: as the decade of the great liners began, a high-level plateau of propulsion technology had been reached. It offered machinery that could drive those ships with technical and economic success on the North Atlantic. The boilers were universally of the cylindrical (or "Scotch" type), tended by gangs of sweating humans who fed the furnaces with coal, and raked out the ash and clinker. Steam pressure was typically in the neighborhood of 200 pounds per square inch, with no superheating. The engines were steam reciprocators, triple or quadruple expansion, exhausting their steam to surface condensers at pressures in the neighborhood of 5 pounds per square inch absolute. These components were manifestations of a mature technology that would serve for roughly another half century while gradually being replaced by distinctly different machinery components that would soon prove their superiority. Although their time had not come, they were already on the scene in 1900.

One of the innovations, the diesel engine, could even be called a radical innovation in that it lay beyond the realm of steam technology. The diesel would eventually come to dominate marine propulsion, but it plays no part in our story, for it needed another half-century before its dominance was to begin.

Other innovations, all lying within the steam realm, might have been applied with some consequence to the Cunard and White Star liners. One of them—the steam turbine—appeared in all of these ships. Since it did, and ranks as a major innovation, it is the principal subject of two later chapters. We'll say no more about it in this chapter.

Two other major innovations—petroleum fuel and the watertube boiler—might have been applied, but were not. We'll give them some discussion in this and the following chapter with the objective of illuminating the whys and why-nots of the choices made in their cases.

Liquid petroleum fuel (usually designated by the simple term "oil") offers overwhelming advantages over coal. As has been noted briefly, those advantages had been recognized well before 1900. On the other hand, the technology of burning it in a marine furnace had not matured at that time, and the infrastructure that was necessary to make it a readily available fuel on the North Atlantic was still years away. Nonetheless, by the time the last of our five ships was launched, examples of oil firing were ubiquitous and numerous. An estimate[8] published in 1907 put the number of oil-fired steamers, world wide, in the 1200 to 1500 range. Examples are easy to find. Two mid-sized oil-fueled passenger liners were built in Japan in 1905, *Tenyo Maru* and *Chiyo Maru*.[9] These were twin-screw, turbine-engined transpacific liners of 13,454 gross tons, with about 20,000 horsepower. The well-remembered American coastwise liners[10] *Harvard* and *Yale*, entering service between Boston and New York in 1907, were converted to oil fuel in 1910. On the West Coast of America the thoroughly utilitarian steam schooners that brought lumber from the Northwest to California were often oil-fired. Even in the United Kingdom, a fortress of coal in these years, oil-burning ships were being built for foreign operators. An example is the Canadian Pacific Railway steamer *Princess Alice*[11] (a modest-sized short-trip passenger steamer) built in 1911 by Swan Hunter on the Tyne, the yard that had built the *Mauretania*.

Oil fuel has several distinct advantages over coal. Its heating value (British thermal units per pound) is at least 30% higher than that of even the best coal, a difference that readily translates into 30% less fuel to be carried for a particular voyage. Since coal comes in lumps, a bunker in which it lies is holding a lot of interstitial air; that, with the heating-value advantage, leads to the volume of oil-carrying bunkers being something like 50% smaller than that of coal bunkers. Further, tankage for a liquid can occupy small and remote places that would never do for a solid. The space between inner and outer bottoms of the hull is a common example of space that is used for carrying oil fuel, but which would be near impossible for the carriage of coal.

Oil is, of course, a liquid, and hence so much easier than a lumpy solid to load aboard a ship, to move from bunker to boiler, and to introduce into a furnace. With respect to loading fuel, it is interesting to note that after the *Olympic* was converted to oil in 1920,[12] taking aboard a full load of fuel was reduced to six hours from the previous 4½ *days*.

When a piping system brings the fuel from tanks to furnace front, and another pump sprays it into the fire through a nozzle, the jobs of several hundred firemen and trimmers on a ship of the size we are treating are eliminated. When the fuel is a liquid, control of the fuel and air flows is easier, and is more easily accomplished, all with good matching of the ratio of air to fuel, and with good matching of fuel flow to the steam demand of the engines.

The door to a hand-fired coal furnace is continually being opened, letting in a gust of cold air at each opening. That air is inevitably heated as it passes through the furnace, all to the wastage of heat that should have gone into the making of steam.

These differences are so large and so significant that a rapid and universal switch from coal to oil might have occurred during the first decade of the 20th century, but there were several factors that obstructed a widespread change-over until after World War I. The first was price. A comparison published in 1909 cites a price of 60 shillings per ton in Britain, a price of 10 to 20 shillings for coal. That was the overriding factor for the North Atlantic. Meanwhile petroleum was being pumped from the ground in California in the face of a rather modest demand. Oil fuel was, therefore, quite cheap, while the distance and the great mountains of the West lay between California and large sources of coal. This circumstance largely accounts for the variety of the examples of oil-fired ships just cited. The conversion of *Harvard* and *Yale* to oil fueling occurred after a change of ownership in 1910 took them to the California coast.

If price was the major strength of coal in the Atlantic, the presence worldwide of its mature infrastructure was also a factor in the success of its rear-guard actions. Just as the absence of coaling stations along most of the world's shorelines was sustaining the fading sail fleet 50 years earlier, now the scarcity of oil-fueling ports was retarding the spread of the new fuel.

As it was, the designers of the Cunard liners almost certainly gave no consideration to the petroleum option for *Lusitania/Mauretania*. Four or five years later, the White Star designers may have considered that option for *Olympic* class, or at least for the second or third ship of that class. Cunard may have considered it for the *Aquitania*, the last of its pre-war ships. But no, all of these ships were to burn coal, though the ones that survived World War I were converted to oil before re-entering peacetime service.

So the discussion of oil fuel is a dead end—except for the "what if" speculations it may provoke. Would the distinctly shorter fueling times allowed by oil have placed the *Titanic* somewhere else on the night of the iceberg? The *Lusitania* somewhere else on the day of the submarine? (Or maybe *Olympic* or *Mauretania* would have been the ships to be at the wrong places at the right times instead?) Or would the wing fuel tanks so typical of later-day liners have saved either *Titanic* or *Lusitania* from the fatal consequences of their encounters?

Notes for Chapter 5

1. The first recognized use of steam propulsion occurred in Scotland in 1788. The first Atlantic crossing under steam occurred in 1838. In recognition of the 100th anniversary of the first event, and the 50th anniversary of the second, a commemorative paper was published by the Institution of Naval Architects in 1888: Professor Henry Dyer, 1888, "The First Century of the Marine Engine," *Transactions*, Institution of Naval Architects, Volume 29. His paper serves as a major source for this chapter.
2. The link between higher pressure and higher engine speed can be demonstrated with the aid of two simple formulas. The formula for power of a reciprocating engine is W = P L A N, where W is the power, P the mean pressure in the cylinder, L the piston stroke length, A the piston face area, and N the number of power strokes per unit of time. Piston speed, always a limiting parameter for any reciprocating engine, is proportional to the product N L. Exercise for the reader: double the pressure while keeping power and piston speed unchanged, and then show how an increase in speed N can follow.

3. The cylindrical boiler is often called the *Scotch* boiler. That name is sometimes not capitalized—scotch boiler—just as the proper noun *Diesel* becomes diesel when designating a type of engine.
4. Boilers are often fitted with heat-recovery devices such as economizers (to preheat the incoming water) and air heaters (to preheat the incoming combustion air), and since those devices reduce the heat being discharged to the atmosphere, they are properly regarded as contributing to boiler efficiency. Since they were less commonly used with the cylindrical boilers of 1900, the marine boilers of the late 20th century were usually reported as being more efficient than their predecessors. The boilers themselves, however, were not much different in that very significant parameter.
5. Speeds and powers of ships had not reached the levels that made vibration and cavitation problems of concern. Efficiency—the fraction of received torque power converted to thrust power—was recognized as important under almost any circumstance, but sophisticated design can add very little to a propeller that has correct size and pitch. Professor Dyer (see Note 1) quotes several authorities of his era as making declarations to this effect.
6. The potential of liquid petroleum as a marine fuel attracted the notice of marine engineers as early as the 1860s; papers on its possible use were published by the Institution of Naval Architects in 1868, 1869, 1870. An example is Jasper H. Selwyn (captain, Royal Navy), 1870, "On the Progress of Liquid Fuel," *Transactions*, Institution of Naval Architects, Volume 11.
7. The early fuel-consumption figures come from estimates by Professor Dyer, Note 1.
8. The estimated number of oil-fired ships appears in the doctor-ingenieur thesis of Ernst Foerster, 1907. It is summarized in *Engineering*, 1902, Volume 84.
9. The *Tenyo Maru* and *Chiyo Maru* are described in *Engineering*, 1911, Volume 92.
10. Information about the *Harvard* and *Yale* comes from David Crockett and Edwin Dunbaugh, 1997, *Eastern Steamship*, The Steamship Historical Society of America.
11. The *Princess Alice* is described in *Engineering*, 1902, Volume 74.
12. The post-war refitting of the *Olympic* is described in *Engineering*, 1920, Volume 109.

CHAPTER 6

The Watertube Boiler

The watertube boiler eventually swept away that entrenched standard of seas, the cylindrical or Scotch boiler that was the near-universal favorite in 1900. Even the Liberty Ships of World War II, the last stronghold of the reciprocating steam engine, were fitted with watertube boilers. In 1900, however, the eventual triumph of the watertube concept was still far over the horizon, beyond the immense turmoils of World War I. Nonetheless, it was already in the process of becoming dominant in naval construction, where its merits were more important than in commercial work. And even in the commercial sphere the watertube innovation was beginning to find applications. For example, the notable American transpacific steamers *Minnesota* and *Dakota*, placed under building contract in 1900, were fitted with watertube boilers.[1]

What kind of boiler was in the five ships that are of interest here? No watertube boilers were installed in any of the five; their boilers were 100% of the tried and trusted cylindrical fire-tube type. Why, then, is this chapter here? The essence of the answer is this: in a step that might be called revolutionary, the designers of the Cunard liners *Lusitania* and *Mauretania* chose turbine engines for those ships, leading to the thought that they might have chosen watertube boilers as well. Speculation on why they did not appear in a later chapter; this chapter fills in a background for that material.

As is usually the case when an established concept is confronted by an innovative challenger, points for and against the new concept are argued in private and in publications. So it was as the watertube concept was promoted against the traditional cylindrical boiler, for relative advantages could be cited for each.[2,3] However, in this instance there was a factor whose dominating importance was not recognized in 1900: the potential of watertube boilers to operate at extremely high pressures—higher than any marine engineer would ever want—in contrast to the limited ability of the cylindrical design in this respect.

Boiler pressures of the era were rarely above 200 pounds per square inch, and the temperatures were those of saturated steam. While the educated practitioners of marine design knew that the Carnot principle demanded higher temperatures, they also knew that higher temperatures would present grave difficulties for the designers of reciprocating engines. Higher pressures that necessarily accompanied higher temperature would also present grave difficulties for the designers of cylindrical boilers. What use, then, of higher pressure (they might ask) if we can't build engines and boilers to take the advantages offered?

The turbine was soon to liberate engines from pressure and temperature limitations, but in their early applications those innovative machines were furnished with the same steam that was common to their reciprocating contemporaries. They could not begin their advance while still linked to the cylindrical boiler because that type was severely limited in the pressure it could withstand.

The key to the cylindrical boiler's pressure ceiling lies in that very word: cylindrical. The shell is cylindrical, and because of the necessity for furnaces and tubes to penetrate at least one end of the cylinder, that end must be flat, and with one end flat, the other end may as well be flat also. Internal pressure is a severe test for the flat ends of any cylinder. To allow the popular cylindrical boiler to withstand even modest pressures, reinforcing staybolts that ran from end to end had to be fitted. This expedient, in effect, used the internal pressure on one end to resist the internal pressure on the other end. Trouble was, the higher the pressure, the more numerous the staybolts, and at some point the bolts began to compete for space with the boiler tubes.

The furnaces of cylindrical boilers were also cylindrical, since they must have that shape to resist the inward force placed upon them by the surrounding water. Although they had no ends for staybolts to reinforce, the inward pressure on the cylindrical surface was hazard enough, and hence that pressure had to be limited.

Watertube boilers (See Figure 5-1) have their cylindrical surfaces as well: the boiler tubes and the drums wherein water and steam are separated, but the pressures are internal, and even the drum is of much smaller diameter than the shell of a cylindrical boiler. In addition, the ends of the drum can be hemispherical because no penetrations of tubes or furnaces

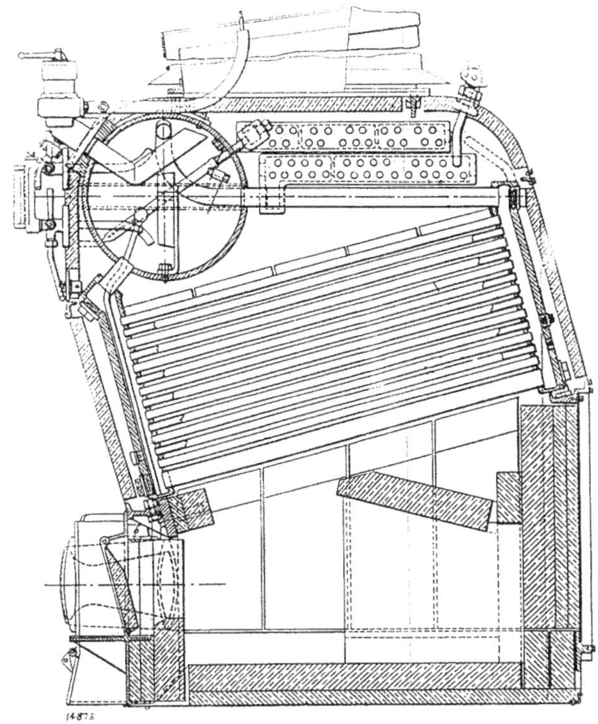

FIGURE 6-1 Cross-sectional view of an oil-fired, water-tube marine boiler, circa 1910. The bank of slanting water tubes is quite evident. Closer inspection shows the headers and pipes connecting the tubes to the drum. (*From an early edition of* Steam, *a publication of Babcock & Wilcox*)

demand flatness. True, some of the early watertube designs featured the convenience of "headers," which were rectangular-shaped conduits that connected groups of tubes to the drum. These placed a fairly low limit on the boiler's pressure. Headers allowed all tubes to be straight, and handholes in their sides opposite to the tube connections could be opened to give access to the tubes for internal cleaning. In later days, the use of artfully bent tubes allowed them all to connect directly to drums, and advances in water chemistry obviated the need for mechanical cleaning of tube interiors. Headers were no longer necessary. Pressure limits may indeed have remained for the steamships of the late 20th century, but those limits were not imposed by their boilers.

The other major advantage of the watertube concept—and one that *was* recognized in 1900—is its reduced weight. With its pressures all being internal, pressure walls can be thinner, and there is the blessed absence of staybolts. On the other hand, since the furnaces were not submerged in the boiler water as in the cylindrical, the watertube furnaces were usually lined with firebrick, a negative factor so far as any weight advantage was concerned. However, the weight of water contained within the boiler was the dominant factor. With the water being inside small-diameter tubes rather than filling all of the space outside the tubes, the wet weight of a watertube boiler was significantly lower than the weight of a comparable cylindrical boiler. Table 6–1 shows typical comparative weights for Scotch and watertube boilers.[4] The data tabulated in Table 6–1 are merely a comparative illustration; the numbers do not come from the ships in this study.

Such a distinct difference in weight was especially attractive to naval designers, and well before 1900 watertube boilers of various designs were common in fast naval vessels such as torpedo boats. They were being used in steam yachts, and a few instances of application to commercial ships can be cited. For example, boilers of the watertube Belleville design powered the passenger ships *North Land* and *North West,* built in the early 1890s, on the Great Lakes. The notable transpacific American freight/passenger liners *Minnesota* and *Dakota* were mentioned early in this chapter.

Where coal persisted as a fuel well into the age of petroleum, as it did on the Great Lakes of North America, its firing was often accomplished by mechanical stokers when the boilers were of the watertube design. In 1900 mechanical stokers were coming into use ashore, and a few marine applications had been tried.[5] A certain imperative drove the trials: the difficulty with hand firing of getting a satisfactory thickness of fuel bed, and getting it uniformly over the entire furnace. Regardless of the type of boiler, a thin fire allowed an excess of air to pass through the combustion zone; a thick fire restricted the air flow to the point where combustion was incomplete. Even one who has never stoked a coal fire can probably picture that proper thickness of a coal bed, and its uniformity, is easier for a shoveling human to maintain in a small furnace than

TABLE 6–1 *Comparison of a Scotch Boiler to a Watertube Boiler*

	Scotch Boiler	Watertube Boiler Babcock & Wilcox Type
Dry Weight (Tons)	564	380
Weight of Water (Tons)	276	77
Total Weight (Tons)	840	457

in a large one. Now, the boilers of the Cunard ships had four furnaces each (three each for the White Star boilers), whereas comparable watertube boilers would have had but one furnace each. And that one would have been about four times as large as the cylindrical furnaces. Given the likely unavailability of stokers with watertube experience—and given the still nascent state of mechanical stoking—this factor was probably a strong negative in any proposal to use watertube boilers where strong incentives for their use was absent.

The steam turbine made possible (and desirable) the use of steam at pressures far beyond the capability of the cylindrical boiler. The watertube boiler thus was to become essential for ships built in the mid-20th century and beyond; it could be built for pressures far higher than any marine engineer could desire, while the cylindrical boiler could not be taken much beyond the pressures it had reached by 1900. At the time of our interest, however, this great benefit could not be applied, leaving the great appeal of the watertube concept to be its comparatively light weight. That virtue would be of significant advantage to a commercial ship that would be loaded to its marks by dense cargo, but it meant very little to a passenger ship. It is also even more significant in the naval domain, and it was there that the marine watertube boiler developed over several decades before it became common in the commercial field.[6]

The renowned and revolutionary battleship of the Royal Navy, *Dreadnought*, whose keel was laid in October 1905, was fitted with 18 watertube boilers of the Babcock & Wilcox design, the type favored by The Committee on Naval Boilers.[7] The ship was at sea little more than a year later, roughly a half-year before the *Lusitania*, the first of the Cunard ships, was launched. The *Dreadnought* was the naval pioneer in big-ship turbine propulsion; the *Lusitania/Mauretania* pair was similarly the commercial pioneer in big-ship turbine propulsion. On the other hand, Cunard opted for cylindrical boilers.

Because design and building periods overlapped, and because the Royal Navy had leverage with Cunard through its provision of subsidy money for the new liners, we might expect that two decisions for turbine propulsion were actually one decision. On the other hand, the two entities disagreed on choice of boilers. Since *Lusitania* and *Mauretania* were eminently successful without a radical boiler plant, we must admit that the disagreement was justified.

The turbines were almost imperative if the Cunard ships were to reach their contract speed, while watertube boilers were not. In addition, what was good for a navy may not have been good for a commercial fleet; a navy ship spends most of its time in port, where leisurely servicing and repair may be done, while its transatlantic counterpart is forever harried to maintain its published schedule. In its early years, before the necessary water chemistry was developed, the watertube boiler did experience a heavy burden of tube cleaning and replacement. Whatever the cause, the watertube boiler would not have its day in the transatlantic fleet until after World War I.

Then there is the question of boiler explosions. The watertube boiler is much less likely to suffer an explosive failure than its fire-tube counterpart, and because it contains much less water, an explosive pressure release will contain much less destructive energy. Of course, since no watertube boilers were fitted aboard the ships in question, comparative points such as these are not really pertinent. The truly pertinent question is whether the second explosion that seemingly sealed the fate of the *Lusitania* was steam related. A similar question accompanies the *Titanic* event: were the loud rupturing noises accompanying its final moment afloat the sounds of exploding boilers? We leave those questions to exploration in other chapters.

Mechanical draft (supply of air to the boilers) was an innovation circa 1900 that seemed minor when compared to the coming dominance of turbines, watertube boilers, and petroleum fuel. Eventually, it became as universal as those other innovations in steam propulsion technology. It might not be regarded as important as those first three, but it nonetheless filled an important role. From the beginning, and for most of steam's first century, natural draft—flow of air to boilers induced by heated air rising through tall stacks—supported the combustion in the boiler furnaces. During those early days, however, engineers recognized that even with the fuel supply depending on the exertions of human stokers, the rate at which air could be delivered to the furnaces was often the factor limiting the output of boilers. Forcing air into the furnaces by mechanical means (i.e., by fans) was in such cases a means of increasing boiler output. This expedient was not a difficult concept to effect, but what would happen when a fireman opened a furnace door? The pressure inside the furnace would be higher than atmospheric, and out would shoot combustion gases, with serious consequences for the fireman. Induced draft (i.e., drawing the air through the furnace by a fan in the uptake) was a solution, since it put the furnace below atmospheric pressure. That alternative was used, but it put the fans into a very hot and testing environment, a circumstance that promoted mechanical problems. Pumping the air into the furnaces was to be preferred.

English marine engineer James Howden[8] investigated the benefits of mechanical draft in the 1880s, and devised a satisfactory mechanism for automatically shutting off the flow of air into a furnace when its door was opened. The "Howden system" of forced (i.e., pumped in) draft was soon accepted by his contemporaries, and within a few years seemed to have lost its aura of being an innovation. The Cunard designers, doubtless identifying it as a means of pushing their boiler output to the level needed in recapturing the speed prize, applied it to the *Lusitania* and *Mauretania*. However, a few years later their White Star counterparts, not striving for extreme speed, were satisfied with natural draft for their *Olympic* ships.

Notes for Chapter 6

1. The *Minnesota* and *Dakota* are described in great detail by Charles R. Hanscom, 1903, "Description of the Design and Building of the 21,000 Ton Steamships *Minnesota* and *Dakota*," *Transactions*, Society of Naval Architects and Marine Engineers, Volume 11, New York.
2. A very thorough explication of relative merits was published by W. A. Fairburn, 1902, "The Water-Tube Boiler in the American Merchant Marine," *Transactions*, Society of Naval Architects and Marine Engineers, Volume 10, New York.
3. A lengthy and thorough testing of relative merits for naval applications was made by a Royal Navy committee in 1900 through 1903, based on shipboard trials of cylindrical boilers and of several types of watertube boilers. The report of The Committee on Naval Boilers (its formal title) was published in full in *Engineering*, Volume 78, August 5, 1904, and interim findings appeared in earlier numbers of that publication. *Engineering* is the source used here.
4. Boiler weights cited are taken from W. A. Fairburn. See Note 2.
5. An example is the American transpacific steamship *Dakota*, commissioned in 1905. Four of its 16 watertube boilers were fitted with mechanical stokers as an experiment. Their owners' intention was to gain experience with this novelty before converting the remaining 12 boilers to stoker firing. The results were discouraging, and in consequence the conversion was not made, and indeed, the experimental outfit was soon abandoned.
6. Circa 1900 a marine engineer interested in applying the watertube concept was forced to choose among many competing designs, several of which were popular, yet doomed to oblivion within a few years. Two French designs, the *Belleville* and the *Niclausse* (both named after their developers) were used by the U.S. and Royal Navies, and by a few non-French commercial ships as well. Both types proved to be troublesome to operate and maintain in service, and by 1910 were finding favor only in their homeland.
7. Boiler Committee. See Note 3.
8. See James Howden, 1884, "On Combustion of Fuel in Furnaces of Steam Boilers," *Transactions*, Institution of Naval Architects, Volume 25.

PART II

The Ships

The two great Cunard liners, *Lusitania* and *Mauretania*, went to sea in 1907, and immediately established themselves as the fastest ships on the North Atlantic. One was sunk 1915 in consequence of an encounter with a German submarine. The other survived until 1935 as a premier passenger-carrier on the North Atlantic. The *Mauretania* held the "Blue Riband" as the fastest passenger ship on the North Atlantic for 20 years until it was wrested away by the German liner *Europa* in 1927. Three great White Star liners, *Olympic*, *Titanic*, and *Britannic*, went to sea in 1911, 1912, and 1915, respectively. Two were sunk, one by collision with an iceberg, the other by contact with a German mine. The survivor of the three had a long and distinguished service in competition with the *Mauretania*, until replaced by the new liner *Queen Mary* in 1936. The *Olympic* was known as the "Old Reliable."

In Part II we say little about the events that conferred such great interest upon these ships. Here the story is simply descriptive: first in Chapters 7 and 8 the ships, generally; then in Chapters 9 and 10 discussions of their pioneering propulsion machinery—steam turbine for the Cunard ships, compounded reciprocating-plus-turbine for the White Star ships.

The *Mauretania* is shown with her hull painted white in New York in June 1935 near the end of her career. Shortly after this picture was taken, she sailed to Southampton for the last time and was laid up before being sold for scrap.

CHAPTER 7

The Lusitania *and* Mauretania[1]

In 1839 the British government offered a subvention (contractual payment) for the steamship carriage of transatlantic mail. Samual Cunard, a Canadian immigrant in his early forties, won the contract, and in so doing founded the bellwether transatlantic passenger and mail line that bears his name. His winning ploy was to negotiate first among the Clydeside ship and engine builders for suitable ships, then to present to the government the soon-to-be fait accompli: three ships designed for the desired service that would be ready to begin that service in 1840. Influential friends and contacts helped, and after the mail contract was signed in May 1839, the necessary start-up capital was easy to attract.

The three ships were built and ready as contracted. The first ship, the *Britannia*, sailed 4 July 1840 from Liverpool, reached Halifax on the 17th, and completed her westward passage to Boston the next day.

Given the many investors who participated, the company was never "owned" by Samual Cunard or his descendents. In the 1840s, for example, there were 33 partners; Cunard's share was £61,000 out of a total capital of £300,000. In 1880 the company was incorporated as The Cunard Steam Ship Co., Ltd. A second generation of the Cunard family (two brothers) held one third of the shares; the Burns and McIver families, who had been prominent partners since the beginning, each held one third. During the early 20th century days when the building of *Lusitania/Mauretania* was being negotiated and undertaken, the corporation chairman was Lord Inverclyde (a Burns).

Cunard was such a pioneer in developing transatlantic navigation that some of the measures it introduced became maritime doctrine. For example, Cunard developed the now universal lighting scheme for ships: a green light to designate starboard, a red light for port, white for range and stern lights. It established "safe tracks" across the Atlantic to separate westbound and eastbound traffic. On the other hand, until its early 20th century ships, it was not a leader in ship design. It was still building wooden ships when its competitors had turned to iron; it was late in advancing from paddlewheels to screw propellers; it favored simple engines for a while even after compound engines had demonstrated their merit. Of course, pioneering often has its price, and Cunard may have benefited for a while from letting others take the risks. Competition forced it to at least go along, and in the early 20th century, to leap ahead.

By 1900, that competition had intensified, particularly from the nascent German program to capture the Atlantic immigrant (in particular) trade with ships that were bigger and faster than any Cunarder. For example, the *Kaiser Wilhelm II* of North German Lloyd, 20,000 GRT and capable of at least 23 knots outranked existing Cunarders in those important characteristics.

Coal rose sharply in price early in the new century, but several alternatives offered possibly offsetting advantages. Petroleum fuel could bring major benefits, even though its price per ton was still much higher than that of coal. Superheating of steam would improve thermal efficiency, though its application was still in the experimental stage. Turbine engines had arrived, and were apparent successes in cross-channel ferry service. Watertube boilers were becoming dominant in naval service.

Whatever decisions might arise from the pressures of competition and the opportunities of new technology, Cunard needed new ships that were bigger and faster than their predecessors. Moreover, the Germans had now replaced France as Great Britain's rival at sea and had introduced luxurious, technically advanced, and fast liner service on the North Atlantic. Two of their ships, the *Kaiser Wilhelm der Grosse* and *Deutschland* captured the Blue Riband, symbolic for the fastest ship afloat. However, the financial resources of Cunard did not seem adequate to compete with this new threat to British supremacy on the Atlantic.

A further complication was the shadow of J. P. Morgan and his North Atlantic shipping trust, the International Mercantile Marine. The competition for the emigrant trade had become very intense by the early years of the 20th century. This American financier had monopolized much of this business, obtaining controlling or major interest in every major shipping line in the Atlantic trade, with the exception of the French Compagnie Gènèrale Transatlantique and the British Cunard Company. Morgan began negotiations to acquire controlling interest in

Cunard soon after he had obtained control of the White Star Line in early 1902. Lord Inverclyde, Cunard chairman, was alarmed, but he lacked the security of a controlling interest in the company. Nevertheless, he decided to approach the government of Prime Minister Arthur Balfour with a proposal to build two superliners that would be large, fast, and the most luxurious liners in the North Atlantic. The Admiralty, which had a vested interest in some of the Cunard ships as potential auxiliary cruisers, might help repel the American takeover. That powerful agency took up Lord Inverclyde's concerns, ostensibly because of the commitment of existing Cunard ships to the Royal Navy in return for mail subventions.

In an effort to forestall Morgan's takeover of the Cunard Company, the Admiralty proposed that the British Government provide a subsidy for the mail and emigrant trade, sufficient in size to allow Cunard to compete with Morgan's shipping empire. Negotiations were initiated between the Cunard Line and the British Government. Although the Cunard Line wanted to secure a loan repayable over 20 years at low interest, the British Government would grant the loan on the basis that the ships would be built to Admiralty specifications. In addition, Cunard had to agree that the two ships could be taken over in wartime and made into auxiliary cruisers when needed to safeguard British trade and intercept the maritime trade of an enemy. Parliament approved the

The *Lusitania* is shown entering New York harbor in 1907. Her ventilators are much smaller than her sister *Mauretania*. She was to hold the Blue Riband for a short period of time before her sister took that honor away. Her stacks are painted with the Cunard colors, red with black at the top. During World War I the funnels were all black.

money, via construction loans at low rates of interest and increased mail subventions.

The intervention of the Royal Navy generated a construction loan of £2,600,000 at a below-market rate of 2.75% and an annual subsidy of £136,000 as part of an agreement with Cunard to build the two ships with a speed of not less than 24.5 knots in moderate weather. In return, the company was to remain purely British. The two ships would seem to have justified this outlay: the *Mauretania* would set a record at 26.06 knots; the *Lusitania* would reach 25.65 knots.

They were great ships—fast, luxurious, and about as handsome as a large passenger liner could be for their times. They were popular with travelers, too, even into the age when the *Mauretania* had to compete with ships like the *Bremen*, *Europa*, and *Normandie*, ships that overshadowed the mature Cunarder.

The design of the *Lusitania* and *Mauretania* was prepared under the joint auspices of Cunard and the Admiralty. Leonard Peskett was the naval architect. A design speed of 24.5 knots was specified and the ship's structure was designed to be capable of supporting an armament of twelve 6-inch guns when used as an auxiliary cruiser. The ships were classed by Lloyds and built to Admiralty specifications. As a result, these ships were the first to be built with high tensile steel on their sides and decks. They would be the first passenger ships to be propelled by steam turbines on all *four* shafts that would develop the requisite horsepower of 68,000. The Admiralty agreed to provide construction loans at a low rate of interest. A yearly operating subsidy of £75,000 was also agreed upon. Finally, it was agreed to build the two ships to the characteristics shown in Table 7–1.

The early designs had specified a beam of 78 feet on a waterline of 760 feet, but tests at the Admiralty's tank at Haslar showed that the ship was not stable for the hull depth and beam chosen. It was agreed that the beam had to be increased to 87.5 feet to permit the luxurious accommodations that were planned by Leonard Peskett. The final design featured a rather full midship section, but with a very fine hull. To provide protection for the machinery spaces, the coal bunkers were placed outboard of a longitudinal bulkhead, despite the experience of the Cunarder *Oregon* after a collision in 1882 off Ambrose Light that had shown such a practice to be dangerous when these spaces flooded.

Publicity posters based on early designs show only three funnels. The final four-funnel design was apparently based on a desire to enhance the intended image of high speed.

TABLE 7–1 *Principal Characteristics of* Lusitania *and* Mauretania

	Lusitania	*Mauretania*
Length overall, feet	785	790
Length between perpendiculars, feet	760	760
Breadth, feet	88	88
Depth, feet-inches	60–4½	60–6
Draft, feet-inches	33–6	33–6
Gross register tonnage	31,540	31,940
Displacement tonnage	38,000	38,000
Passengers, first class	552	563
Second class	460	464
Third class	1186	1138
Boilers	Cylindrical; 23 double-ended, 2 single-ended; 192 furnaces; 158,350 sq ft heating surface; 4,048 sq ft grate area	Cylindrical; 23 double-ended, 2 single-ended; 192 furnaces; 159,000 sq ft heating surface; 4,060 sq ft grate area
Indicated HP* (total, 4 shafts)	68,000	68,000
Rated speed, knots	25	25
Rated propeller rpm	180	180

*"Indicated power" is a meaningless term in turbine technology, but the figure given here is "what the power would be if the engines were reciprocating," and is usually quoted so that the Cunard ships can be compared to their numerous competitors powered by reciprocating engines. In 1907, the most powerful of those was the North German Lloyd liner *Kronprincessin Cecilie* with 45,000 ihp. The actual output (shaft or brake power) of *Lusitania* and *Mauretania* is usually quoted as being 70,000 hp. On trials in July 1907, the *Lusitania* recorded 76,000 shaft hp at a mean propeller speed of 200 rpm, 25.7 knots, 32 ft 9 in. draft. Curves showing this and other trial data were published in *Engineering*, Volume 85, April 10, 1908.

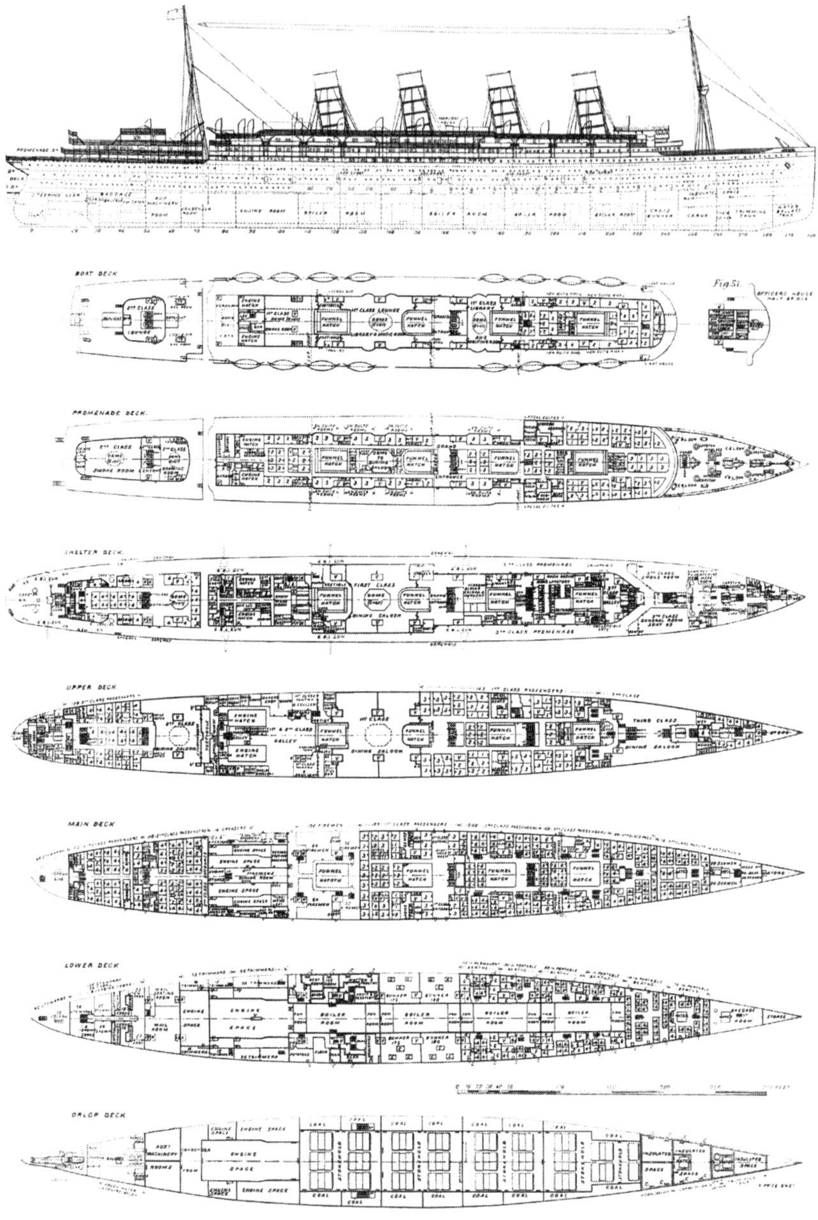

Profile and deck plans for *Mauretania* (from *Engineering*, vol 84, 1907)

FIGURE 7-1 Profile and deck plans for *Mauretania*. (*From* Engineering, *vol. 84, 1907*)

Significant dates in their careers were:

Announcement of a subsidy contract with the British government	July 1903
Laid down at John Brown, Clydebank: *Lusitania*	February 1905
Laid down at Swan, Hunter, Wallsend-on-Tyne: *Mauretania*	May 1905
Launched, *Lusitania*	June 6, 1906
Launched, *Mauretania*	September 1906
Maiden voyage begins, *Lusitania*	September 7, 1907
Maiden voyage begins, *Mauretania*	November 16, 1907
Lusitania sunk by German submarine, *U-20*	May 7, 1915
Mauretania maiden voyage, post-war	September 21, 1919
Mauretania maiden voyage after conversion to oil fuel	March 1922
Mauretania, last departure, New York	September 1934

The data of Table 7–1 show that the two Cunard ships were not identical, though they were close enough to be called "sister ships." The most evident external difference lay in a different type of ventilators mounted on Bridge Deck on either side of the funnels. The *Mauretania* was equipped with the "macaroni elbow" shaped ventilators while the *Lusitania* had a pipe type with hinged covers. A greater curvature in the forward end of the Deckhouse, a modest (approximately 20 inches) athwartship overhang of the Promenade and Boat Decks beyond the hull, and a longer forecastle for *Mauretania* can be recognized by ship *cogniscenti*. These differences increased the second ship's internal volume slightly, and hence its greater gross register tonnage.

Machinery. The propulsion machinery outfit of both ships was identical in its essential parameters. A careful examination of all particulars will show numerous differences in detail, due (especially) to the ships being built by different shipyards, and with different vendors for some of the machinery elements. None of these are significant.

The *Mauretania* is shown underway shortly after her completion at the Swan Hunter & Wigham Richardson Shipyard in Newcastle. She would hold the Blue Riband from 1907 until 1929 as the fastest ship in the Atlantic passenger trade.

The veritable heart of each ship was its boiler plant: 25 cylindrical boilers distributed through four boiler rooms, occupying approximately 42% of the hull's length (length between perpendiculars), and providing 192 furnaces to challenge the approximately 400 stokers that each ship carried[2] (192 furnaces: 23 boilers, doubled-ended, with 8 furnaces per boiler; 2 boilers, single-ended, with 4 furnaces per boiler). Save for the number of boilers and the space they occupied, the two liners were nearly identical in their boiler outfit to the three White Star ships that would follow in a few years. Two differences might be noted: the *Lusitania* and *Mauretania* boilers were fitted with eight furnaces per double-ended boiler and with Howden's forced-draft system; the boilers for the *Olympic*-class ships were fitted with six furnaces per double-ended boiler and operated on natural draft. Differences aside—or included—we are speaking of boiler plants entirely conventional for steamships of the early 20th century.

The turbine propulsion engines were definitely *not* conventional. True, they were successfully powering smaller ships such as cross-channel ferries, and were being chosen by the Royal Navy for the revolutionary battleship *Dreadnought* during the time that Cunard's powering decision was being debated by a special committee. The committee recommended the turbine machinery, but the decision was delayed until two other new Cunard ships could enter service in 1905. One, the twin-screw *Caronia*, was fitted with quadruple-expansion reciprocating engines; the triple-screw *Carmania* was fitted with Parsons steam turbines. The turbine ship not only operated successfully in transatlantic service, but reached a speed one knot higher than that of its near-sister, and with the same coal consumption.

The *Lusitania/Mauretania* turbines thus had precedent, but nonetheless were a leap forward from convention in their size and power, and in the size of the ships they were to drive.[3] These were nonetheless machines from the dinosaur age of steam turbines, but with only minor upgrades during their lifetimes, they powered the *Mauretania* throughout her 27 years. To be sure, steam turbine technology had advanced during those years, and would advance further through the years of World War II, but the biggest change affecting turbine technology was the advent of successful speed-reducing gears, typically with two stages ("double reduction") of gearing that allowed a turbine to turn at something like 50 times the rpm of the propeller it drove. When that innovation had become common, turbines impressed an observer by how *small* they were for the power produced. With the turbines fed by steam from watertube boilers at pressures perhaps five times that of these early Cunard ships, the last of the transatlantic liners might have as much as

The *Lusitania* is shown in dry dock after undergoing a change in her propellers. There were severe vibration problems with both ships so the outboard three-bladed propellers were replaced with four-bladed ones. Note the bossings for the inboard and outboard shafts. (*Courtesy of the Mariners Museum*)

70,000 hp on a single shaft, or the total power of *Lusitania* or *Mauretania* on four shafts.

An observer familiar with the later ships would, on looking back, be impressed at how *large* the early turbines were. The rotor diameters of the Cunard ships, high pressure, low pressure, and astern, respectively, were 10 ft, 13 ft 6 in., and 9 ft 4 in. Their casings, including external ribbing, were about two feet greater in diameter. In length, including their bearings, they were 45 ft 8 in., 48 ft 2 in., and 30 ft 1 in. It is not surprising that in spite of the well-justified principle that turbines are lighter than the equivalent reciprocating engines, Cunard estimated that their choice of turbines gave their new ships only about a three percent displacement saving.

Efficient turbine action depends largely on the ratio of steam jet speed to the linear speed of the blade that it impinges upon. The same is true of a propeller, where the "jet" is inflowing seawater. Unfortunately, the steam jet velocity is much higher than that of the seawater jet, so that with turbine and propeller at opposite ends of the same shaft, the turbine diameter must be very large to produce a high linear speed at a low rpm, or the turbine must be divided into an impracticably large number of stages to produce a lower jet speed to match that of the propeller. With the speed-reducing gear, the problem is greatly mitigated. Without gearing, best efficiency was unattainable at both ends of the shaft, but very large turbine diameters of *Lusitania* and *Mauretania* were a step toward reducing the consequences of a problem unsolvable without the gearing expedient.

In the late 20th century, "cross compounding" was a common way of maximizing the number of turbine stages with a machine of reasonable length: the machine was divided into two side-by-side units through which steam flowed in series. Their shaft outputs were joined by the accompanying speed-reduction gear. A similar arrangement was employed in the Cunarders' propulsion plant, though in the absence of gearing, the high- and low-pressure outputs could not be combined within the ship. The combination was made instead by having the two units drive separate shafts, thus the arrangement of high-pressure turbines driving the two outboard shafts, low-pressure turbines receiving exhaust from the other two units to drive the inboard shafts.

Reversing the direction of propeller thrust is an essential prerequisite for maneuvering. Three ways of doing so are in common use: swinging the propeller drive unit through 180 degrees (think of a small outboard motor), reversing the pitch of the propeller blades, or reversing shaft rotation. Reversal of reciprocating steam engines is a simple matter, and hence shaft reversal was the only way familiar to marine engineers in the early 20th century. The advent of the turbine therefore seemed to present an insurmountable problem because turbines, in any practicable sense, cannot be reversed. The problem could be bypassed, however, by mounting a "reverse turbine" on the propulsion shaft. This turbine might be identical to the ahead turbine, but simply mounted backwards so that its sense of rotation was opposite that of the ahead unit. This was, in effect, the solution used throughout the turbine era, with exception that the astern unit usually had fewer stages, and might be smaller in diameter, than the ahead because required maneuvering power is typically much smaller than that required at sea, and because efficiency of propulsion is less important. The use of the astern turbine, and its smaller size, is evident in the description already given for the *Lusitania* and *Mauretania* machinery.

High-pressure turbine driving the outboard shaft . . . low-pressure turbine driving the inboard shaft . . . a necessity for the shafts to be operated independently . . . an astern turbine on the inboard shaft . . . a single condenser serving all three turbines . . . all of this required a complicated array of piping and valves. Description is complicated, but is greatly facilitated by a sketch. See Figure 7-2, and comments that follow:

- Observe large piping size. All valves are power-operated by steam cylinders.
- Overspeed trip on any of the three turbines closes the bulkhead stop valve.
- Maneuvering valve is interlocked so that the ahead steam and astern steam cannot be applied simultaneously.

In its ultimate form, the steam-turbine propulsion plant was usually "regenerative" in that liquid water returning from condenser to boiler was preheated by steam that was extracted from one or more intermediate stages of the propulsion turbine.[4] *Lusitania* and *Mauretania* did not feature extraction from propulsion turbines, but were amply provided with a similar regeneration by exhaust steam from pump-drive engines, such as those driving the condenser cooling-water pumps, and the boiler feed pumps. This steam was used as the heating medium in a first-stage feedwater heater. The electric-power generators were driven by steam turbines; their exhaust steam was used as the heating medium in a second-stage heater.

In the pre-turbine, pre-electrical age of steam propulsion, the many pumps, fans, and deck power units (e.g., anchor windlasses, cargo winches) were driven by individual reciprocating steam engines. We are faced here

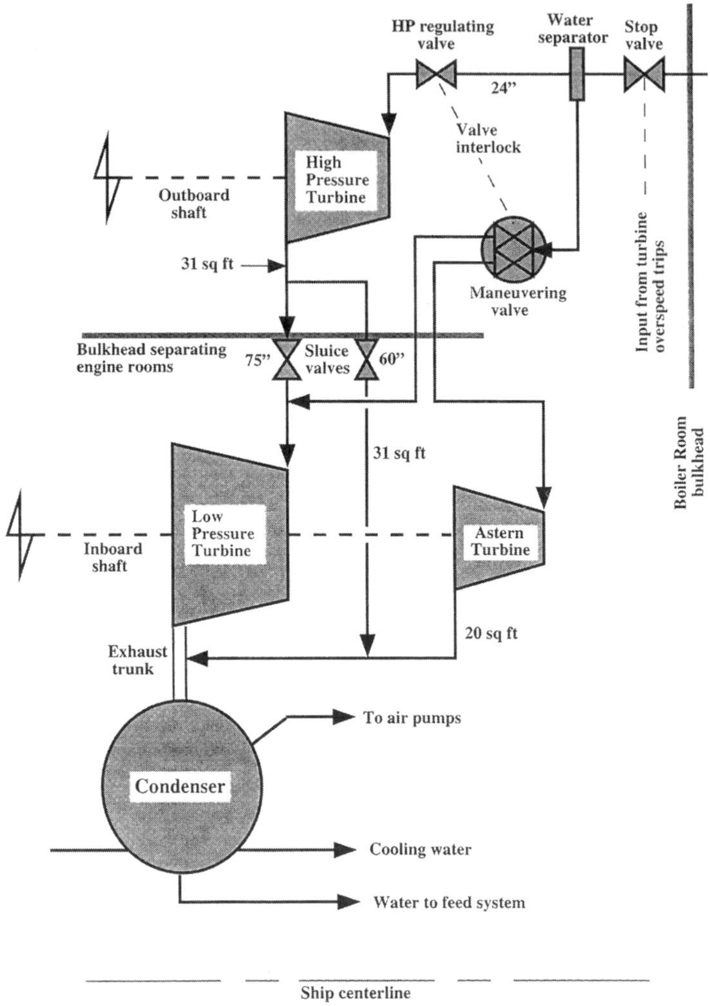

FIGURE 7-2 Schematic arrangement of the *Lusitania/Mauretania* propulsion machinery. (*Sketch by the authors*)

with ships early in the transition to steam turbines and electric motors for essentially all such functions. At that time many of the ships' auxiliary power services remained the domain of reciprocators. However, the Cunard ships' electric-power generators were driven by steam turbines. (The White Star ships soon to follow took a step backward by using reciprocating engines for this service.) The fans for the Howden forced-draft system were driven by electric motors. Ventilating fans, passenger and baggage hoists, boat winches, and one of a pair of refrigerating machines, were also motor-powered. Steering gear and anchor windlasses were of the conventional (for the time) reciprocating steam units, as were the cargo-handling winches.[5] On the whole, these ships can be called pioneers in the use of electricity in power applications, though not so strikingly advanced as they were with their turbine propulsion engines.

The fuel was coal, shoveled into those 192 furnaces by a hard-working "black gang." Coal was the fuel of all transatlantic liners of the era, as well as of a great majority of ships worldwide. This fact is commonplace to the point of scarcely being worth more than a sentence or two. But the bunkers—those dark chambers alongside the boiler rooms whence the fuel was drawn—proved to be a thing of hazard that might have been a key element in the foundering of the *Lusitania*. The ship's arrangement plan at the machinery level shows the boiler rooms flanked by the bunkers, with the separating bulkheads being nominally watertight. Here was a good idea—it seemed—because these ships were built to Royal Navy specification, and that navy had found by testing that such bunkers, when filled with coal, provided a protection to the boilers from enemy gunfire. As events were to prove, the same could *not* be said with respect to an enemy's torpedoes, since the doors which allowed coal to flow out onto the firing aisle of the boiler rooms rendered the bunker bulkheads nonwatertight. Supposedly the doors would be closed in time of danger, but they were inevitably clogged with coal, and it would take time to clear the coal away. Almost certainly there would be insufficient time after a torpedo ruptured the bunker. And if the coal doors could be closed in time, the flooded bunkers might still cause a dangerous list. One authority seriously proposed that such a list could be rectified by blowing down the boilers on the affected side of the ship. Again, not a reasonable countermeasure when the sea is pouring into the boiler room through the bunker doors. These matters we shall examine at greater length in Part IV.

The electric generating plant consisted of four turbo-generators with total capacity of 1500 kW. Power was generated at 120 volts, direct current, and fed to a two-wire distribution system through switchboards in the auxiliary machinery rooms. Feeder circuits were protected by circuit breakers, branch circuits by fuses. By means of shunt trip-coils on the boards, breaker trip timing was arranged so that the fuses responded to moderate overloads, the breakers to fault currents.

During the acceptance trials of the *Lusitania* and *Mauretania*, heavy stern vibrations were detected during their full speed runs. John Brown reconstructed the stern framing of the *Lusitania*, reducing the vibrations somewhat. This was a very expensive alteration because the second-class cabins in the area of reconstruction had to be removed and then reinstalled once the structural modifications were complete. Stern vibrations remained a concern, and the outboard three-bladed propellers were replaced with four-bladed propellers in June 1908 in an effort to improve the situation.

The *Mauretania* suffered from a similar vibration problem. One serious manifestation of the problem was deck vibration in the navigating bridge, a vibration of such magnitude at 20 knots that bridge personnel could not stand comfortably. Swan Hunter and Wigham Richardson engineers traced the problem to the propellers. However, the immediate remedy was strengthening of stern structure, as had been done with the *Lusitania.* Although ameliorated, vibration problems continued during the maiden voyage, and afterward. The final correction came in 1909 with the replacement of the outboard three-bladed propellers with four-bladed ones.

Passenger Accommodations. Accommodations (including those for crew and officers) and public spaces were heated, ventilated, and cooled by an array of 53 thermo-tanks,[6] these being compact assemblies of fans and tanks in which steam coils and water sprays adjusted temperature and humidity of the air before it passed into distribution ductwork. Although the contemporary literature speaks of cooling, the word apparently means "cooling of spaces by unheated outside air," since the tanks were not fitted with cooling coils.

In terms of arrangement, both ships of the class were essentially identical, but the décor of the passenger spaces differed between them. James Millar did the interior design of the *Lusitania* in lighter colored woods, while Harold Peto did the interiors of *Mauretania* in darker woods. Cunard chose different architects for the two ships, apparently with the objective of giving them individual personalities in the minds of transatlantic passengers. Throughout the first- and second-class spaces, the principal basis of décor was wood paneling, as was the case with the great majority of passenger ships until much later in the 20th century. Given the many varieties of woods, and the many ways that it could be applied, a different "feel" was given to each of these ships.

Passenger spaces were luxurious throughout the first-class public area. The same can be said of the second-class public spaces, though richness of décor is much more evident in photographs of the first-class area. Third-class spaces receive scant attention in the literature of the day. We shall nonetheless assume that Cunard did as well by third class (emigrants, largely) as any of its competitors.

Passenger staterooms varied in size and amenity according to class, and in first class, at least, according to the desires of a passenger to pay for luxury. Most rooms were French-based design resembling some of the rooms in the Palace of Versailles. However, different periodic styles were used to decorate the first-class staterooms. This gave the impression that all the rooms were different and individual. There were two regal suites, a combination of two bedrooms, a sitting room, parlor, dining room, private bath, lavatory and pantry.

Next were twelve *en suite* accommodations, each consisting of a bedroom and a sitting room. Six of these had private bathrooms. The regal suites and *en suite* rooms were on Boat Deck and Promenade Deck. The more ordinary first-class staterooms were on Boat, Promenade, Upper, and Main Decks. Most were intended for either two or three persons, with a few single rooms on each of the lower three of these decks. All of these rooms tended to be small; a typical two-person room was approximately 75 square feet, with two lower berths, hanging locker, chest of drawers, and a vanity with sink. No other bathing facility or toilet was provided in any of these rooms.

In the early 20th century, personal plumbing on even the best of ships and in the best of hotels was woefully deficient by later-day standards. But given that the ships we speak of were built as "luxury liners" (for first class, at least), and accepted as such by the travelers of the time, our listing in the next paragraph of facilities is not intended as criticism. We merely observe facts that may be interesting as curiosities to readers of a later century. Further, you may note that the chapter describing the passenger outfit of the *Olympic* class makes no mention of bath or sanitary facilities. The omission does not imply that Cunard was either better or worse than White Star. Both classes of ships were built to essentially the same standards in this kind of outfit.

In first class below the *en suite* level in the *Lusitania* and *Mauretania*, the ratio of passengers to water closets seems to have been 10-to-1. The ratio of passengers to bathtubs (no showers save for a few "needle baths") was about 30-to-1, all in public "gents" and "ladies." By the '30s, the transatlantic traveler was expecting private facilities, and the lack of them aboard the *Mauretania*, and the surviving White Star liner, *Olympic*, as well, appears to have been a significant factor in marking these ships for the scrapyard.

Second-class staterooms were approximately the same in size and amenity as first class, but typically were fitted to accommodate twice the number of passengers; each single bed of a first-class room usually became a double-decker affair in second class. None of the rooms had its own plumbing beyond a sink. Ratio of passengers to public water clos-

This is the first-class Dining Room of the *Mauretania*, located on the Upper Deck; there is a cutout in the Shelter Deck to create a balcony effect for this space. A dome was positioned over the cutout in the Shelter Deck.

This is the first-class Lounge in the *Mauretania,* located on the Boat Deck between the third and fourth stacks. A dome was arranged over the center of the room.

This is the first-class Lounge and Music Room on the *Lusitania*. The dome over this space indicates that it is one of the public first-class spaces on the Boat Deck. (*Courtesy of the Mariners Museum*)

Like the *Titanic*, the *Lusitania* followed the tradition of having a Smoking Room for the gentlemen to use after dinner to be parted from wives to discuss matters of politics, shipboard life, or business. The Smoking Rooms for first and second class were outfitted in an Italian 18th century style. A stained glass barrel-vaulted dome done in three panels can be seen in the ceiling of the first-class Smoking Room. There were two fireplaces in the room, seen at the rear that actually burned coal.

ets was about 15-to-1; the bathtub ratio was about the same as in first class.[7]

Third class, also known as "steerage," was berthed in staterooms, but they appear to have been mean places indeed.[7,8] The berthing arrangements varied from two persons to eight. Most of these rooms, with berths for four, were given only about 40 square feet. Water closet ratio was about 18-to-1; bathtub ratio about 375-to-1. All of these meager sanitary facilities were on Upper Deck and Shelter Deck. Sleeping accommodations were on Main Deck and Lower Deck, one and two decks down, respectively, from Upper Deck.

These sleeping and personal-care spaces may have deserved such mild encomiums as "nice . . . tasteful . . . attractive" (save perhaps in third class) from those who traveled in them. At least, Cunard seemed happy to have photographs of them published. However, the real appeal to the traveler was intended to be the public rooms, especially in first class. These passengers entered on Main Deck; from the entrance hall, a Grand Staircase and two elevators carried them successively through Upper, Shelter, Promenade, and Boat Decks. On the three highest decks the stair hall was known as the Grand Entrance, apparently because it was the principal access to and from the open decks that were available to first-class passengers at these levels.

The principal public space for each of the three classes was its dining saloon. All of the saloons were on Upper Deck, with the first-class room being by far the largest and most ornate. It echoed the splendor of the French King Louis Seize in the *Mauretania*, while in the *Lusitania* it was decorated in the style of Louis XVI. It consisted of two levels, the second being in the form of a balcony on Shelter Deck that overlooked the main floor, i.e., Upper Deck. Above, a dome extended skyward through Promenade and Boat Decks. The dome contained segments depicting the four seasons re-enacted by Cherubim. Beneath the grand dome the circular opening in the Shelter Deck Corinthian columns provided support to the deck structure. The dome was done in cream and gold in *Mauretania* with white and gold used in the *Lusitania*. The elegance of mahogany paneling gave the space a distinctive look. Potted palms were arranged on the tables.

Other first-class public spaces were on Boat Deck. The Library and Writing Room, the latter used mostly by women, was directly forward of the Grand Entrance. The Lounge, Library, and Music Room were directly aft. The Lounge was the most ornate showpiece with its stained glass ceiling lit by skylights. The first-class Lounge and Music Room in the *Lusitania* was completed in Georgian style with a yellow and green color scheme over a mahogany paneling. There was a fireplace at one end of the Lounge. The same space in the *Mauretania* was done in the style of an 18th century French Salon.

The *Lusitania* and *Mauretania* followed the tradition of having a Smoking Room that was used after dinner for the gentlemen once they had separated from their respective partners to discuss daily activities, business, or politics. The rooms were done in an Italian 18th century motif. The first-class Smoking Room on the Boat Deck had a stained glass barrel-vaulted dome. Two fireplaces in the room actually burned coal.

There was a Verandah Café. The interior designers originally wanted a gymnasium instead of the Verandah Café but Cunard decided the issue. This space, like all of the Boat Deck public rooms, was well-endowed with exterior light from overhead domes.

First-class spaces, both staterooms and public rooms, occupied the mid-length of the ship. Second class was aft, third class forward. The second-class entrance hall was on Shelter Deck, with a modest staircase (but no elevators) leading down to the Dining Saloon, and up to Drawing Room and Smoking Room on Promenade Deck, then to the Lounge on Boat Deck. The second-class Dining Saloon was one of the grandest rooms afloat for its class. Huge pillars had been placed in the dining room to support the deck structure that had a cut for a skylight. A Smoking Room was provided for the second-class gentlemen on the Promenade Deck. They enjoyed sitting under a barrel-vaulted stained glass skyline. The Drawing Room, also on the Promenade Deck, was furnished in the style of Louis Seize of France. Different shades of gray and rose complimented the satinwood furniture. All of the second-class public rooms were fitted with domes or skylights.

The third (or steerage)-class entrance hall was also on Shelter Deck, but the staircase (no elevators) led only downward. The third-class galley and most of the toilets for this class were on this deck, as were the Smoking Room and General Room for this class. The Dining Saloon was directly below, at forward on the Upper Deck. There were no domes or skylights for third class, but the open deck at Shelter Deck level was available for those who favored sea and sky over the press of humanity indoors. Although third-class spaces were not as elegantly decorated as the other two spaces, the immigrants who used them believed that they were.

Armament. Published arrangement plans show one unusual and controversial feature: 12 six-inch guns. The outlines of the guns are presented as dashed lines, thus not representing fixed armament, but rather the places where the guns would be mounted if the ships were taken into

The *Lusitania* is shown at speed after her departure from Liverpool. At the time of her completion in 1907, she was the largest ship in the world. Both *Lusitania* and *Mauretania* were designed also to be auxiliary cruisers in times of war, hence the speed requirement. Neither ship saw that service during World War I.

the Royal Navy as auxiliary cruisers. If actually mounted, the Admiralty's auxiliary variant of the *Lusitania* would have been a more heavily armed ship than most light cruisers of the day, with a projected armament of 12 six-inch guns in single mounts. However, as an auxiliary cruiser, the ship lacked even the modest armor protection afforded light cruisers.

Six circular foundations for gun mounts were installed, four on the forecastle (Promenade Deck forward of the deckhouse) and two on the poop (extreme aft Shelter Deck). Six more foundations were installed on the Shelter Deck, distributed three to a side between frames 80 and 175, approximately. The foundations were behind a special bulwark whose plating and stiffeners could be unshipped when the guns were mounted. Portable stanchions could be placed at each removed bulwark. The guns, if mounted, would have limited arcs of fire; one pair of forecastle guns could be trained through 150 degrees, the other pair through 135 degrees. Arcs for the Shelter Deck guns were 120 degrees, except for the poop guns with arcs of 150 degrees. The latter arcs included aiming directly aft. The guns were never fitted. However, all that was necessary for installation was to lower each gun mount onto the foundation ring and bolt it in place with twelve castellated bolts.

A forward ammunition magazine was located between frames 245 and 260 in the reserve coal bunker in Cargo Hold No. 2 to supply the forecastle guns and the forward-most two on Shelter Deck. Ammunition was to be brought up through the main hatch for the cargo hold. An after magazine was fitted into the Mail Room on the Lower Deck between frames 31 and 50. Electric hoists in the baggage hatches brought the shells and powder to Shelter Deck, whence it was carried to the six aft-gun positions. Shell and powder stowage facilities were part of each of the magazines.

In spite of the apparent ease of fitting out the *Lusitania* as a merchant cruiser, the Admiralty did not invoke its right to do so. The amount of coal that would be required to keep a turbine-driven ship at sea for lengthy periods[9] was the likely explanation. There is one later event that may have helped to influence the Admiralty's decision not to proceed with the auxiliary-cruiser conversion of the *Lusitania*. This was the loss of the German armed merchant cruiser *Kaiser Wilhelm der Grosse*. This converted passenger ship with six 105-mm guns and two 37-mm machine guns had no armor protection but was among the fastest of the ocean liners of that period with a speed of approximately 22.5 knots. On 26 August 1924, the British light cruiser HMS *Highflyer* intercepted her. In a running gun battle that lasted 75 minutes, the German vessel sustained severe damage to her shell plating that caused her to list to port and sink within 45 minutes.

Structural modifications pertinent to its role as a potential warship were reportedly made to the *Lusitania* while she was in the Cammell-Laird Shipyard in Liverpool during 1913. The shell plating (sheer strake) between the Upper and Shelter Decks (a width of 14.5 feet) was alleged to be double plated. (This will be confirmed when the wreck is evaluated during an expedition to the wreck site sometime in the future.) The Shelter Deck plating was doubled in way of the gun foundations. Upon completion of her shipyard stay, the ship returned to her North Atlantic run to New York.

When war broke out on 4 August 1914, the *Lusitania* was about to leave New York. Upon her arrival at Liverpool she was handed over to the Admiralty for her role as an armed merchant cruiser. Other British passenger ships such as the *Aquatania*, *Carmania*, *Ivernia*, and *Transylvania* were armed by mid August 1914. She was taken to a shipyard in Liverpool, where all third-class passenger accommodations that amounted to 176 berths on "F" Deck over the boiler rooms were removed. Access to that deck was closed off. Some of her armament of 6-inch guns installed by 17 September. However, seven days later the Admiralty announced to

Cunard that it no longer intended to use the ship as an armed merchant cruiser and that they wished to discuss her future employment with Alfred Booth a soon as possible. On 3 October a meeting was held between Chairman of Cunard Alfred Booth and the Secretary of the Admiralty, Sir William Graham Greene to discuss how the liner would be used. Cunard was to operate the ship between Liverpool and New York, but the Trade Division of the Admiralty would use some of her cargo space, particularly on eastbound voyages. As part of the agreement, all the forward passenger accommodations on and below the Shelter Deck in way of the forward armament were closed off, requiring the closures of the third-class Smoking Room, Ladies Room, and Dining Room. Third-class passengers had to use the second-class Dining Room. However, Admiralty requests for some of her accommodations were to take precedent over regular bookings. In addition, her deployment at sea would be an Admiralty responsibility. The annual subsidy would be continued and the Admiralty would pay for any cargo and accommodation spaces it used.

The structural changes for the gun armament added weight high in the ship, decreasing its transverse stability. The vertical center of gravity of the ship was already high, but at a value that would make ship motions most comfortable for the traveling public. It would have been necessary to carry some form of ballast in the full load departure condition to compensate for these changes. This development made *Lusitania*'s stability suspect in a loading condition with the coal low in supply. Even the addition of salt water ballast may have not been enough to compensate for the depleted coal.

Although the guns never were installed, their ghostly presence entered the realm of controversy in 1915 after the *Lusitania* was torpedoed. The question of whether it was a warship or an innocent passenger boat became highly pertinent. The guns were *not* mounted, but were they stowed somewhere aboard? If so, did that circumstance make it a warship? This book does not address these questions, but the point does arise again in later chapters.

Stability and Subdivision. An inner bottom, five feet (six feet in way of the engine room) above the baseline along the centerline, extended the full length of the ships, and was carried to the eight-foot level at the turn of the bilge. Thirteen transverse bulkheads, watertight to Main Deck (several forward were watertight to Upper Deck), subdivided the ships sufficiently for two-compartment survivability. Longitudinal watertight bulkheads separated the two wing engine compartments from the compartment containing the inboard shafts and their machinery. Forward of the engine rooms, the four boiler rooms extended 332 feet, flanked outboard by coal bunkers. The coal in that location supposedly protected the boilers from enemy gunfire, an important consideration for ships that were intended to be auxiliary cruisers in wartime. To make this protection effective against underwater hazards (icebergs, the bows of other ships, torpedoes), the longitudinal bulkheads separating boiler from bunker had to be watertight. Each boiler room had five subdivisions in the coal bunker spaces outboard. Vertical sliding watertight doors were necessary for access to the coal. It is questionable whether these doors could remain watertight because their openings were inevitably clogged with coal. Watertight or not, any athwartship barrier water flooding as a consequence of a rupture of the ship's shell will produce a list. The list, if of low magnitude, may not be fatal unless it leads to secondary flooding, e.g., water pouring in through open portholes that are submerged by the list. Such an event is believed to have occurred on *Lusitania*'s fatal day, and hence is to be discussed at appropriate points in this book.

During the design of these ships, Leonard Peskett found to his dismay when doing a damage stability analysis that when the coal bunkers on one side in one subdivision were flooded that the subsequent list caused the waterline in the damage condition to rise to the level of the portholes on "F" Deck. These portholes were subsequently sealed and watertight doors were fitted on "F" and "E" Decks in way of the boiler spaces below. These doors were remotely controlled by a switch on the Bridge. There was a manual override mechanism fitted for emergencies.

Hull. The hulls were largely conventional, both in constructional details and in general aspect. A profile view (see Figure 7-1) shows the vertical steam and the fantail stern in conformity with almost universal practice of the early 20th century. A cross section of the hull, showing framing and plating details, appears in an illustration here, and must be judged quite conventional. Fastening was by riveting, of course. Some of the riveting was done by hydraulic machines.

A profile view does show features of note around the stern. The rudder, although it was of the plate type common to the era, in profile it appears similar to the "balanced" rudders of later ships.[10] The arch just forward of the rudder was included because model tests showed that it would enhance flow of water to the inboard propellers.

The hull feature most noteworthy of comment is the use of high-strength steel for parts of the hull structure, with the prime objective of saving weight through lighter scantlings.[11] Figure 7-3, reproduced from *Engineering*, shows the extent of silicon steel for hull plating, and for

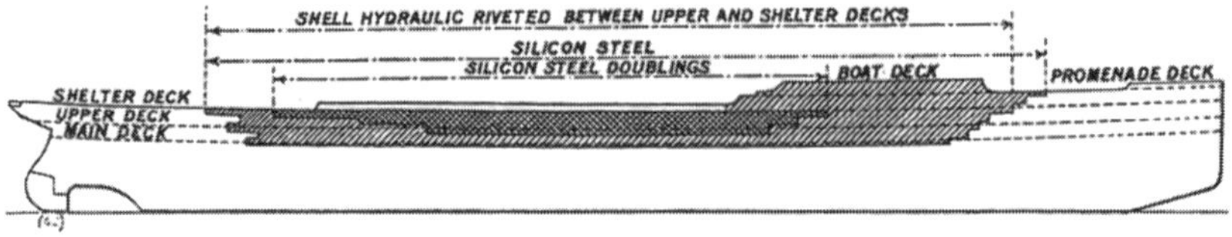

FIGURE 7-3 Sketch showing the use of high-strength steel in the *Mauretania* hull. (*From* Engineering, *vol. 84, 1907*)

doublers added to that plating. This steel was also used for deck plates and their doublings on Upper Deck, between the sides and the boiler casings. High-carbon steel was used between inner bottom and Upper Deck for transverse and longitudinal bulkheads. Where used, these special steels allowed reductions of about 10% in scantlings.

This special steel was chosen for construction after tests were conducted to determine to what extent the strength of the upper portion of the hull could be improved with the use of high-strength steel.[12,13] The tests, including explosion tests of heavy charges of dynamite against the plates, demonstrated that high-strength steel was the best choice because its 36% superior ultimate strength permitted the scantlings of the upper structure to be reduced 10%, a substantial savings in weight for ships with a low metacentric height (GM). This reduction improved the stability of these ships and shortened their erection process because thinner plates could be more easily riveted.

Lloyds ruled that rivets of high-strength steel would not be required in the high-strength steel plates. Rather, rivets could be made from a mild ingot steel with an ultimate strength of 26–30 tons per square inch with an elongation of 20% and an elastic limit between 13 and 15 tons per square inch. This was controversial, since naval practice required high-tensile steel rivets. However, no wrought iron rivets were used, in contrast to the partial use of that material for the *Olympic*-class liners. Shearing forces on the rivets were minimized through use of a special tool that rounded and reamed the drilled holes.

All the watertight doors on the *Mauretania* and *Lusitania*, whether they were the vertical sliding type on the main transverse bulkheads or those watertight coal bunker doors in the longitudinal bulkheads were closed by hydraulic pressure on the Stone-Lloyd principle. That pressure was applied at the door frame; that is, the cylinder operating the door was adjacent to the frame, so that any distortion of the bulkhead did not interfere with the operation of closing the door. These doors were closed by an electric current through an actuation device on the Bridge or manually operated from within the space.

Inside the coal-bunker doors were iron screens that were there to prevent the coal from falling down into the door opening, but where the hydraulic doors were fitted, the coal on many occasions was allowed to come to the angle of rest from a full bunker. The door, nevertheless, did close down through the residue. According to Leonard Peskett's testimony at the 1912 Mersey Inquiry the system was never known to fail. The doors, according to Peskett were constantly working so that no build-up of coal dust and clinkers was possible. As a precaution, all the bunker watertight doors were to be closed at 12 o'clock as a matter of precaution.

Lifeboats. As per Board of Trade Regulations, the *Lusitania*, as completed, was equipped with 16 wooden lifeboats. After the tragic loss of the *Titanic* on 15 April 1912, the number of lifeboats carried aboard the two Cunard liners was increased from 22 to 48 boats (22 wooden and 26 collapsibles). The new collapsible lifeboats were stowed on chocks below the wooden lifeboats rigged on the davits. Each of the collapsibles

The *Lusitania* is outward bound from New York on one of her many eastbound voyages prior to World War I. The red stacks and black topping are the Cunard colors, but during wartime her funnels were painted black. (*Courtesy of the Mariners Museum*)

was also fitted with two chains to anchor them to the deck. This would prove disastrous when the ship sank because the chains would have to be released before the boats could be fitted to the davits and swung clear of the ship. When the *Lusitania* was sinking many of the chains were not released, thus preventing the boats from being launched successfully. Many of the lifeboats went down with the ship. The additional lifeboat complement was added to the *Lusitania* during an extensive industrial period that began on 12 May 1913. The number of davits was not increased, and no changes were incorporated to accommodate the added service now expected of these davits.

The *Lusitania* and *Mauretania* were nicknamed the "greyhounds of the sea" because they were the fastest passenger liners on the North Atlantic in their heyday. However, that speed was attained at the price of beam, and with the high stacks and impressive superstructure meant a high vertical center of gravity. The events of May 7, 1915 would show the price these features would exact.

Careers.[14,15] The RMS *Lusitania* sailed on her maiden voyage to New York on 7 September 1907 under the command of Captain James B. Watt. She carried over 3,000 passengers and crew. She arrived in New York on 13 September 1907. Her voyage lasted 5 days and 54 minutes, but she did not recapture the Blue Riband on this occasion, which was a great disappointment for the authorities and owners. A record speed on the return passage could not be attained because of bad weather. Her second voyage in October 1907 was faster. She averaged 23.993 knots crossing the Atlantic in less than 5 days, recapturing the Blue Riband for the British merchant marine in that eastbound crossing. She eventually would become the holder of both east and westbound Blue Ribands, setting westbound crossing records in October 1907, July 1908, August 1908 and September 1909.

Most of her passengers spoke eloquently about the design and accommodations of the new ship. However, there were also problems with vibrations in the stern, causing some discomfort for some of the second-class passengers. This would be corrected in 1909 when the liner was taken out of service to correct the vibration problem.

At sea when World War I broke out, *Lusitania* arrived home safely. Upon arrival in the Mersey, the Admiralty decided that they did not need the ship as an armed merchant cruiser. However, they paid for the ship to remain at Liverpool at their disposal until October 1914. The *Lusitania* was allowed to return to service and made two trips between Liverpool and New York during October 1914 and then began a monthly service on this route. In order to save on coal and labor, Cunard decided to shut down six of the ship's boilers, reducing her maximum speed to 21 knots.

George Booth, cousin of Alfred Booth, chairman of Cunard, went to the United States in October 1914 aboard the *Lusitania*. Captain Dow was made the master after the Admiralty returned the ship to Cunard the previous month. Booth noticed several things about the ship on which he had made many voyages. The stewards were not as helpful as they had been in the past and the ship motions were much more pronounced than Booth had experienced on earlier crossings. For the first time on the *Lusitania* he was seasick the entire voyage. Before arriving in New York, Booth asked Captain Dow why the ship seemed so different from a ship motions aspect. Dow explained that the changes made to the ship in September had removed much of the equipment from "F" Deck and the steel deck of the Shelter ("C") Deck had been increased in thickness. Much of the area forward of the Bridge was entirely in the hands of the Admiralty and closed off from the rest of the ship. Dow even showed Booth the concealed gun rings in the planking of the Shelter Deck. He also went over the changes made in the third-class accommodations that forced the steer-

The *Mauretania* is shown underway in 1907. Her distinguishing feature from the *Lusitania* was the large ventilators. The Grand Old Lady served from 1907 until her retirement in 1935. She was used as a troopship and hospital ship at different times during World War I.

age passengers to use the second-class Dining Saloon. Captain Dow explained that these changes made her trim by the stern and as a result she would "corkscrew as well as roll heavily." Dow also pointed out that eastbound voyages were more comfortable as the forward spaces were filled with some sort of "cargo."

On 7 May 1915, while heading east during her 202nd voyage off the Old Head of Kinsale, Ireland, the *Lusitania* was torpedoed without warning by a *U-20* and sank within 18 minutes. See Chapter 17 for details.

Mauretania. The *Mauretania* made her maiden voyage from Liverpool to New York on 16 November 1907, and set an eastbound record of 23.69 knots on the return leg of that trip. The westbound crossing of that voyage was marred by three days of heavy weather during which a spare anchor broke loose from its restraints. The anchor was finally secured by heading the ship for three hours away from the direction of the high waves. Nevertheless, there was some storm damage. Several large boarding waves broke windows on the upper decks. An iron railing over the wheelhouse was damaged. This was the first of seven consecutive eastbound records that the *Mauretania* would set until losing a propeller blade after hitting a submerged object in May 1908. She was forced to steam on three shafts for the duration of the year until her new propellers were ready. In September 1909 she established a westbound record speed of 26.06 knots with her new propellers. Her last eastbound speed record came in August 1924 when she made 26.25 knots. In all, she held the eastbound and westbound records from 1907 until 1929, until broken by the new German liner *Europa*.

In December 1913 the *Mauretania* returned to Liverpool for her annual overhaul, part of which involved work on the main propulsion turbines. On 26 January 1914, while men were working on the turbine blades, one of the gas cylinders exploded. Four men were killed and six were injured. The resulting fire was extinguished and the main damage was confined to the blades in the starboard turbine. The ship was not ready to rejoin the Atlantic service until March 1914.

Mauretania was enroute to New York when World War I commenced on 4 August 1914. Returning to the United Kingdom, she would be used to bring home American citizens in Europe. After the sinking of the German auxiliary cruiser *Kaiser Wilhelm der Grosse* on 27 August 1914, the British Admiralty decided not to use these ships as auxiliary cruisers. Therefore, the reduced demand for transatlantic passages required that the ship had to be laid up at Liverpool on 26 August 1914.

The *Mauretania* is shown arriving in New York in the late 1920s during the winter period. By this time a number of changes had been made in the ship's interior to reflect the fact that the immigrant trade had vanished. The third-class spaces had been upgraded to attract American tourists, particularly female schoolteachers.

After the loss of the *Lusitania* in May 1915 Cunard planned to return the *Mauretania* to service. Before that could happen the Admiralty decided to requisition her to transport troops during the Gallipoli campaign, in late May 1915. After this, she made several voyages to Mudros Bay Island of Lemnos, the Allied base for operations in the area. On one of these voyages the *Mauretania* was attacked by a submarine but managed to avoid the torpedo, largely due the ship's high speed. At the end of August 1915 the *Mauretania* returned to Liverpool and was fitted out as a hospital ship. Under hospital colors she left Liverpool on 21 October 1915 to assist with the evacuation of the wounded from Gallipoli. The *Mauretania* made several further voyages as a hospital ship and completed her last voyage as a hospital ship on 25 January 1916.

The *Mauretania* was requisitioned as a troopship on 29 September 1916. In October–November 1916 she made two voyages from Liverpool to Halifax carrying Canadian troops bound for France. After this duty she was laid up on the Clyde until March 1918. She was used again as a troopship to carry more than 30,000 American troops to Europe before the Armistice was declared in November. After the end of the war the ship was used in the repatriation of American and Canadian troops. She made her final trooping voyage on 28 June 1919 and was then returned to Cunard to be refitted for peacetime duty at Southampton. The

Mauretania would make her transatlantic voyages from Southampton, Cunard's new principal terminus. With loss of the *Lusitania* during World War I, Cunard used the *Mauretania*, *Aquitania*, and *Berengaria* as a trio in establishing a regular pattern of transatlantic passenger service.

In June 1921, the *Mauretania* was in Southampton undergoing routine turn-around work. Some crew were on board with many on leave while the cleaning crew was working on the carpets in the first-class cabins on "E" Deck. One of the workers was using an inflammable cleaner to remove grease spots. A fire broke out at 1330 on 25 June and spread quickly through this area of the accommodations. What crew was aboard tried to contain the fire with fire hoses, but with only two boilers in operation, there was no firemain pressure. Engineers were notified and began to raise steam. However, land-based fire fighters arrived on the scene and quickly poured water from the port side of the ship through open portholes and doors. However, the starboard side doors and portholes were not open and water from the fire hoses began to accumulated on that side of "E" Deck. After four hours of firefighting effort, the weight of water on the starboard side caused the port side mooring lines to fail. The *Mauretania* took a 15-degree list to starboard. First Officer James Bisset and a volunteer crew managed to open the starboard doors and allow the water to run off. The ship was on an even keel in ten minutes.

While the fire damage was repaired, it was decided to convert the *Mauretania* from coal to oil as part of her refit. The dramatic decrease in emigrants to North America also necessitated changes in her third-class accommodations. She returned to service in March 1922 and continued her transatlantic service through the 1920s. On 25 July 1922 the *Mauretania* broke her pre-war Atlantic speed record with a crossing of 26.16 knots, her best speed. In January 1923 the ship was chartered by an American travel company and made a Mediterranean cruise. In November 1923 she was again withdrawn from service to overhaul her turbines. This included an upgrading in her power rating to 90,000 shaft horsepower (total for four shafts). Because a strike at the Thornycraft shipyard interrupted this work, she was towed to Cherbourg for the repairs to be completed for scheduled return to service in December.

The upgrading of the turbines allowed the ship to make higher speeds. In 1924 the Cowes Harbour Commission complained about the *Mauretania*'s speed as she left the Solent. The heavy wash from her propellers flooded main street of Cowes and caused considerable disruption to this village. A Board of Inquiry decided that the pilot was to blame. Another overhaul in 1928 involved modernization of the ship's furniture and décor. With worldwide depression in 1929–1930, she was used mainly for cruising. This was also necessitated by the fact that the new German liners *Bremen* and *Europa* were now the fastest passenger ships on the North Atlantic service.

On 27 November 1929 the ship collided with a train ferry near Robbins Reef, after leaving New York. Fortunately no one was injured, however, the ships bow was slightly damaged. Repairs were completed within 24 hours. After a winter overhaul she returned to service in February 1930 and during the following years was used mainly for cruises. For this duty she was painted white. The *Mauretania* made her final passenger sailing from Southampton on 30 June 1934, the day Cunard and White Star Lines merged. After two cruises to the West Indies she left New York on 26 September 1934, the day *Queen Mary* was launched, for her final Atlantic crossing. Returning to Southampton on 2 October 1934, she was laid up. She was sold to Metal Industries Ltd. of Glasgow for scrap on 3 April 1935.

Notes for Chapter 7

1. The source for Cunard history is Francis E. Hyde, 1975, *Cunard and the North Atlantic, 1840–1973,* The MacMillan Press, London. The source for *Lusitania/Mauretania* description is the journal, *Engineering*, Volume 84, 1907. The issue of 2 August 1907, pages 129–170 describes the *Lusitania*. The 19 July issue also presents a number of interior photographs. The 8 November issue, pages 633–647, describes the *Mauretania*.
2. Given 70,000 hp and 192 furnaces each requiring 0.2 (an estimate) human horsepower to fire, we see that the great machines of the industrial age were providing a multiplication of human effort by a factor of approximately 2,000. Advances such as mechanical stoking, oil fuel, and internal combustion engines were soon to send this ratio soaring.
3. The battleship *Dreadnought* was powered by turbines that were not significantly larger than those of Cunard, and it went to sea roughly a year before *Lusitania*. But its claim to being the pioneer is weak because it was designed *after* Cunard had made its decision for turbines.
4. Students of steam power-plant thermodynamics will recognize that this seemingly illogical passing-around of steam produces a worthwhile increase in steam-plant efficiency (ratio of shaft work to thermal energy in the steam). In the limit, i.e., with an infinite number of heaters fed by an infinite number of turbine extraction points, this efficiency approaches that of the Carnot cycle.
5. Cargo winches on passenger ships? These ships had a small cargo hold forward, and a baggage room that was loaded/unloaded in the manner of a cargo hold.
6. "Thermo-tank" appears to have been a proprietary name.
7. In many ships of the era, steerage passengers were berthed, barracks-style, in large open spaces, with one such space for men, a second for women and young children.

8. A disagreement here. One of the articles in *Engineering* that describes these two ships contains the statement, "Indeed, the arrangements made for the third-class passengers are *exceptionally fine*, so far as public and private rooms are concerned." Emphasis added!
9. The statement about excessive coal consumption stems from the circumstance that turbines are highly inefficient when operating at low loads, the operational mode an auxiliary cruiser would encounter in, say, convoy escort duty. Warships powered by steam turbines are usually fitted with small "cruising turbines" that are fully loaded when driving those ships at low speeds, and hence avoid excessive fuel consumption.
10. Balanced rudder: a portion of such a rudder extends forward of the rudder stock. When the rudder is turned this portion produces a torque that aids in the turning, thus reducing the torque that the steering engine must produce.
11. Tests made on the high-strength steels proposed for the Cunard ships were summarized in *Engineering*, Volume 84, 1907, pages 129ff. Ultimate stresses for both the silicon and the high-carbon steels were found to be in the 36–37 tons/sq. inch range. Elongations (minimum 20% specified) were in the 22–23 percent range. The mild steels used for shipbuilding at the time typically had ultimate stresses in the 28–32 tons/sq. inch range.
12. *Lusitania*, Souvenir Edition, Patrick Stephens, Ltd., 1986 (reprinted from *Engineering*, 1907).
13. Attwood, E.L. and Pengally, H.S., Theoretical Naval Architecture, 1931.
14. Shaum, John H., Jr. and Flayhart, William H., III, Majesty at Sea, The Four Stackers, W.W. Norton and Company, New York and London, 1981.
15. Simpson, Colin, The *Lusitania*, Little Brown and Company, Boston and Toronto, 1972.

CHAPTER 8

The Olympic *Class*[1,2]

The White Star Line, owner and operator of the *Titanic* and its two sisters, *Olympic* and *Britannic*, began operation in the mid-19th century with sailing ships serving the England–Australia trade. It continued to serve this route into the age of steam, but fell into bankruptcy in 1867. After a brief hiatus, Thomas Henry Ismay, owner of the Oceanic Steam Navigation Company, bought the name and assets to form a subsidiary that would compete in the lucrative North Atlantic passenger service. Thus began the Ismay era that would see White Star become a major factor in the Europe–North America passenger traffic. The service to North America, the vehicle of both its fame and its misfortune, grew under the direction of the founding Ismay, and as it grew, the line became a fierce competitor with other North Atlantic passenger lines. Had it not been for the tragedy of April 1912, and then the loss of the *Britannic* in World War I, the large and luxurious ships that White Star built just before that war might have advanced the shipping line to the dominant position on the North Atlantic.

One of Ismay's financial backers was Gustave Swabe of Liverpool, who in turn was an uncle of Gustav Wolff, a junior partner in the shipbuilding firm of Harland & Wolff in Belfast. Through this connection an arrangement was made by which Ismay was guaranteed the support of Swabe if he built his ships at the Belfast shipyard. Ismay's long-time friend William Imrie also became involved, leading to the formation of Ismay, Imrie and Company to undertake the management of the White Star Line.

The reborn line soon built its first ship to compete on the North Atlantic; the *Oceanic*[3] was completed in March 1871. Her admirers spoke of the new ship as "the ship of the future," and that sobriquet

The *Olympic* and *Titanic* had an identical first-class Grand Staircase with the exception of the woodwork surrounding the clock, which on the *Titanic*, seen here, was much more ornate. Behind the clock were three Otis elevators that served "E" to "A" Deck.

seemed well deserved since the *Oceanic* gained both the admiration of passengers and profit for White Star. Several more ships followed, including a sixth ship built for the North Atlantic, the *Baltic*, destined to set a speed record for the North Atlantic crossing. Over the span of the first six ships, White Star competed vigorously for the speed prize with its principal rival, Cunard, as well as other lines, both British and foreign. But a major shift in strategy for success came in the 1890s, based on a perception that had grown with experience in passenger trades: amenities as well as speed attracted passengers. The White Star Line ships typically had more beam than those of its rivals, a feature that meant more open deck space for passengers at the cost of greater propulsive power to remain competitive in speed. Also, the dining areas and accommodations for first class were centered around amidships where motions that produce the terrors of mal de mer are the least, and those accommodations were concentrated in a single superstructure rather than in separate deckhouses. These design features seemingly made the Ocean Steam Navigation Company preeminent in its continuing Australian service. Emphasizing them on the North Atlantic might also be a successful signature feature for that service as well.

Thomas Ismay retired in 1892 and died in 1896. His sons, James and Bruce Joseph, ages 22 and 28 in 1892, succeeded him as partners in Ismay, Imrie and Company. James never gained a major role in company affairs, and soon left to follow other interests. Bruce Ismay, whose name was to be tarnished beyond repair from the moment he stepped aboard the *Carpathia* on 15 April 1912, became the effective boss. During his tenure the momentous decision was made to abandon speed as the ultimate of ship design in favor of luxury. The decision to switch goals from first-in-speed to first-in-luxury was first evident in the *Cymric,* new in 1895, a ship that was primarily a freighter, but which carried 250 passengers in first class and 1,160 in steerage. Its speed was a non-competitive 14 knots, but it was capacious, comfortable (whether those adjectives applied to steerage is difficult to say), and financially rewarding. That success was attributed to the new design philosophy. The profitable record of the *Cymric* inspired the design of the next new ship, the *Oceanic* (second of that name), a vessel that was described as "an aristocrat of the ocean, a beautiful ship . . . the crowning success of the 19th century in naval architecture and marine engineering."[4] On its maiden voyage in September 1899 it crossed at an average speed of 19.5 knots, far short of the 22.59-knot westbound record set by the new German liner *Kaiser Wilhelm Der Grosse* the year before. Regardless of this undistinguished performance, the *Oceanic* and her near-sisters *Celtic*, *Cedric*, and *Adriatic*, remained the bellwethers of the White Star fleet until the advent of the *Olympic* class in 1911.

Whether Bruce Ismay made the decision to abandon the speed chase is unknown, but that decision was of lasting consequence, for its success with *Oceanic* and its sisters provided the guiding principle of design for the *Olympic* class. The ships of that class were designed for capacity and luxury with no attempt to match the speed of *Mauretania* and *Lusitania,* reigning speed queens of the North Atlantic when the *Olympic* was launched on 20 October 1910. Luxury can be had through suitable investment in quality, but speed can be had only through investment in quantity, as manifested in size of ship and power of engines. When competitors are already straining at the limits in those things, it may be very difficult to best them. The decision to stick with quality meant that White

Star ships could be the products of conservative design—except in the décor that cosseted the first-class passengers.

The Ships of the *Olympic* Class

The new *Kaiser Wilhelm der Grosse* had captured the North Atlantic speed prize with an average speed of 22.59 knots, westbound, in 1898. It was known at the time that the Germans aspired to larger and faster ships in their ambition to capture the emigrant trade. Indeed, the 16,000 ton *Deutchsland* of the Hamburg Amerika Line attained 24.37 knots. Another seeming threat to British maritime supremacy on the transatlantic route was the growing presence of the American financier J. Pierpont Morgan. Morgan built no ships, but he was adept at forming trusts that would monopolize the fields he deigned to invade. Following some successes at home with coastwise shipping, he moved into ocean shipping. An early target was the White Star Line.

The Ismay family resisted the advances of Morgan, but a majority of stockholders accepted the invader's terms, and White Star became a subsidiary of International Mercantile Marine, effective 31 December 1902. However, the White Star house flag still flew, and Bruce Ismay became president and managing director of IMM with full consent of Morgan. William Pirrie, manager and director of White Star's favored shipbuilder, Harland & Wolff, and who had helped arrange the takeover attempt, was also a prominent member of the new management.

J. P. Morgan continued his advance toward monopolizing the North Atlantic. With some of the German shipping lines falling under his control, he monopolized much of the emigrant business. But apparently his appetite remained unsated—Cunard was to be his next prize. He may well have succeeded in this ultimate courtship, but British pride in its merchant marine could not accept its consummation by this wealthy American: the government in the form of the Royal Navy intervened with a subsidy that would allow Cunard to build two ships to regain the speed prize without first sacrificing both soul and ship to the American dollar. The upshot was the building of the *Mauretania* and *Lusitania*, ships that were to be successful challengers for the title "fastest" after their advent in 1907.

The shadow of the new Cunarders lay over White Star in its unending need to compete successfully with foreigners and countrymen alike. The powering of the Cunard ships by steam turbines was unprecedented among the first-tier liners of the North Atlantic, but this machinery was the key to unsurpassed speed rather than to any other quality evident to passengers. However, even if White Star did not care to be provoked into a speed challenge, Cunard's success with turbines presented design alternatives that the former could not ignore. Further, the simple presence of outstanding new ships by a principal competitor meant that White Star must respond, and with ships that would outshine the competition in ways appealing to its ocean-crossing customers.

The announcement of three new German liners, *Vaterland*, *Imperator, and Bismarck* also goaded White Star toward construction of a new generation of ships. These were to be large liners with luxurious accommodations, a large capacity for third class, but moderate speeds of 23–24 knots.

Early in 1907 Bruce Ismay dined with Lord William Pirrie, chairman of Harland & Wolff to explore these concerns. New ships, a class of three,[5] were to rival the new Cunarders in size, safety, comfort, elegance—all things that for the past decade had been elements of White Star success. Reaching for speed prizes was to be forgone, but the ships were to be 50% larger (by gross tonnage) and 100 feet longer than the Cunard ships. They were to carry over 3,400 passengers and crew. Their design speed was to be 21.5 knots. The two men reached agreement, and within three and a half years the first of the *Olympic* class was launched at Belfast. The naval architect in charge of the design, who was also managing director of the shipyard, was Lord Pirrie's brother-in-law, Alexander Carlisle. However, another naval architect and nephew of Lord Pirrie, Thomas Andrews,[6] was the chief overseer of the two ships' design and

This is a starboard view of the *Olympic* in 1911. The sister-ships *Olympic* and *Titanic* were almost identical with the exception of "B" Deck Promenade in *Titanic*, which was enclosed. Note that the lifeboat complement is the same as *Titanic*'s.

construction, while Carlisle devoted his efforts to managing the shipyard. Thomas Andrews was assisted by Edward Wilding, who was responsible for the design calculations. When Carlisle left Harland & Wolff to take a position with the Welin Davit Company, Andrews became the managing director of the shipyard and Wilding became his assistant.

The decision led to the initial maiden voyage of the *Olympic*, in June 1911, but its consequences appeared long before. Both shipyard and ports had to be improved in response to the projected size of the ship. Between 1906 and 1908 Harland & Wolff built a pair of slipways, 840 feet long and 240 feet wide, on which two ships of *Olympic* size could be built, side by side. A gantry, providing trackage for mobile cranes over the ships, was built over the new ways. With an elevation of 214 feet to the highest point of the cranes, it became a dominating fixture of the Belfast skyline. Heavy lifts—the boilers, for example—would be made while the ships lay at their outfitting berths after launching. For

This is a view of the construction slip on 8 May 1909. The tank top of the *Titanic* is completing in the foreground and the plating tank top and the erected aft frames for the *Olympic* are in the adjoining slip in the background. This was one of the rare times in maritime history where two superliners were being built side-by-side. (*Courtesy of the Ulster Folk Museum*)

this task, a floating crane that could lift 150 tons with a reach of 130 feet was ordered from Germany.

The new ships would be based in Southampton, requiring a new deepwater dock at that port. In New York, construction of the new Chelsea Piers was put in train with the intention of having them in service by the time of their first maiden voyage.

The characteristics of the *Olympic* and *Titanic*, as built, are outlined in Table 8–1. The design featured four 22-foot diameter funnels rising 62 feet above the deckhouses, a high cutaway counter stern—reminiscent of sailing ships—nine steel decks, and two large raked masts that supported the wireless aerials.

Many alternatives were considered during the preliminary design

TABLE 8–1 *Principal Characteristics of the* Olympic *Class*

	Olympic	*Titanic*	*Britannic*
Displacement, full load (tons)	52,250	52,310	53,220
Gross register tonnage	45,324	46,329	48,158
Length overall	882' 9"	882' 9"	882' 9"
Length between perpendiculars	852' 6"	852' 6"	852' 6"
Beam, maximum	92' 6"	92' 6"	94' 0"
Depth of strength deck to bottom of keel	65' 3"	65' 3"	65' 3"
Draft, full load	34'	34' 7"	34' 7"
Draft, maximum	36' 1.5"	36' 1.5"	36' 1.5"
Draft, light	27' 3"	27' 3"	27' 3"
Rated shaft horsepower	46,000	46,000	50,000
Service speed	21 knots	21 knots	21 knots
Trial speed	21.7 knots*	—	—
Passenger capacity:			
First class	1,054	905	790
Second class	510	564	830
Third class	1,020	1,134	953
Crew	860	900	950
Total	3,444	3,533	3,523

*June 1911

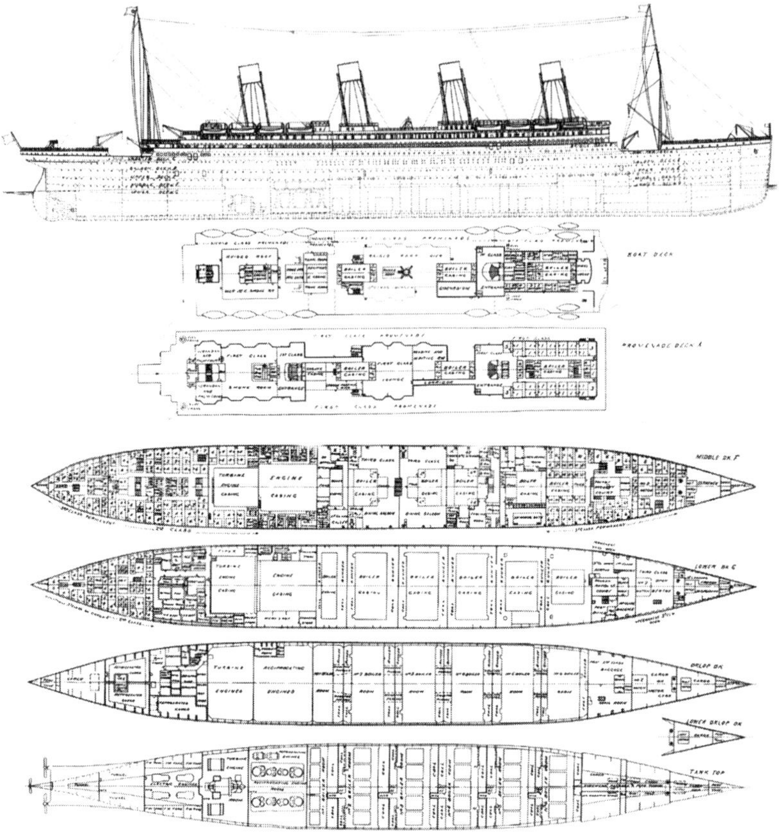

FIGURE 8-1 Profile and deck plans for the *Olympic* class. (*From* Engineering, *vol. 90, 1910*)

process; four significantly different ones can be identified. One of those is worth mentioning here because of the implications it had on the night the *Titanic* went down—the one that featured lifeboat capacity for the entire ship's complement.

The ships were built under the regulatory umbrella of the British Board of Trade. Under rules promulgated in 1894, ships of over 10,000 gross tons were required to carry only 16 lifeboats. Given the typical lifeboat size, this number was equivalent to only about 1,100 persons. Alexander Carlisle and Thomas Andrews, well aware of the enormous discrepancy between that figure and the number of persons who would come into hazard in a sinking, believed that the Board of Trade would update its regulations, and asked the Welin Davit Company of Sweden (vendors for the lifeboat davits) to prepare a design for the carriage of 64 boats, even though 48 might be enough. The requisite boat arrangement is shown by an illustration in the chapter on the lifeboat problem. It required the stacking of three boats at each pair of davits, an arrangement that would clutter the topsides of the new giants, and might even require reduction of public spaces on Boat Deck. A perhaps more telling point was the lack of similar initiatives by other shipping companies. Harland & Wolff and White Star had been caught by the problem of an imminent change in the Board of Trade rules and had endeavored to cover for the most severe case of boats for all being required. As it turned out the Board of Trade was thinking of fewer boats rather than more until the *Titanic* tragedy changed their minds. The White Star Line was reluctant to adopt the arrangement because it did not want to be the first shipping line to break precedent and it also felt that so many lifeboats would spoil the appearance of the ship. Bruce Ismay was shown the lifeboat plans, but later claimed that they were not discussed during a two-day conference on design changes being offered by Harland & Wolff. In all, no one after 15 April 1912 would admit to having made the fateful decision.

The *Titanic* is shown under construction with the gantry overhead. The photograph was taken in way of one of the uptakes. Note the rows of pillars and transverse framing of the deck structure. (*Courtesy of the Ulster Folk Museum*)

The designers of the propulsion machinery were faced with two radically different types of steam engine: the traditional reciprocating engine and the innovative turbine. On the one hand, Harland & Wolff could offer an advanced reciprocating engine of its own design and manufacture that was efficient and essentially vibration free. On the other, the two new Cunard liners had successfully demonstrated turbine engines designed and patented by Sir Charles Parsons, a dominant figure in the development of the steam turbine.

The propulsion machinery adopted a compromise, a combination of reciprocating engine and steam turbine. This failure to opt for the turbine alone, in spite of its having proved itself by several years' service with Cunard, might seem to be a step backward in marine engineering. However, given the lack of need for highest possible power in a reach for the highest possible speed, the choice was an excellent one, since it allowed exploitation of the best features of the two different engine types. Further explication of that point is found in Chapter 10.

The two reciprocating engines of the *Titanic* can be seen completing on the test bed in the erection shop before being moved to the outfitting pier for installation. The four cylinders of the port engine can be seen in the foreground. (*Courtesy of the Ulster Folk Museum*)

That essential adjunct to the propulsion engines, a boiler plant to produce the necessary steam, was based on standard practice for transatlantic liners of the time. It consisted of 29 cylindrical ("Scotch") boilers, hand-fired with coal. Twenty-four boilers were double-ended, the remainder single-ended. These were essentially the same boilers that had been chosen for the Cunard ships; the principal differences were in the number of furnaces per boiler, and in their draft systems. The Cunard boilers had four furnaces at each end, and were supplied with air by the Howden forced-draft system; the White Star ships had only three furnaces at each end, and relied on natural draft.

In sharp and significant contrast to the *Lusitania* and *Mauretania*, coal was stowed in cross bunkers, i.e., coal stowage spaces that ran from side to side between boiler rooms. The ostensible rationale for this scheme was that it placed a source of fuel close at hand even for the boilers located on the centerline. A seeming disadvantage was the lack of protection from collision that the Cunard-style longitudinal bunkers seemed to provide. Indeed, one can reasonably speculate that if the *Titanic* had been given those longitudinal bunkers, and *if* their coal-passing doors could have been closed expeditiously after the collision, the ship would have survived. On the other hand, in the actual event, no bunkers impeded cross-flow of water, and hence no significant list developed to worsen the plight of the ship's people.

The machinery plant and the factors that underlay its design are discussed in greater detail in Chapters 9 and 10.

There were 16 watertight compartments that were bounded by transverse watertight bulkheads (see Table 8-2) that were carried up to at least the height of the "E" Deck. The two forward bulkheads (labeled as A and B) and all the bulkheads aft of amidships (K through P) were watertight up to the underside of "D" Deck. The collision bulkhead (A) extended to "C" Deck, but it was watertight only to "D" Deck. The first two bulkheads forward and aftermost bulkhead (P) had no openings in them. All the other bulkheads had openings that were fitted with watertight doors. The transverse bulkheads within the machinery plant (bulkheads D through O) each had a vertical sliding watertight door at the level of the floor plate (in the engine and boiler rooms for the use of the engineers and firemen). There was one watertight door at the level of the Orlop Deck (bulkhead N) that was used for access to the refrigerator rooms. There were no watertight doors in the transverse bulkheads on "G" Deck. On both "F" and "E" Decks almost all the transverse bulkheads were fitted with watertight doors to allow communication between the different blocks of passenger accommodations. All the doors outside the engine

TABLE 8–2 *Heights of Watertight Bulkheads*

Bulkhead No.	*Deck Extends to*	*Height Above Keel*	*Space Forward of W.T. Bulkhead*	*Space Aft of W.T. Bulkhead*
A	C	65.08	Fore Peak Tank	Cargo Hold No. 1
B	D	54.58	Cargo Hold No. 1	Cargo Hold No. 2
C	E	45.08	Cargo Hold No. 2	Cargo Hold No. 3
D*	E	45.08	Cargo Hold No. 3	Boiler Room No. 6
E	E	45.08	Boiler Room No. 6	Boiler Room No. 5
F	E	45.08	Boiler Room No. 5	Boiler Room No. 4
G	E	45.08	Boiler Room No. 4	Boiler Room No. 3
H	E	45.08	Boiler Room No. 3	Boiler Room No. 2
J	E	45.08	Boiler Room No. 2	Boiler Room No 1
K	D	54.08	Boiler Room No. 1	Reciprocating. Eng. Rm.
L	D	54.08	Reciprocating Eng. Rm.	Turbine Room
M	D	54.08	Turbine Room	Electric Plant
N	D	54.08	Electric Plant	Refrigerated Cargo
O	D	54.08	Refrigerated Cargo	Cargo Hold
P	D	54.08	Cargo Hold	Aft Peak

*There was another watertight door at the after end of the watertight passage through the bunker immediately aft of D bulkhead. This door and the one on the D bulkhead formed a double protection to Boiler Room No. 6.

rooms and boiler rooms were horizontal sliding doors workable by hand both at the door and at the deck over.

The twelve vertical sliding watertight doors in the boiler and engine rooms could be closed from the bridge. A warning bell, rung from the bridge, preceded the door closing. Once these vertical sliding doors were closed by electric power, they could not be opened until this power was turned off on the bridge. The doors could be opened only through a manual mechanical operation at each door. They could, however, be individually lowered again by operating a lever at the door. In addition, an opened door would automatically close if water should enter the compartment. A float device that was activated when water flowed in through the door opening would activate these vertical doors.

By their dimensions, the three ships were clearly to be the largest on the North Atlantic. Elegance . . . luxury . . . these are more subjective in-the-eye-of-the-beholder qualities, quite at the mercy of contemporary taste. Commentators of a later century, especially those who are mere engineers, should refrain from passing judgments on White Star's accomplishments in the aesthetic domain. Readers who would like to form opinions in this regard should consult sources[7] that treat such matters, and make comparisons between ships. Here we shall merely outline features that were especially remarked on, circa 1910.

First-Class Spaces

The superiority in speed of the *Lusitania* and *Mauretania* (as well as the German liners) was conceded—elegance and comfort would draw passengers to the White Star liners. Even on first boarding, passengers would be awed by a magnificent Grand Staircase, soaring through seven decks, and capped by a glass dome. By its size and elegance the staircase might well constitute the signature of the *Olympic* class in the eyes of its first-class passengers.

This main access featured engraved live oak paneling, richly trimmed rails, and carpeted stairs and decks and may be compared to multi-deck atriums in modern cruise ships. The Grand Staircase extended through seven decks, being 60 feet in height and 16 feet in width. The overall décor suggested the English decade of William and Mary, contrasted by wrought iron scroll work reminiscent of the Louis XIV era. Paneling in the staircase featured a clock framed in an elaborately carved Honor and Glory motif.

Three electric elevators flanked the Grand Staircase, serving six decks. Although the *Lusitania* and *Mauretania* had pioneered this convenience, elevators aboard ship were still regarded as innovative.

The after staircase, also paneled in oak, extended through five decks. It was fitted between the intakes and uptakes for the third funnel and the machinery casings to reciprocating and turbine engine rooms. In the last stages of the sinking, the *Titanic*'s hull fractured in this vicinity, indicating a possible weakness influenced by this large opening. This matter is discussed in a later chapter.

A swimming pool (30 feet by 14 feet), Turkish bath, and electric baths on "F" (Middle) Deck were innovative features. Pools were later to become essential features of all transatlantic liners—and of cruise ships, of course—and were always located on the centerline of the ship. The *Olympic* pools were placed off-center to starboard. However, there was to be no known consequence of this arrangement in the sinkings of the *Titanic* and *Britannic*.

This is the a-la-carte Restaurant and the Palm Court outboard on the *Titanic*. (*Courtesy of the Mariners Museum*)

This is the luxuriously furnished first-class Smoking Room on "A" Deck of the *Titanic*, where men spent many social hours aft or before dinner. Note the motif of the stained glass window and the height of the space. It was equipped with a raised roof. (*Courtesy of the Mariners Museum*)

The Turkish bath included steam, hot, temperate, and cooling rooms, as well as a space for shampooing. The cooling room was decorated in an Arabian style of the 17th century. Portholes in this room were covered by an elaborately carved Cairo curtain through which light created an atmosphere of the Mysterious East.

A gymnasium on the starboard side of the Boat Deck, and squash court above the aft end of Cargo Hold 3 attracted the athletically inclined passengers. Neither were firsts for transatlantic liners, but the Dining Saloon was the largest afloat, extending the full width of the ships and 114 feet lengthwise. The space was paneled in oak in a Jacobian English style of the 17th century. Natural lighting came through groups of four and six sidelights behind stained glass that was fitted into a bay window arrangement, giving the effect of dining in a large mansion. The room was furnished as a restaurant might be, with a large table and small tables on the centerline, and a few tables of medium size outboard to port and starboard. The arrangement allowed groupings for private parties. Overall, only the elegance and size (300 feet long) of the Versailles Court first-class Dining Saloon of the *Normandie* surpassed the attractions of the *Olympic* class.

An a la carte restaurant, 60 feet long and 45 feet wide, was located on "B" Deck at the aft end of the deckhouse. Like the Dining Saloon, it was accompanied by a reception room. Styling was Louis Seize with paneling in French walnut stained to a light tan color. A bandstand was provided at the after end.

A Verandah Café and a Palm Court were located at the after end of each side of the first-class Smoking Room on "A" Deck. Entrances from the open deck were made through draft-proof revolving doors. The motif was Treillage of the Louis Seize period with live ivy and climbing plants trained to grow up a trellis panel of French walnut.

"A" Deck, located one deck below the Boat Deck, had several notable first-class public rooms. It featured a long and irregular deckhouse, extending nearly the whole length of the deck. This deckhouse was divided into three structures by two expansion joints, one located between the two forward funnels and a second located between the third funnel and the main entrance for the second-class passengers. Thirty-four first-class staterooms were located at its forward end. Aft of these staterooms were the Grand Staircase and several first-class public rooms, a Reading Room, Lounge, the Smoke Room, and the Verandah and Palm Court. Outside the deckhouse was a closed promenade for first-class passengers. In the *Titanic*, the forward end of the "A" Deck promenade on both sides of the ship, below the forward group of boats and for a short distance further aft, featured a steel screen, 192 feet long with large windows to protect against the weather. This was an exterior recognition feature between the two ships.

The first-class Smoke Room was intended to be one of the finest afloat. Its dimensions were 65 by 63 feet, with a 12.5-foot ceiling height. Styling was early Georgian. Paneling was mahogany, set off by carvings in mother-of-pearl. A large open fireplace was located at the aft end of the room; over it hung the painting "Approach to the New World" in the *Olympic*, "Plymouth Harbor" in the *Titanic*. Thomas Andrews is believed to have chosen this place to wait after hope was gone for rescue on the dark morning of 15 April 1912. He was last seen there, sitting with arms folded, a life jacket on the table behind him.

The first-class Lounge was located on the Promenade ("A") Deck just aft of the Grand Staircase. Its styling was adapted from that of a Louis Quinze apartment at Versailles. The room was intended for conversation, reading, card games, and afternoon tea. An open fireplace was located at the forward end, a bookcase with popular reading material at the aft end. A reading and writing room styled in 1770–1780 Georgian, was adjacent to the Lounge. The paneling and ceiling were in white. A large bay window gave the passengers a wide view over the Promenade and the sea beyond.

"B" Deck featured a number of first-class staterooms and cabins with an a la carte restaurant and the Café Parisien at the aft end. There were more first-class staterooms on "C" Deck.

The main first-class Dining Room (or Saloon) was located on "D" or Saloon Deck with a reception room located between it and the Grand Staircase. Forward of that were more first-class staterooms.

Second-Class Spaces

The clientele intended for second class were the tourists of the British middle class. They were treated to an atmosphere less ornate than that of first class, but White Star intended that they nonetheless have the finest accommodations.

Second-class passengers boarded the ships at an entrance abreast of the fourth funnel.

The second-class Dining Saloon, located on the Saloon ("D") Deck aft of the main galley, extended the full width of the ship for a length of 71 feet, in all a capacious room that could seat 394 diners. It featured long rows of tables with swiveling chairs, giving a level of comfort that matched that of first class on many other ships. Paneling was solid Eng-

This is a view of the port side of the *Olympic*'s Promenade Deck ("B" Deck) looking aft. The windows on the left are for first-class cabins at the forward end of the deckhouse. The forward expansion joint can be seen under the sign, "Emergency Station A." The joint is about two inches wide and covered with a leather strap on the house side and bulwark. The deck had a brass plate covering the gap in the deck structure with one end free.

lish oak, that noble wood that had sustained Old England at sea for many centuries.

The second-class Library was a spacious and carpeted space, 40 feet in length and 58 feet in width, on Shelter ("C") Deck. Paneling was a combination of sycamore and mahogany; furniture was mahogany. Overall, the styling was that of Colonial Adams.

The second-class Smoking Room was placed on "B" Deck, directly over the Library. Décor was that of the Louis Seize period with paneling of oak, a linoleum deck covering, and oak furniture with dark green leather.

Third Class (Steerage) Spaces and Crew Spaces

Although third-class ("steerage") accommodations were austere and primitive by standards extant in the late 20th century, they were a vast improvement over what was provided in contemporary ships. And well they might be, since the revenue from the carriage of immigrants (the mainstay of third class) was a vital factor in the profitability of the transatlantic lines of the early century. Nonetheless, the cabins were small compared to the third-class (typically labeled "tourist") cabins of the last transatlantic liners, such as the *United States*, *France*, and *Queen Elizabeth II*. The single men were berthed forward in a common room; single women and families were berthed aft, though in small cabins that in many cases were fitted with six berths. Heating, ventilation, and electric light were provided. Common spaces were a smoking room, dining room, and a lounge, and two open-deck promenades. Although the promenades were in the well decks, one forward, one aft, and obstructed by cargo hatches and cargo-handling cranes, they were furnished with deck chairs and tables suitable for all weather conditions to be encountered on the North Atlantic.

The dining facility was divided between two compartments amidships on Middle ("F") Deck, seating 473 diners. A galley and pantry were located aft of the dining space. The entire facility was directly over the boiler rooms. Though this location might be considered poor in comparison to those of the other classes, its low and midship location minimized the ship motions that so often trouble the seaborne diner. The food was also plain, but good and plentiful. By the standards of the time, both food and accommodations were a wonderful value.

The third-class General Room and Smoking Room were placed far aft, in the Poop Deckhouse aft of the Well Deck (which was also the third-class promenade). The General Room, paneled in pine with a white enamel finish, was to starboard; the Smoking Room, to port, was paneled in oak with teak furniture. The third-class promenades, each about 50 feet in length, were located on "C" Deck nestled between the forecastle and deckhouse forward and the poop and deckhouse aft.

The forward end of the Boat Deck was the Wheelhouse where the steering wheel and steering compass were located. The Chart Room was located immediately aft. On the starboard side of the Wheelhouse and funnel casing were the navigating room, the captain's quarters, and some of the officers' quarters. On the port side were the remainder of the officers' quarters. On the centerline aft of the forward funnel were the wireless telegraphy rooms and the radio operators' quarters. The top of this deckhouse formed a short deck. The connections from the Marconi aerials were made from this housetop and two of the collapsible boats were also placed on it at the aft end. Aft of the officers' house was the Grand Staircase, the first-class passengers' entrance, and stairways, and other adjuncts to the passengers' accommodation below. The Gymnasium was also located on the Boat Deck.

At the aft end of the forecastle here were the crew's galley and the seamen's and firemen's messroom. Also located here on each side of the ship, were the entrances to the third-class spaces below. On the port side, at the extreme after end and opening on to the deck, was the Lamp Room. A break in "B" Deck between the forecastle and the first-class passenger quarters formed an open space about 50 ft in length which was used as a third-class promenade. This area contained two hatch openings, one for the No. 2 hatch and the other for the reserve coal bunker hatch. The latter of these hatchways also provided access to the first- and second-class baggage hold, the Mail Room, and to the lower hold, which was used for cargo or coal stowage.

The forward end of this deck provided accommodation for 108 firemen who served in two separate watches. Lavatories were fitted at the aft end of these quarters on each side of the ship. Immediately aft of these firemen's quarters and on each side of the centerline, there was a vertical spiral staircase leading down to the tank top to an enclosed Firemen's Passage that extended to Boiler Room No. 6. This allowed the firemen to go to and from their work without mingling with passengers. Just aft of this area on "D" Deck was the third-class enclosed promenade that was covered by "C" Deck. From this promenade, third-class passengers could go down to their quarters on "E," "F," and "G" Decks.

At the forward end of "E" Deck there were accommodations for three watches of trimmers, in three separate compartments, each holding 24 trimmers. Aft of this, there were accommodations for 44 seamen on the port side. Further aft, and also on the starboard side of it, were the lavatories for crew and third-class passengers; further aft again came the forward third-class lavatories. Immediately aft of this was a passageway right across the ship that led to stairwells to decks above and below and port and starboard gangway doors. This passage was 9 ft wide at the sides, increasing to 15 ft at the center of the ship.

From the after end of this cross passage main alleyways on each side of the ship ran right through to the after end of the vessel. That on the port side was about 8 ft wide. It was the general communication passage for the crew and third-class passengers and was known as the "working passage" or "Scotland Run." This passage ran 475 ft of "E" Deck with a slight jog around the Reciprocating Engine Room casing. Direct access was made from this passage to the third-class Dining Saloons on "F" Deck. On the port side of the working passage there were accommodations for the petty officers and most of the stewards. An Engineers' Messroom was located outboard of the engine casing. From this passage access was also obtained to the Reciprocating Engine Room, the Turbine Room, and all the boiler rooms and the engineers' accommodation, some third-class lavatories and also some third-class accommodations at the after end. There was another cross passage at the end of this accommodation about 9 ft wide, terminating in gangway doors on each side of the ship. The port side of it was for third-class passengers and the starboard side for second class. A door divided the parts, but it could be opened for any useful purpose or for an emergency. The second-class stairway leading to the Boat Deck was in the cross passageway.

The forward end of "F" Deck had accommodations for 15 leading firemen and 30 greasers. The next watertight compartment contained third-class accommodation in 26 rooms for 106 persons in third class. The next watertight compartment contained the first-class baggage room, the post office and registry, a racquet court, and seven third-class rooms for 34 passengers. From this point to the after end of the boiler rooms the athwartships spaces forward and aft of the stokeholds were used for the stowage of coal.

Alongside the reciprocating engine room were the engineers' stores and workshop. Abreast of the turbine engine room were some of the ship's stores. In the next watertight compartment aft of the Turbine Room were the chill and freeze stores for perishable stores. The next two compartments were appropriated to 186 third-class passengers in 60 rooms; this deck was the lowest on which any passengers or crew accommodations were located.

Lower Decks. Below "G" Deck were two partial decks, the Orlop and Lower Orlop Decks, the latter extending only through the fore peak and No. 1 hold; on the former deck, aft the turbine engine room, were some storerooms containing stores for ship's use.

Below these decks again came the inner bottom, extending fore and aft through about nine tenths of the vessel's length, and on this were placed the boilers, main and auxiliary machinery and the four dynamos that provided the ship's electric current. In the remaining spaces below "G" Deck were cargo holds or 'tween decks, seven in all, six forward and one aft. The Firemen's Passage provided direct access from their accommodation to the forward boiler room by stairs at the forward end of Cargo Hold No. 2. It also contained the various pipes and valves connected with the pumping arrangements at the forward end of the ship, and also the steam pipes that carried steam to the windlass gear forward and exhaust steam pipes leading from winches and other deck machinery. The Firemen's Passage was made watertight on its sides and overhead throughout its length, and at its after end was closed by a watertight vertical sliding door of the same character as other doors on the inner bottom. Special arrangements were made for pumping this space out, if necessary. The pipes were placed in this tunnel to protect them from possible damage by coal or cargo, and also to facilitate access to them.

Hull Structure. The *Olympic*-class ships were transversely framed with a frame spacing that varied. In the forepeak area it was 24 inches to the collision bulkhead. That spacing increased to 27 inches in the forward end of Cargo Hold No. 1 and 30 inches in the after part of that space into Cargo Hold No. 2. In Cargo Hold No. 2, the frame spacing increased to 33 inches over the forward part. Finally the spacing increased to 36 inches in the middle portion of that hold. This frame spacing was maintained through all of the mid-length of the ship. That spacing was stepped down over the stern end until it was 24 inches in way of the aft-peak area.

The double bottom structure of the *Olympic* class was a very stout structure. It was 63 inches deep[8] in all parts of the ship except in way of the Reciprocating Engine Room where it was increased to a depth of 76 inches. There were four longitudinal girders with a center vertical keel. Six longitudinal girders would be used in the *Britannic* because of her greater beam. A bar keel, 19.5 inches by three inches in thickness, was joined to the flat plate keel that ranged in thickness from 1.2 to 1.5 inches. This structure was the last to fail during the hull failure of the *Titanic*.

For a very appropriate instance, all the steel rivets in the *Olympic* ships were hydraulically riveted.[9] These rivets would be found between the keel and the margin plates and the three strakes of plates below "B" Deck in the amidships 3/5ths length with wrought iron rivets found elsewhere. The use of steel rivets was expanded in the construction of the *Britannic*. That 3/5ths midlength was the realm where everyday bending stress would be the highest, but the forward portion of the hull was where harsh contact between hull and ice would ravage the seams of the *Titanic*'s hull. It also would be the same area where a mine devastated the bow structure of her sister *Britannic*. There will be discussion of that point in a later chapter. In contrast, all the rivets in the *Lusitania* and *Mauretania* were steel.

The other choice of possible significance was the option of using hydraulic riveting, i.e., riveting by a machine that developed the rivet-forming force hydraulically rather than by hammering. This technique was used in building both the Cunard and the White Star ships, with the principal incentive of speeding the process. The John Brown Shipyard in Glasgow pioneered this process in the construction of the *Lusitania*. During the period the *Olympic* class was under construction, the

John Brown and Harland & Wolff Shipyards signed a cooperative agreement. John Brown would build the turbines for the *Olympic* and *Titanic* and also would provide the hydraulic riveting technology to the Harland & Wolff Shipyard. The fact is that this technology was used only in the driving of steel rivets in the *Olympic*-class ships. Hydraulic riveting offered a faster and more economic way of constructing such large ships just as the use of automated welding would a century later. But we do believe that there may have been some deleterious consequences in the case of the *Titanic*. This will be explained in a later chapter.

The design was approved on 31 July 1908, signified by a contract between the Oceanic Steam Navigation Company (White Star) and Harland & Wolff for the construction of the first two ships. An option for construction of a third was also included. The financial terms were cost-plus, with White Star to specify vendors. With their customer to meet all costs, the shipyard had no incentive to purchase substandard equipment or materials, and no one can say that it did.

Beginning of construction was marked by the keel laying of the *Olympic* (Harland & Wolff's hull 400) on 16 December 1908. The keel laying for the *Titanic* (hull 401) followed on 31 March 1909. Construction of two large liners simultaneously was a first. The offset of three months (which had stretched to ten months when the *Titanic* was delivered) between the two was a deliberate measure to preclude the delivery and processing of large machinery items simultaneously. The *Olympic* was launched into the River Lagan on 20 October 1910, and was followed by her sister on 31 May 1911. The *Olympic* made a successful maiden voyage from Southampton, beginning on 14 June 1911.

The *Titanic* was completed in April 1912, and after brief sea trials, steamed to Southampton to prepare for the maiden voyage sailing on 10 April. Numerous small changes had been incorporated in this second ship of its class. Most notable interior change was the addition of the Café Parisien, a reproduction of a French sidewalk café, added where the restaurant promenade had been on the *Olympic*. With this change and the alteration of the forward end of "A" Deck, there were additions to the ship's internal volume, making its gross tonnage some 1,000 tons greater than that of the *Olympic*. In addition, the forward end of the first-class promenade on "A" Deck was enclosed to protect passengers from wind and spray coming over the bow. Changes such as these added to the ship's internal volume, making its gross tonnage some 600 tons greater than that of the *Olympic*. The title of "world's largest ship" thereby devolved upon the *Titanic*.

After the Cunard Line ordered the *Aquitania* and the German Hapag Lloyd announced its decision to proceed with the building of the 50,000-ton *Bismarck*, the White Star Line announced that it would exercise its option for a third *Olympic*, though with some modifications. This ship, to become the *Britannic*, was ordered on 20 June 1911, just as the *Olympic* was completing her first passage to New York.

Of the three *Olympic*-class sisters, only the lead ship successfully crossed the Atlantic. The *Titanic*, as the world knows, never completed her maiden voyage, or even the westward passage. *Britannic* was completed during World War I, and immediately went into service as a hospital ship. Her wreckage lies on the seabed near the Greek Island of Kea in the Aegean, the consequence of contact with a German mine on 21 November 1916. When the post-war renaissance of transatlantic shipping began in the early 1920s, only the *Olympic*, known by then as "Old Reliable," was there to resume service to New York. With the restrictive revision of American immigration laws, effective in 1923, the bottom fell out of the once lucrative third-class market. With the advent of worldwide depression in the early 1930s, first- and second-class trade suffered as well. Having no government subsidy, White Star tottered toward bankruptcy, only to be saved by a merger with Cunard in 1934 into what was known for a while as Cunard-White Star. *Olympic*, grown old, and facing competition from the many ships built in the 1920s (and from the coming *Queen Mary* and *Queen Elizabeth*) was kept in service for only a few years; during 1935 she was towed to the scrap yards.

Within a few more years there was no more White Star; after World War II Cunard-White Star was once more just the Cunard Line.

About *Britannic*

Are "sister ships" ever identical? Where sisterhood is extended to mass-produced ships like the Liberty Ships of World War II, then pairs (or groups) of truly identical ships might be found. Where the ships are large and complex, are built with some time lag between ships, and especially when they are built at different yards (as were *Lusitania* and *Mauretania*), the answer almost certainly has to be negative. Hull and machinery of the first two White Star ships may have been identical, but differences in passenger accommodations gave *Titanic* a gross tonnage 1,000 tons greater than the *Olympic*, and with that the title of World's Largest Ship. *Britannic* was yet larger, and in overall hull dimensions as well—her beam was 18 inches greater than that of the first two ships (reasons to be discussed in Chapter 18). Displacement tonnage and gross tonnage were correspondingly greater. Boilers were slightly bigger, and rated power was 50,000, up from the 46,000 of the first two ships. Lifeboat outfit of *Britannic* was radically different as a consequence of her design well after the *Titanic* event (description in Chapter 27).

The completion of *Britannic* after the outbreak of World War I surely had an impact on her internal décor, for all that she otherwise would have been near-identical to that of the first two White Star ships. In any case, her short career, ending as a sunken hospital ship, makes discussion of such matters of only trivial interest.

Notes for Chapter 8

1. Descriptions of the *Olympic*-class ships can be found in the contemporary engineering literature such as *Engineering* and *The Shipbuilder*. A convenient source is a 1988 reproduction, *Olympic and Titanic*, Patrick Stevens, Welingborough, Northamptonshire.
2. A source for a complete treatment of White Star history is Roy Anderson, *White Star*, 1964, T. Stephenson & Sons, Prescot, Lancashire, England.
3. With its first ship in 1871, the *Oceanic*, the reconstituted White Star Line began the tradition of choosing ship names ending in -ic.
4. This quotation appears in the Roy Anderson book, note 2.
5. Shipping lines serving the North Atlantic passenger trade thought it important to offer a sailing with ships of a particular class on the same day of the week, every week. Given the distance to be covered, the reasonable speed of ships, and the time required in port for turnaround (especially while coal was still the fuel), adherence to this precept required three ships of near-identical characteristics. This ideal was rarely attained.
6. A biographical sketch of Thomas Andrews, based on a eulogy written by Sean Bullock, appears in Appendix 3.
7. Many books have been written about the transatlantic liners. Two among the several that treat the décor of the first-class spaces are John Maxton-Graham, 1972, *The Only Way to Cross,* MacMillan, New York, and Lee Server, 1996, *The Golden Age of Ocean Liners*, Smithmark Publishers, New York.
8. The dimensions of the plate and all rivet spacing and sizes in this book pertain to the RMS *Titanic* from the official document, "Form of Comparison of Scantlings of Unclassed Iron and Steel Ships with the Rules of Lloyds Register for 1885." The form was signed by surveyor, F. Carruthers.
9. This statement was extracted from the Shell Plating Plan of the SS *Olympic*.

CHAPTER 9

The Steam Turbine for Cunard

The near-standard propulsion outfit of 1900, reciprocating steam engine plus cylindrical fire-tube boiler, was threatened in that era by several innovations—watertube boilers, petroleum fuel, and turbine engines being the principal ones. All of these were eventually to triumph over the legacies of the 19th century, leading to steam propulsion plants that were distinctly lighter and more compact, and that typically consumed no more than half the fuel.[1] These new concepts had also come into use in 1900, but none had clearly passed beyond the experimental stage, and for none did its eventual dominance seem certain. In particular, watertube boilers and petroleum fuel were not to be found on a major transatlantic liner during the decade before World War I. Because they were not used by the ships treated here they receive only cursory attention in our discussions.[2] However, Cunard was soon to be a major innovator by powering *Lusitania* and *Mauretania* by turbine engines, and White Star to use them in combination with reciprocating engines in its *Olympic* class. Hence the more detailed discussion of the turbine innovation in this chapter.

The turbine reached the status of an acceptable marine engine quite late in the 19th century, but the concept of an engine that rotates without any reciprocating parts had surfaced early in the steam age. Reciprocation . . . expanding steam pushes piston, piston pushes connecting rod, connecting rod applies torque to a crank shaft: in that way the force of expanding steam creates the steady turning of a propeller shaft. The combination works well, but all that "running gear" adds weight, occupies a large part of the engine's volume, and requires numerous bearings to be lubricated (and watched carefully for overheating). But why can't the steam push the shaft around directly? That question was a near-obvious one early in the 19th century, setting many inventors to their tasks. They found that the why-can't-it question could be answered positively in several ways, allowing an assortment of rotary steam engines to be developed during that century. Many, perhaps all, *worked*, but never satisfied anyone that their virtues outweighed their inherent problems.

Then there was the turbine engine. It, too, is rotary, and thereby dispenses with the need for mechanical conversion of linear motion to rotation. However, it was successful when its round-and-round precursors were not because of one significant innovation: it separates the expansion of the steam from the rotation of the shaft,[3] and thereby allows continuous flow of the steam, and without valves that must open/close with every turn of the shaft. The expansion takes place through a nozzle, which is usually stationary; the high velocity jet from the nozzle then interacts with a bladed wheel. This continuous process permits a much higher rate of the steam flow than does the batch operation of a cylinder, and in consequence a small turbine can produce the same power that would require a monstrously large reciprocating engine, or its rotary equivalent. To be sure, the blades (also known as vanes or buckets) are usually precision components that require sophisticated machining operations in their manufacture, but by the late 19th century the necessary manufacturing techniques had been mastered.[4] Now, the turbine did have some drawbacks that kept it from quickly replacing the reciprocating engine, but once the essential manufacturing processes were reduced to practice, the turbine was ready to begin its steady march to dominance over reciprocation.

Almost inevitably, there was professional inertia, a concatenation of many conservative minds that stood in the way of the turbine's adoption by the marine engineers. All of that was shattered by the bold stroke of Charles Parsons, a renowned turbine pioneer, in 1897. In addition, a serious technical problem hindered widespread marine application: that of matching the speed of the turbine to the speed of the propeller. In simple terms, this matching problem can be described as a dichotomy between the demands of the Carnot Principle and the dictates of a propeller's interaction with the water. The Carnot Principle states that the fraction of mechanical energy extractable from a heat source (e.g., steam) is a function of the temperature of the source and of the sink (usually the sea in the case of marine engineering). The sink temperature provided by the sea falls within a narrow range, and is obviously beyond the powers of an engineer to change, so that in essence the Principle simply demands that the source temperature be as high as practicable. However, the source temperature is always limited by the capabilities of the technology used in exploiting the heating value of the fuel. In the present case, the burning coal might release its heat at temperatures over 2000°F, but those limited capabilities had set marine boiler pressures at about 200 pounds per square inch.[5] Without superheat (quite uncommon in 1900), the steam temperature at that pressure is 388°F. A calculation from elementary thermodynamics shows that if this steam expands through a nozzle to a pressure of 1.0 psia, the jet velocity is about 4100 ft per second. A calculation from elementary dynamics shows that maximum transfer of energy from the jet to a moving object (turbine blade) takes place when the object speed is about one-half the jet speed, thus about 2050 ft per sec for the linear speed of the point where jet and turbine blade interact. Now, the velocity of water approaching the propeller (also a turbomachine) is likely to be about 20 ft per second. The linear speed of the object accelerating this water aft, by the same principle used for the turbine blade, should thus be about 40 ft per second.

In an illustration that would incorporate all the real-world complexities of turbines and propellers, this simple comparison would be clouded by many assumptions that would greatly expand the magnitude of the computations and of our explanations. However, the simple case just sketched shows the required linear speed of a 1900 turbine blade to be about 50 times that of the propeller blade it might drive, a difference that no contribution from practical realities can change much. The upshot is that a turbine (whether steam or the later-day gas turbine) is fundamentally a much higher speed machine than a ship propeller. Given the reasonable diameters of the two machines, this linear-speed difference translates into an rpm difference of about the same magnitude.

The speed-matching problem was recognized even before the first attempt was made at turbine ship propulsion. One approach that was technically feasible was multi-staging of the turbine, perhaps suggested by the multiple-expansion reciprocating engines that were becoming common during the early development days of practical turbines. Turbines can also be multi-staged, i.e., exhaust from one turbine can be the input to a second turbine, mounted on the same shaft, whose exhaust becomes input to a third, and so on. For example, if the turbine just cited expanded the steam to 50 psia, the resulting velocity would be about 2300 feet per second; when the steam continued its expansion to 1.0 psia in a second stage, the velocity there would be about 3300 feet per second. These velocities are still much too high for the propeller, but additional turbine stages provide still smaller velocities. The possibilities are not infinite, to be sure, but staging can be (and commonly is) carried far beyond the triple and quadruple expansions used with the reciprocators. On the other hand, mechanical considerations do preclude using enough stages to make the best propeller rpm coincide with the best turbine rpm.

A second speed-matching remedy is to make the diameter of the turbine so large that the required linear blade speed is obtained at an rpm that suits the propeller. Photos of the turbines being built for the Cunarders *Lusitania* and *Mauretania* show this principle in practice: the turbine rotors appear monstrously large in diameter. To viewers familiar with the ship-propulsion turbines of the post-World War II era, those same photos also make obvious the many stages included in the pioneering Cunard engines.

By providing turbines with large diameters and many stages, their rotational speeds could be brought into a range suitable for ship propulsion, but turbines never attained a speed-matching with propellers that was common with reciprocating steam engines. The problem would be reduced to triviality, however, if speed-changing gears could be fitted between turbine and propeller. This solution to a vexing problem was recognized from the early days of the turbine, but its application had to wait until the technology of manufacturing acceptable gears caught up to the demand. That was not to occur until the 1920s.

An alternative method of speed reduction is electric drive, effected (usually) by a synchronous generator and a synchronous motor with different numbers of poles. However, this technique, like that of geared speed reduction, was not reduced to practice until well after the ships of interest here were built.

Meanwhile, the loss in propulsive efficiency inherent in the engine/propeller speed mismatch was accepted in applications requiring high power, and especially in those requiring high power without excessive weight and dimension of engine.

Turbines are afflicted with one other inherent defect, one that was of significant magnitude in their early days, but which was to be reduced to insignificance as the technology matured. That defect is leakage of steam past the end of the blades. Clearance between the fast-moving blade tips and the stationary casing is essential, and given that pressure must be lower in each succeeding stage, some of the steam is going to flow through the clearance without having done any of the work intended. The comparable problem in a reciprocating engine is eliminated by piston rings, but the high velocities at the tips of turbine blades makes the piston-ring solution impracticable. Nonetheless, partial sealing between blade tips and casing was gradually improved as turbines became increasingly common. In the early days, however, the leakage problem was serious enough to be cited frequently in published turbine versus reciprocator comparisons.

Now the professional inertia problem will be addressed. Inertia as a *physical* problem never troubled the steam turbine, but bureaucratic inertia was a hurdle for the turbine in a way that it never had been for the reciprocating engine. In the early 1800s the advantages of steam propulsion over sail and oars were so apparent that the concept sold itself. In contrast, the turbine of the late 1800s was initially just another kind of steam engine that would have to contend against more traditional machinery. But even before it went to sea, there was one place where it was recognized as offering an overwhelming advantage over reciprocation: the powering of small, high-speed craft such as torpedo boats.

The torpedo had reached the stage of practicality, to the potential bane of large armored ships, but its method of delivery was a matter of development. True, torpedo tubes were to be part of the armament of some 20th-century dreadnought battleships, but with the ranges at which battleships (and even lesser warships) were supposed to fight each other, the torpedo was a highly uncertain weapon. Submarines and aircraft would some day demonstrate its devastating utility, but at the time in question they were the concepts of Jules Verne and the like. The vehicle of choice for delivery in the 1890s was the torpedo boat, a small, low, swift craft beneath a belching cloud of coal smoke that would dart toward an enemy to launch the deadly weapon from comparatively close range. Such craft could be (and were) built with reciprocating engines, but they really needed turbines. The inability of the turbine to match well with good propeller speed might be a serious flaw to a commercial ship, since the resulting inefficiency of propulsion meant unacceptable coal consumption. What mattered most to the torpedo boat, however, was that few minutes as the rapier of the sea, thrusting at maximum speed through the geysers from the frantic enemy's near-miss shellfire to deliver the weapon against which there was no defense if it were to be aimed truly from close enough range. Speed was of paramount importance, and speed came from power, and much more power could be fitted in the form of a turbine. A lavish expenditure of coal was of trifling concern.

Charles Parsons, renowned in later years as the man who developed the marine steam turbine to the point of practicality, thought that way, but he was not welcomed when he went to London to convince the conservative denizens of the Admiralty. But that inertia problem, which manifested itself mainly in keeping the doors to the Admiralty (and to the minds within) shut, was a serious barrier to the innovator. Won't work, won't work . . . authority scoffed, and its members vied with each other in finding the reasons why. Oh, but they were to be sorry when Parsons embarrassed them in front of the Queen—*the* Queen, Victoria herself.

The occasion was Victoria's Diamond Jubilee, celebrated in 1897 by a naval review. Parsons was not invited—why should he be?—but he was there with his *Turbinia*, a craft something like a torpedo boat without the torpedoes, powered by Parsons turbines. The boat was capable of 34 knots, a speed that had never been approached by any watercraft, and maybe twice the maximum speed of the patrol boats that attempted to head him off from the approaches to the review. In war his hull might have been riddled by gunfire, but in a crowded and peaceful harbor the water-bobbies were completely frustrated. Parsons swept by the stern of the Royal Yacht, watched in surprise by Her Majesty[6] and in dismay by her attending admirals. We don't know what Victoria said to her keepers of the sea, or how they managed to explain themselves to her. No matter—the hero of the hour was soon to be knighted and all torpedo boats were soon to be powered by steam turbines. The *Turbinia* herself, after serving a few more years as a demonstrator and test bed, was retired into the marine museum of the University of Newcastle where she now can be seen.

That was 1897, and by 1900 the steam turbine had found a niche in the marine propulsion field, and much more would be heard from it within a few years. But beyond the realm of very specialized craft such as torpedo boats, the multiple-expansion reciprocating engine remained the day's standard. The biggest powering question facing the ship designer was "shall it be triple expansion or quadruple expansion?" With that settled, the key question remaining was the combination of stroke and cylinder bores that would produce the required propulsive power.

However, the canon was soon to be upset. Very soon: the great German liners (e.g., *Deutschland*, which captured the North Atlantic speed prize in 1900 at 23.51 knots, 36,000 indicated horsepower). British sea-pride was severely damaged. In addition, mundane commercial considerations demanded a response that was to manifest itself in the Cunarders *Lusitania* and *Mauretania*. Power that would be a significant step upward from any existing ship was essential to produce the speed that would trump the Teutonic intrusion, and the designers decided (quite successfully, indeed) that expansion of the torpedo boat solution into the realm of Atlantic liners might be the only reasonable way to get it. The company of Sir Charles Parsons built turbines totaling approximately 70,000[7] in horsepower, and the two Cunard liners set speed records that were not challenged until after World War I.[8]

That newer, larger, and more powerful ships would eventually replace the *Mauretania* as North Atlantic speed champion was inevitable, but it was a tough fight. After the Cunarder's boilers were converted to firing by oil, they could produce steam for 90,000 hp (apparently the shaft power rather than the inappropriate "indicated"). To match this new potential, the ship was taken out of service for seven weeks in late 1928 so that supposedly secret adjustments could be made to the turbines to take advantage of a higher steam flow.[9] This enhancement allowed a record (for the *Mauretania*) crossing of 26.9 knots in 1929, only about 0.9 knot less than the then-extant record.

Meanwhile, White Star deigned not to compete with Cunard or with upstart foreigners by speed, either when its prize ships were new, or in the post-war years. The White Star cynosure remained size and luxurious accommodation. Following roughly five years after *Lusitania* and *Mauretania*, its *Olympic* class came at a time when power plants that featured a series combination of turbine and reciprocating engine could offer economy and weight-saving advantages (outlined in Chapter 10). The White Star designers chose this combination, employing reciprocating engines on two propulsion shafts that exhausted steam into a low-pressure turbine on the third. The propulsion plants of both liner classes that we are to treat were, therefore, innovative by the standards of the early century. However, the innovation for the *Olympics* lay in a scheme for combining the best features of turbines and reciprocating engines, not in obtaining a maximum of power. Additional insight also appears in Chapter 10.

Notes for Chapter 9

1. We speak here of specific fuel consumption, e.g., pounds per horsepower-hour or kilograms per kilojoule-hour.
2. *Mauretania* and *Olympic*, the two survivors of that war, were converted to oil-burning before returning to peacetime service, but that circumstance is of minor consequence to our story.
3. One can argue that in a *reaction* turbine this separation is not present. The point can be disputed, but there should be no disagreement that in any kind of turbine the flow is continuous whereas in a rotary (or reciprocating engine) it is not.
4. Sophisticated blade shapes are not *required* for a turbine that will work; witness the simple flat blades of the traditional farm windmill (a "wind turbine"). The principal requirement for optimal extraction of energy from the moving fluid is correct speed ratio between fluid and blade. Aerodynamic sophistication adds a few points of efficiency.
5. It is assumed here that the 200 psi is "gage pressure," or pressure above atmospheric.
6. Story-telling improves a bit on historical fact at this point. Ill health kept the

Queen from attending the review. Parsons made his point with important persons nonetheless.

7. The figure 68,000 *indicated* horsepower is often quoted for the *Lusitania/Mauretania* turbines, an inappropriate parameter for turbines. See definition in Appendix 1.
8. Another German passenger ship, the post-war *Bremen*, took the Blue Riband in July 1929 with an average crossing at 27.83 knots.
9. The details of the turbine improvements may have been secret at the time, but they were surely based on an increased nozzle area, injection of additional steam at some turbine stage other than the first, or a combination of both.

CHAPTER 10

Machinery for the Olympic *Class*

Cunard was a pioneer in application of steam turbine propulsion to transatlantic liners: first its step forward with *Carmania,* then the great liners *Lusitania* and *Mauretania.* White Star was less bold in that it remained loyal to the reciprocating engine for its new ships. From another viewpoint, it was even bolder than Cunard in that it gave the *Olympic* class (and the immediately preceding *Laurentic*) a combination of reciprocating and turbine machinery, and that combination came close to being a unique application—very little before, very little after. It needs considerably more explanation than the steam turbine itself, for the turbine-alone innovation was to be made familiar by thousands of applications before the diesel engine became dominant late in the 20th century.

In the first year of the 20th century, the reciprocating steam engine, in either its triple- or quadruple-expansion form, was the standard propulsion engine for the Transatlantic liner. It was to drive the great German liners of new century—*Kronprincessin Cecilie, Kaiser Wilhelm II, Deutschland*—to record speeds across the Atlantic. It drove outstanding French liners such as *La Provence*, as well as all English liners before the advent of the *Lusitania* and *Mauretania.* These engines were large, to say the least; the low pressure cylinders (the largest among the ships cited) of *Kronprincessin Cecilie* were about 9½ feet in diameter, and moved through a stroke of about six feet at 82 rpm. When Cunard first faced up to the challenge of building ships with a decisive speed edge over the Germans, its early designs called for reciprocating engines that would have to produce 50% more power than those of the *Cecilie.* Fortunately for the Englishmen's project, the steam turbine had advanced to the point of in-

This is one of the reciprocating engines of the *Titanic* being assembled in the erection shop at Harland & Wolff Shipyard before it was taken to the ship for installation. This engine would be disassembled and then moved to the outfitting pier where it would be lowered into the ship in sections. These were the largest of their type and the size can be compared to the man standing in the lower right of the photograph.

spiring such confidence among ship designers that it was a possible alternative. In a pioneering move, the Cunard designers opted for the turbine. Within a few years, it was to become the standard for major ships.

White Star was to follow with its *Olympic* class about four years after the two Cunarders. The new ships were to be distinctly bigger than their turbine-powered predecessors, but the design speed was several knots less. Power required was only about two-thirds of the 70,000 hp fitted to *Lusitania/Mauretania*. The near-imperative to select turbine engines did not confront the White Star designers, but on the other hand, the turbine had accumulated several years of successful service on the North Atlantic before they needed to make an irrevocable choice of machinery. Their choice was a compromise—a combination of reciprocating engine and turbine.

Now "compromise" might be an inapt word, since the choice was more that of a unique combination of reciprocator and turbine than that of a please-everybody brutalizing of the contemporary steam technology. The two disparate engines were joined in a compound arrangement for a very good reason: together they minimized the principal handicap borne by each.

The *Olympic*-class reciprocating steam engines were triple-expansion, four-crank,[1] of 15,000 ihp each as power for two of the three propulsion shafts. Citing a few pertinent numbers can illuminate the handicap inherent in these engines (and all reciprocating steam engines).

These engines were furnished with steam at about 200 psi, saturated. The steam expanded to a pressure of 9.0 psi absolute (i.e., to a pressure about 6 psi below atmospheric). In so doing, its volume increased by a factor of approximately 20; it was this volume increase that moved the pistons that turned the crankshaft that turned the propellers. The volume increase was, therefore, an essential factor in converting thermal energy to mechanical power, but it was in turn a limiting feature of the reciprocating engine—limiting because it determined the size of the engine. As will be seen, the size could reach a daunting magnitude, one that might outstrip the capabilities of engine builders and might exceed the space and weight that ship designers cared to assign to propulsion machinery.

The 15,000 hp of each *Olympic* engine required a flow of steam that had become approximately 100,000 cubic feet per minute by the end of its expansion. In so doing, it had traveled in succession through three cylinders—high pressure, intermediate pressure, and low pressure (hence the term "triple expansion")—with each cylinder being larger than the preceding to accommodate the increasing volume. Such was the volume by the time that the low pressure cylinder was reached that two cylinders in parallel were required, making four in all, and accounting for the four in the term "four crank." Even with the expedient of dividing the volume burden between two cylinders, each of those cylinders was quite large. The bore (internal diameter) of the four (two per engine, four per ship) low-pressure cylinders was 97 inches, the stroke was 75 inches, and the full-power rpm was 75. The engines were double-acting, meaning that steam was admitted alternately to both the underside and topside of the piston, so that two cylinder volumes were presented to the steam upon every revolution of the crankshaft. The resulting volume, multiplied by 75 and by 4 (for the number of cylinders) is approximately the 100,000 cubic feet cited.

Now, a cylinder 97 inches in diameter is a large one indeed.[2] But the designers of the machinery might wish that it still could be larger. If, for example, the steam could be expanded further to 1.0 psia, the power produced would be approximately 50% greater, and that without any increase in the mass rate of steam flow or in the coal consumption. However, the volume of steam would be found to have increased by a factor of 170 instead of the much more modest 20. The size of the low-pressure cylinders, or the number of them, would be beyond the realm of practicality. It is quite feasible with vacuum pumps and steam condensers to produce the vacuum of 13.7 psi that the 1.0 psia exhaust represents, but the size handicap of reciprocating technology precluded its exploitation by a practicable engine. The steam turbine, with its immensely superior ability to pass large volumes of vapor, offered a solution that was adopted by the *Olympic* ships: exhaust the steam from the reciprocating engines to a (single) turbine at 9 psia; let the turbine exhaust at 1 psia after producing the extra power that was beyond the capability of the reciprocator. The flow *area* for the exhaust leaving the turbine would be in the range 100–200 square feet—large, but quite practicable.

The handicap inherent in the turbine is leakage of steam past the ends of the blades. A clearance between blade ends and turbine casing is obviously necessary, and because there must be a stage-to-stage pressure drop, some of the steam inevitably leaks through the clearance. When steam pressure is high, the steam is dense and in consequence the between-blade flow passages are small. The blade-end clearance area, compared to the flow-passage area, is not insignificant. Steam that leaks through the latter area, therefore, constitutes a loss that is not insignificant. On the other hand, down in that pressure domain below 9.0 psia, the opposite is true: between-blade flow passages are much larger, leakage flow is insignificant, loss consequent to the leakage is insignificant. The leakage at high pressures was greatly reduced in later years by improved inter-stage seals, but during the days before World War I it was often cited as a significant handicap to the turbine concept. However, the combined-unit machinery chosen for the White Star ships neatly reduced the problem to insignificance because the steam never entered the turbine until its pressure was very low—below atmospheric, in fact.

The preceding chapter has discussed the speed dichotomy between turbine and propeller. With steam entering at 9.0 psia and exhausting at 1.0 psia, the most efficient turbine speed was much closer to that of the propeller than would have been the case with an all-turbine propulsion plant.

The combination of engines used by White Star is illustrated by Figures 10-1 and 10-2 that shows the connections among reciprocating en-

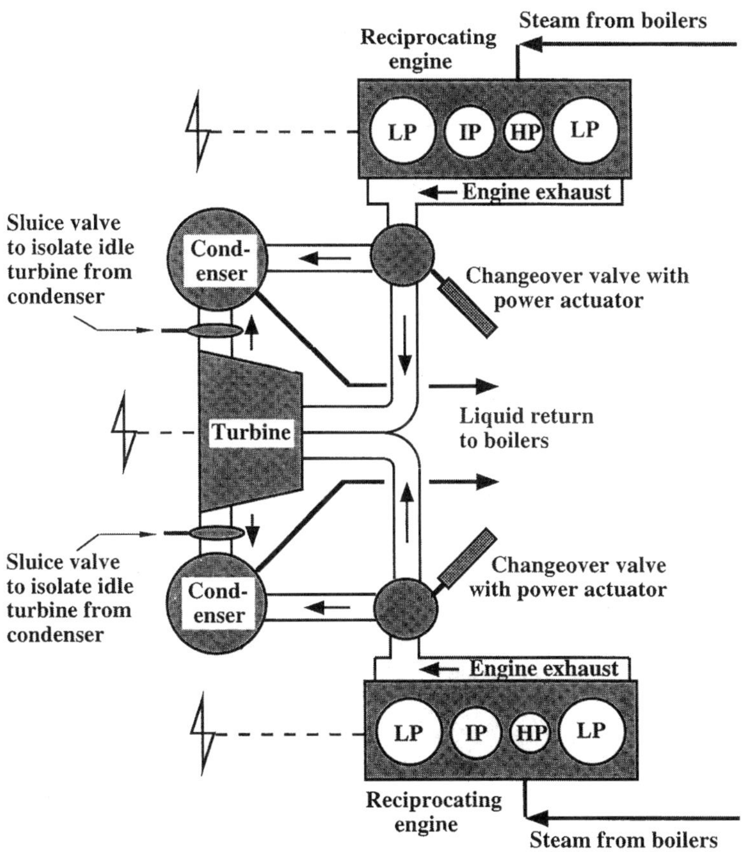

FIGURE 10-1 Schematic arrangement of the *Olympic*-class propulsion engines. (*Sketch by the authors*)

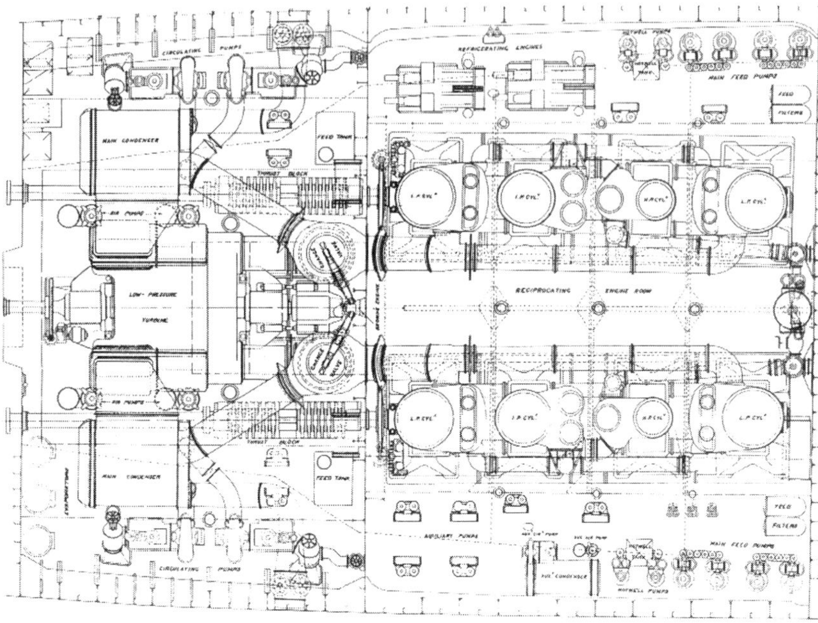

FIGURE 10-2 Arrangement of the *Olympic*-class reciprocating and turbine engine rooms, particularly intended to show the exhaust change-over valves. (*From* Engineering, *vol. 88, 1910*)

gines, turbine, condensers, and exhaust change-over valves. Note that the reciprocators are reversible, but the turbine is not, so that the steam that normally exhausts from the former to the latter must be diverted directly to the condenser when the ship is to go astern—hence the changeover valves.

Like the turbine itself, the combination of the new machine with the old was largely the work of Sir Charles Parsons. It was first publicized in a 1908 paper[3] by Parsons and R. J. Walker in which the Royal Navy destroyer *Velox* of 1900 is credited with being the first ship to use turbines and reciprocators in a compound arrangement, and the New Zealand inter-island ferry *Otaki* is named as being the first commercial ship to do so. Otherwise, the three White Star ships constitute the only notable example. Like Cunard in its *Carmania* and *Caronia* pair that constituted a test of turbine versus reciprocator before *Lusitania/Mauretania* went to sea, White Star did make its own comparative test with *Laurentic* and *Megantic*. The *Laurentic* of 565 feet overall length and 15,000 gross tons was intended for the line's England-Canada route.[4] She was fitted with the combined reciprocating-turbine machinery that was destined for

Olympic, *Titanic*, and *Britannic*. The *Laurentic* entered service in 1909; her more conventional sister, the *Megantic*, was put in service about a year earlier with an all-reciprocating propulsion plant.[5]

The *Laurentic* machinery was nearly identical (except in power rating) to that fitted to the larger ships. Boiler pressure was 215 psi, steam exhausted from reciprocator to turbine at about 10 psia, and exhausted to the condenser at about 1.0 psia (i.e., at about $\frac{1}{15}$ atmospheric pressure). Propeller speeds were 83 rpm on the wing (reciprocating engines) shafts, 220 on the centerline (turbine) shaft. Steam consumption was found to be 13 pounds per horsepower hour during the builders' trials of the ship; the steam rate for the *Megantic*, driven by reciprocating engines only, was estimated to be about 16.

Fuel consumption estimates would have been more meaningful, and in the later age of liquid fuel its consumption per horsepower hour (or per kW hour) became the universally quoted ship-engine propulsion efficiency indicator. However, handling of coal by the shovel full made its consumption rates difficult to measure except in such overall quantities as tons per voyage. In any case, the consumption of coal should have been proportional to the steam consumption rate. Of course, the reduction in steam rate meant a reduction in the number of boilers as well, and with that direct benefit were associated such savings as reduction ship size and reduction in the size of the boiler room crew. For instance, if the $\frac{13}{16}$ ratio was extrapolated to the *Olympic* class, then a ship of their size and power fitted solely with reciprocating engines might have required five or six more boilers.

The White Star designers did not make so radical a step forward as their Cunard counterparts had done with their all-turbine choice a few years earlier. On the other hand, we might justly call it radical, since the Cunard choice turned out to be a step forward along the mainstream of marine propulsion development, while the White Star choice turned out to be a short-lived aberration. In any case, the *Olympic*-class designers must have reasoned something like this:

- The turbine propulsion plants of the *Lusitania* and *Mauretania* appear to have been successful, but we have reasons for choosing differently, as we list here.
- Our ships are to be different from the Cunarders in significant ways. They are not intended to compete for any speed records, and so are not to be enlarged *Turbinias* for which turbine engines are almost essential. No, economy of propulsion is of much greater concern to us.
- The combined reciprocating-turbine propulsion plant that we are in-

stalling in the *Laurentic* provides the highest efficiency possible. It also obviates the need for excessively large low pressure cylinders that an all-reciprocating plant of the power we contemplate would have. If there are any unrecognized problems with the concept, we will have our test ship at sea to correct problems long before the new ships are ready.

- An all-turbine propulsion plant will be burdened by additional units for reversing. With the combined plant the reciprocating engines will handle reversing duties. The only complication in this regard will be the fitting of power-operated valves to divert engine exhaust directly to the condensers so that the turbine won't continue driving ahead when the reciprocating engines are reversed.
- The combined plant has the imprimatur of Sir Charles Parsons, who is its originator, and who advocates its use for applications like ours.

The designers at Harland & Wolff reasoned well, for 1908. In the 1920s they would have been confronted with matured technology for matching engine speed to propeller speed, as well as advances in turbine technology, and would never have made the decision they did. But we might note that the *Olympic*, surviving and serving into the 1930s, was a successful ship whose propulsion plant was certainly outmoded in old age, but which never reflected upon the wisdom of choices made for her two decades earlier.

Concerning the role of Sir Charles Parsons: he, with his collaborator, S. S. Cook, was to be granted a patent on the reciprocator-turbine combination in 1911.[6] The most significant difference between the patent and the machinery as fitted to the White Star ships is the omission by the ships of the speed reducing gear between the turbine and its propeller.[7] This "reduction gear" was to become the universal handmaiden to turbine propulsion machinery after World War I, but in 1911 it had not graduated from the experimental stage.

As something of a nonpertinent, but an interesting sidelight to the reciprocator-turbine compound machine, is its later-day counterpart: the turbocharged diesel engine! The diesel is a reciprocating engine, and it is limited in output by the inability to expand its working fluid (mostly air) to atmospheric pressure without an unacceptably long piston stroke, i.e., without the same enormous cylinder volume that plagued the reciprocating steam engine. The remedial expedient used in the diesel's early days was "turbocompounding," exhausting at an above-atmospheric pressure through a turbine whose output was geared to the engine shaft. Around 1920 the greater benefit of using the turbine to drive a supercharger was recognized, and thereby was born the now familiar "turbocharging," which is simply a particular form of turbocompounding. Indeed, the early form of turbocompounding—turbine output added to engine output—was revived in the 1980s, a time at which turbochargers had become so efficient that they couldn't use all of the exhaust energy in compressing combustion air.

We need to describe two nonpropulsion auxiliary systems, the seawater pumping system and the electric plant, because they were called on to serve during the sinking of the *Titanic*.

Seawater Piping Systems

Every seagoing ship is fitted with piping systems that distribute seawater to a variety of applications aboard: engine cooling, refrigeration cooling, condenser cooling (for steamships), fire-fighting, toilet flushing, ash ejecting (for coal-fueled ships), deck washing, and ballasting are common enough to be near universal elements of a ship's machinery outfit. Then there is the system that pumps water *out* of the ship—the vital bilge pumping system.

Every human contrivance that floats upon water, and has an open interior to carry cargo or humans, has a "bilge" (or bilges), a low point to which liquids that escape from internal sources, or force their way in from the exterior, must drain. A sponge or bailing can is sufficient for keeping the bilge dry in a small boat, but from modest-sized craft to the largest ship, a bilge pumping system is essential. The typical ship system consists of one or more pumps in the engine room, with main and branch suction pipes running to other spaces. The pipes are open-ended, at low level, and there are several of them in each space because different lists and trims will concentrate loose water in unexpected locations. A strainer keeps extraneous solid matter from being sucked into the pipe. A valve is fitted near the open end, and is normally closed to prevent water from a flooded compartment from finding its way in from adjacent compartments via the bilge piping. These valves can be manipulated locally, but can also be opened or closed from high in the ship by means of "reach rods" that are, in a way, extensions of the valve stems.

The *Olympic* ships were fitted with three bilge pumps (see Figure 10-3), each with a capacity of 150 tons per hour; one pump was in the turbine engine room, the other two in the reciprocating engine room. The latter two served the piping system that ran to suction points in all inner bottom-level compartments forward of the engine rooms.

As is common practice, the bilge pumping system was supplemented by connection to a ballast system whose principal function was to move seawater ballast to and from the inner bottom tanks and the two peak tanks at the extreme ends of the ship. The five ballast pumps, each of 250 tons per hour capacity were also connected to piping by which they could take suction from the bilges of the space in which they were installed. (The system was, therefore, properly known as the "bilge and ballast system.") Two of these pumps were in the reciprocating engine room, the remainder were installed, one each, in three of the boiler rooms.

Ash from the boiler furnaces was sent overboard by a jet of seawater from ash ejector pumps in each boiler room. In the three boiler rooms without a bilge/ballast pump, the ash ejector pumps were fitted with bilge suctions for use in emergency. The intention of the designers apparently was to provide each boiler room with at least one pump that could evacuate the bilges under control of the local personnel. Although the capacities of the ash ejector pumps is not known, they must have provided only a minor addition to the total pumping capacity (1,700 tons per hour) provided by the other pumps.[8]

Steamships of a later day were required to have their main condenser seawater circulating pumps fitted with suction piping connections to the engine room bilge. The two main circulating pumps of the *Olympic* ships had a combined capacity of approximately 12,000 tons per hour. Such a rate might have slowed the sinking of the *Titanic* significantly. However, the machinery plans of the ship do not indicate the necessary piping to the circulating pumps. Moreover, the flooding of the engine room occurred far too late for pumping of any magnitude to change the march of events.

Figure 10-3 shows in schematic form the elements of the bilge pumping system of the *Olympic* ships. Please note that it is truly schematic and does not show details. For example, each boiler room had six suction points, but the illustration does not attempt to show all six.

Piping systems that can take suction from the bilges are essential for housekeeping, and are also essential safety systems since they may counteract leakage that will sink the ship. But note the word *leakage*, which implies a rate of inflow at least an order of magnitude less than that likely from a major collision or serious war damage—piping and pumps that can handle events of such magnitude are impracticable. In the case of the *Titanic*, the initial inflow of water to Boiler Room No. 5 was at such a rate that pumping could counteract it. Not so for the more severely holed Room No. 6, nor for Room No. 5 once water began to find its way in from forward.

In the early stages of the *Titanic* sinking the flooding rate is estimated to have been 400 tons per minute (24,000 per hour), a rate that is an order of magnitude greater than the total capacity of the pumps. The pumps, therefore, were far from adequate as a means of saving the ship. Indeed, pumps have never been regarded as a means of salvation in the event of a major accident because an exploding mine or torpedo, or the bow of a colliding ship, will make an opening far beyond the ability of any practicable pumping outfit to counteract. And the *Titanic* incident wasn't even "major"—the total area of holing has been estimated to be only about 12 square feet.

Regardless of the hopeless circumstances, the *Titanic*'s engineers kept the bilge system running as long as steam came to its pumps, and this action must have delayed the sinking by some amount. But it could not hold back the sea until the *Carpathia* came.

Overboard Drains

In the days before environmental awareness brought significant complication to waste disposal, ships' plumbing drains led directly overboard, and functioned by gravity flow (exceptions would have to be made, of course, for any drains originating below the waterline). *Olympic Class* overboard plumbing connections were at "C" Deck level, and were

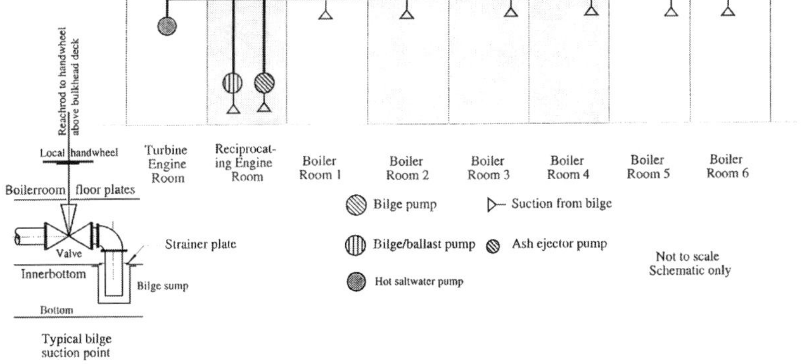

FIGURE 10-3 Schematic arrangement of the *Olympic*-class bilge and bilge-and-ballast systems, based on descriptions in the Mersey report. (*Sketch by the authors*)

not fitted with check valves to prevent backflow. The "bulkhead deck" was "D" Deck, so that unprotected hull openings at "C" Deck should not have been a hazard. However, once the forward part of the hull had dipped below the margin line, backflow through these drains may have contributed significantly to the rate of flooding.

Electric Plant

The *Olympic* electric system produced direct-current power that was generated, distributed, and used at a modest 100-volt level. Distribution was by a single wire; all electrical equipment was grounded to the steel structure of the ship, and through that steel the current found its way back to the generators.

The main generating plant consisted of four generators, each rated at 400 kW output, each driven by a three-cylinder reciprocating steam engine. These units were installed in a compartment just aft of the turbine engine room, and were fitted around the three propulsion shafts that perforce passed through the generator room. The necessary switchboards through which the current was distributed were installed on the deck just above the generators.

Steam came to the generator (or "dynamo") room from the auxiliary steam main, which was so arranged that it could receive steam with any two of the boiler rooms flooded. This feature (see Figure 10-4) was consistent with the overall "unsinkability" standard of the ship, a standard that required the ship to remain afloat with any two watertight compartments flooded.

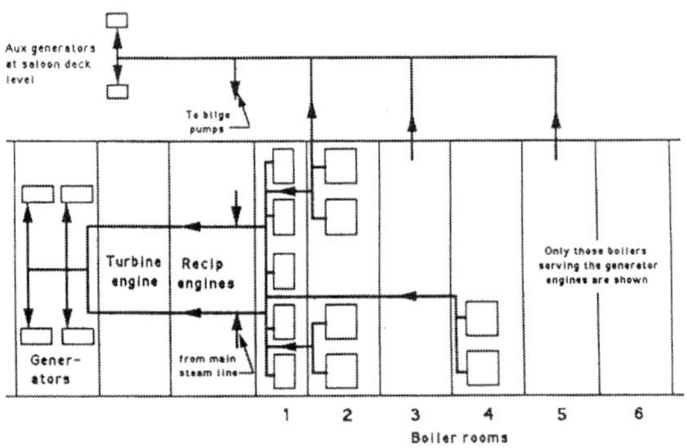

FIGURE 10-4 Schematic arrangement of the *Olympic*-class steam supply to the main and emergency generating engines, based on descriptions in the Mersey report. (*Sketch by the authors*)

A further disaster-related feature of the electrical plant was the provision of emergency generators on "D" Deck, which was above the level that water could reach in any flooding incident that was insufficient in extent to sink the ship. There were two (steam) engine-driven generators in this plant, though each was rated at only 30 kW. They were not intended to keep all lights burning by any means, but only to supply emergency lighting and certain other loads, e.g., boat winches, that might be essential in an emergency.

The emergency generators were located directly above the main electrical plant. However, the steam supply did not come via the machinery spaces, but by a pipe that rose directly from the boiler rooms to "E" Deck, and then ran aft along that deck. The intention was clearly to protect the steam supply from any accident that might flood the aft end of the ship. Like the supply to the main generators, this emergency supply of steam could be maintained with any two of the boiler rooms flooded.

It might be said of the *Titanic* electric plant "give me steam and I will give you light." How that steam was supplied so late in the sinking cannot be known, given that the humans involved in the act of supplying it did not survive to tell of it. We shall discuss the matter further in Chapter 26.

Notes for Chapter 10

1. "Four crank" refers to the crank shaft, and implies that the engine had four cylinders.
2. In 1995 the largest marine diesel engine had a bore of 980 mm (39 inches). With 12 such cylinders its rated output was 92,000 hp.
3. Charles Parsons and R. J. Walker, 1908, *Transactions*, Institution of Naval Architects, Volume 49.
4. *Laurentic* description. *Engineering*, 1909, Volume 87.
5. *Megantic* description. *Engineering*, 1909, Volume 87.
6. The *Parsons/Cook turbine/reciprocator* patent is as follows:

> Patent 16,346/10 granted to C. A. Parsons and S. S. Cook, July 12, 1911. From page 444, Sept 29, 1911, of *Engineering.*
>
> The invention consists of a three-shaft installation, having two of the shafts thereof driven by reciprocating engines, which exhaust into low pressure turbines driving through gearing to the third shaft. The reciprocating engines are preferably arranged on the two side shafts, the turbines thus driving the central shaft, while means may be provided to enable the reciprocating engines to exhaust directly into the condenser when going astern or under other circumstances if desired. The two side propeller shafts are a and b, each driven by one of the reciprocating engines, c and d, respectively, while in the space between the side shafts a and b, two independent low pressure turbines f and g are mounted on separate shafts h and k, and are suit-

ably geared to the central propeller shaft m by gearing n. The installation is thus composed of two parts, each comprising a reciprocating engine and a low pressure turbine. A pipe connection p and q from the exhaust of each reciprocating engine c and d carries the working fluid exhausted therefrom to the inlet of the corresponding low pressure turbine f and g, and after being exhausted from the latter, the fluid passes to the condenser r and s, one of which is provided for each part of the installation. In addition to the pipe connection t between the reciprocating engine, say c and the side shaft a, and the turbine b, another pipe connection allows engine c to exhaust directly into the condenser r, and thus, if desired, to be run alone, a similar connection u being provided between engine d and condenser s. In operation, the high pressure working fluid is supplied to each of the reciprocating engines c and d, from which the fluid thereafter passes by the pipe p and q to the corresponding low pressure turbine f and g and the condenser r and s, each turbine thus being in series with a reciprocating engine. It will be understood, of course, that the connections between the turbine and the engine are similar on each side of the ship. When the ship is going astern, the reciprocating engine on each side of the ship is directly connected to its respective condenser by the connection t and u, respectively, the turbines f and g geared to the central shaft m thus being idle. The provision of a special reversing turbine is thus rendered unnecessary by this arrangement. For the purpose of maneuvering, the reciprocating engine c and d, respectively, on each side shaft can be operating independently, either both ahead or both astern, or one ahead and the other astern, as desired.

7. Note that the Parsons patent application required gearing because it showed two exhaust turbines connected to the centerline shaft.
8. An interesting sidelight: in an emergency, the ash ejector pumps could supply seawater to the boilers.

PART III

The Events

The "Events" are the descriptions of what happened: the *Titanic*'s collision with an iceberg, the *Lusitania*'s falling victim to a torpedo, *Britannic*'s fatal contact with a mine. Eight chapters in Part III tell their stories. Six of the eight chapters are concerned with the *Titanic*. This is scarcely an evenhanded treatment, it would seem, but there are reasons for the seeming imbalance. For one, if we eschew (as we will) the conspiracy legends accompanying the fate of the *Lusitania*, then the story of the *Titanic* sinking has much greater general interest than the other two. For another, the implications of the *Titanic* event for the dubious doctrine of unsinkability, the faults in ship design and construction that were implied, and the effects on future passenger ship safety, are important concepts that will be discussed later in this book. Overall, the space given to the *Titanic* in Part III is justified by the significance of this ship and her impact on maritime design and history.

CHAPTER 11

The Advent of the Titanic[1]

By the calendar, *Lusitania* was the first of the five liners considered in this book: her maiden voyage was the first to take place among the five, occurring in 1907. Her sister *Mauretania* followed soon after, and the first of the White Star ships, *Olympic*, began her transatlantic service in June 1911, shortly after the launch of the *Titanic*. However, the events that give these ships their special place in history began with what was, chronologically, the fourth ship—the *Titanic*—and hence our story shall also begin with that most famous and most fated of all passenger ships.

The first two White Star ships arose side by side on paired building berths, especially constructed for them at the Harland & Wolff Shipyard in Belfast, Northern Ireland. They might have arisen simultaneously, but scheduling one approximately a half-year later than the other was of great advantage to the builders of major equipment units, and also for the yard's own labor resources. So *Olympic* was launched on 20 October 1910, *Titanic* on 31 May 1911.

The launch day of the *Titanic* was warm and sunny, perhaps more comfortable than *Olympic*'s November day. In any case, over 100,000 people were estimated to have watched from vantages within Harland & Wolff and along the adjacent banks of the river Lagan. The event itself may seem strange when viewed from a century later, for there was no christening ceremony with the traditional champagne bottle. A portent of misfortune, as declared by so much of the ancient lore accompanying seagoing (e.g., the beginning a voyage on Friday)? Probably not, since no-christening launchings were a traditional procedure with the White Star Line.

Another incongruous element was the American Stars and Stripes, paired with the United Kingdom's Union Jack, and both displayed from the peak of the shipway's gantry crane. Appropriate, however: ultimate ownership of the White Star Line lay in the moneys of the American financier, J. P. Morgan. Morgan was present, as was the Lord Mayor of Belfast, Bruce Ismay, managing director of the White Star Line, and Thomas Andrews, the naval architect who was to guide the ship through its outfitting to completion. Lord Pirrie, principal owner of Harland & Wolff, was there as well, and not only as a dignitary, but as overseer of the event.

Except for a workman who removed shores beneath the ship and suffered a fatal accident, the launch was uneventful and successful. Tugs moved the new ship to the outfitting piers, where she floated side by side with the *Olympic* for a few hours.

The elder ship was delivered to her owners that very day, and at 1500 *Olympic* sailed on her delivery voyage to Liverpool. J. P. Morgan, Bruce Ismay, and other dignitaries from the launching party were aboard. After preparations for the maiden voyage were carried out at Liverpool, the *Olympic* transferred to Southampton. On 14 June, and with appropriate fanfare, the big ship sailed to begin her first transatlantic passage, carrying 1,316 passengers and 850 crew.

The *Titanic* is shown on Belfast Lough on 2 April 1912 about to undergo her brief trials before sailing to Southampton to begin her fateful maiden voyage on 10 April. Her trials were completed without incident, but those items remaining would be completed during her maiden voyage. (*Courtesy of the Ulster Folk Museum*)

The voyage to New York from Southampton took six days and almost 22 hours to complete, despite some rough weather enroute. The most difficult task was docking the *Olympic*. It took almost an hour to align the ship side ports with the Pier 59's gangways.

The maiden voyage of the *Olympic* was a great success except for a noise at the aft end of the first-class Dining Saloon. This was traced to the water being thrust against the hull by the wing propellers. Bruce Ismay, who was aboard, wrote that everything on board worked "most satisfactorily" and that the propulsion machinery worked excellently without any mechanical problems. Ismay made copious notes during the voyage, however, that would make the *Titanic* even better. During the remainder of the summer of 1911, the *Olympic* continued carrying passenger and mails with the *Oceanic* and *Majestic*.

The *Olympic* collides with HMS *Hawke*. On 20 September 1911, while outward bound from Southampton on her fifth voyage, the *Olympic* encountered the 7,350-ton Royal Navy cruiser HMS *Hawke* in the Solent.

The starboard tailshaft for the *Titanic* is being fitted prior to her launch. The port tailshaft is already installed, as is the centerline rudder, which resembles the types used on wooden sailing ships.

The two ships were on parallel courses off the Isle of Wight, the *Hawke* at 16 knots while the *Olympic* was gradually overtaking and passing the cruiser. As she did so, the "suction" that develops between ships passing close aboard drew the *Hawke* onto an inescapable collision course, and her ram bow punched two holes abreast of the liner's mainmast, one above the waterline, the other below the waterline.

The bow of the cruiser was demolished by the impact, with the failure of structure being characteristic of ductile steel. In contrast, the plating of the *Olympic* exhibited a combination of buckling and brittle-like fractures near riveted seams. Hundreds of rivets failed (some torn loose from their holes), including those in a plating seam that was just above the waterline. There was also damage to the starboard bossing as the cruiser's damaged bow slid aft along *Olympic*. As a result of this damage, the two after cargo holds flooded. The failures of these wrought-iron rivets from impact upon the hull structure would be repeated when the hull of the *Titanic* was struck by ice seven months later. In 1996 and again in 1998 forensic investigators of that ship re-examined the case of the *Olympic* wrought iron rivets in attempting to analyze the initial flooding of her sister.

Fortunately there was no loss of life involved in the collision, despite

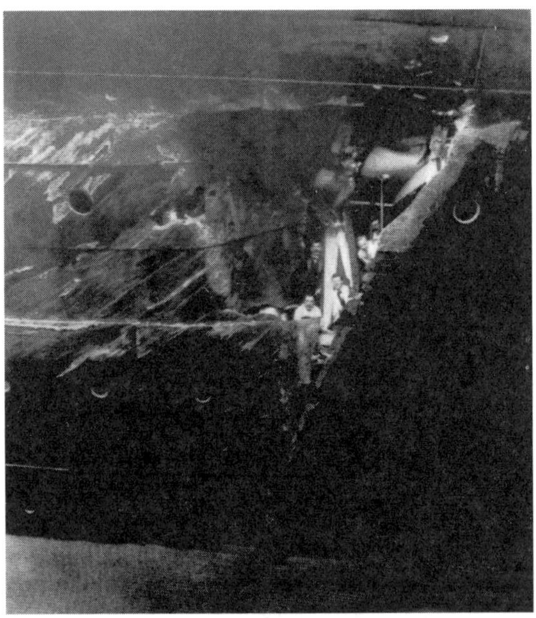

This exterior view of the *Olympic* shows the type of damage that she sustained in her collision with the cruiser HMS *Hawke* on 20 September 1911 in the Solent. This is a view of the damage above the waterline. (*Simon Mills Collection*)

This is another exterior view of the collision damage to the *Olympic*. This is a view of the damage to hull below the waterline from the ram bow of the HMS *Hawke*. Although the plates exhibit a brittle-type fracture, it is evident that the plates and stiffeners had a significant amount of ductility. Also note the number of missing rivets. (*Author's Collection*)

This is a view of the internal damage looking to starboard and aft. A number of rivets can be seen missing. The large pipe structure in the foreground is to hold the wood patch in place over the hole. (*Courtesy of the Ulster Folk & Transport Museum*)

the fact that the *Hawke* had collided in an area of several second-class staterooms. The passengers in these spaces were at lunch when the mishap occurred.

Following the *Hawke* collision, the *Olympic* dropped anchor in Osborne Bay where the damage was assessed. Recognizing that the ship needed repairs and could not be taken to sea, the ship headed to Cowes where her passengers were taken ashore. She returned to Southampton where she was made seaworthy by the installation of a temporary wooden patch, so that she could return to Harland & Wolff for permanent repairs. The demand for resumption of her scheduled service as quickly as possible meant that workmen had to be taken from the *Titanic* to expedite the repair. Some material (the starboard propeller shaft, for example) also was requisitioned during the six-week repair. In consequence of these measures, the completion of the newer ship was delayed. The *Olympic* resumed her transatlantic service on 29 November 1911. The maiden voyage of the *Titanic*, which was scheduled at the time to begin 20 March 1912, was rescheduled for 10 April.

The interaction phenomenon between two ships close aboard was poorly understood in 1911. Authorities of a later century might conclude that the lack of knowledge of such physical phenomena was the ultimate cause of the *Hawke–Olympic* collision. A Court of Inquiry ruled otherwise; it faulted the *Olympic*, noting that the *Hawke* had been drawn into the *Olympic* by the suction of her great hull. Nonetheless, White Star steadfastly supported the captain, E. J. Smith. Had they not, would Captain Smith not have received the honor of commanding the *Titanic* on her maiden voyage?

The Olympic *Loses a Propeller Blade.* On 24 February 1912, while eastbound for Southampton, the *Olympic* lost a propeller blade after hitting an uncharted obstruction some 750 miles east of Newfoundland. Repair required dry docking at the Harland & Wolff Shipyard in Belfast, where she arrived on 1 March 1912. The *Titanic* was then in the Thompson Dry Dock, but again the imperative of scheduling of an in-service ship preempted resources of the shipyard and its facilities. The *Titanic* was taken from the dock for five days. The consequence was further delay in outfitting the new ship. This time, however, the maiden voyage was not rescheduled. The lost time was regained by cutting short the time allowed for tests and trials. Some of these were made up on the delivery passage to Southampton. Other tests were to have been carried out on the maiden voyage.

When the *Titanic* was lost, the *Olympic* was 650 miles east of her ill-fated sister and on her way to Southampton. Although Captain Haddock tried to make a rescue attempt, he was outdistanced by the closer *Carpathia*. She resumed her voyage to Southampton. While preparing for the next transatlantic voyage on 24 April 1912, her engine room complement of greasers, trimmers, and firemen went ashore a few hours before departure time. They complained that the 24 additional canvas collapsibles that the White Star Line had provided in the wake of the *Titanic* disaster were unseaworthy. The men insisted that the White Star Line provide wooden lifeboats. Unable to sail, the voyage was canceled two days later. Tests done at Spithead with the collapsibles did prove that some of these were not reliable craft. When the ship was ready to resume transatlantic service on 15 May 1912, wooden lifeboats were installed instead.

The *Olympic* continued in service until September 1912. At that time it was decided to return her to Harland & Wolff's Belfast shipyard where she would undergo a very expensive transformation. For a period

This is a view of the *Olympic* (left) entering the Thompson Dry Dock at Harland & Wolff Shipyard in March 1912 to undergo repairs to a propeller blade. The *Titanic* (right) was already moved from the dry dock to a pier to continue her outfitting. The repairs to the *Olympic* would force *Titanic*'s maiden voyage to begin on April 10 instead of earlier. (*Courtesy of the Ulster Folk and Transport Museum*)

of six months her interiors in way of the propulsion plant were gutted and a four-foot double bottom extension was carried up to "E" Deck. Several forward bulkheads also were extended a deck height as a result of the recommendations of the *Titanic* Board of Inquiry. The number of lifeboats was increased to 48. These changes convinced the White Star management that the *Olympic* was the safest ship afloat.

The *Olympic* returned to transatlantic service in April 1913 and continued in passenger service until the outbreak of World War I. When World War I began on 4 August 1914, the *Olympic* was enroute to New York. The cruiser *Essex* accompanied the ship to New York harbor where she discharged her passengers. The subsequent return voyage was canceled so that the crew could paint part of the ship grey and black out windows. Passengers scheduled for the return voyage to the United Kingdom were put aboard the *Adriatic*. She began her return voyage to the United Kingdom on 8 August 1914. For several months she was kept in service returning Americans stranded in Europe by the war.

During a voyage on 27 October 1914, the ship was steaming off the coast of Northern Ireland, headed for Belfast. Ahead of her was the Second Battle Squadron with the new battleship *Audacious* on their way for gunnery practice in Lough Swilly. Around 0800 the battleship exploded a mine abreast of her port engine room and eventually lost all power. Unable to maneuver, the *Audacious* began to slowly sink by the stern. When the *Olympic* arrived on the scene, after safely passing through the minefield, the stricken battleship was wallowing in swells. Captain H. J. Haddock volunteered to tow her to safety. All but 50 of her crew were removed and a towline was secured to the *Olympic*. Unfortunately the cruiser *Liverpool* ran over the towline and severed it. Another tow line was rigged, but the battleship had been taking on water and began to roll heavily and the seas began to rise. It was concluded that the *Audacious* was too big a tow for the large liner so the cruiser *Liverpool* took over instead. The battleship later sustained an internal explosion and sank. In the meantime, the *Olympic* went to Lough Swilly where she would remain for a period of a week while the Admiralty debated how to keep the loss of the *Audacious* a secret after so many of the passengers and crew had witnessed and even photographed the rescue attempt. Captain Haddock was unaware of the minefield and was criticized for risking his ship in the rescue attempt. However, he was judged to have made a courageous attempt to save the battleship and the utmost in seamanship.

Once the *Olympic* was completed in bringing home American citizens from war-torn Europe, the *Olympic* was requisitioned as a troopship to take troops to the Dardenelles campaign. During 1915 the ship returned to Harland & Wolff for an overhaul, a dazzle camouflage scheme, and an armament of six 6-inch guns. She would carry Canadian and later American troops to the war in Europe. The *Olympic* was a participator in several notable incidents during World War I. One was the ramming and sinking of a German submarine during World War I on 18 May

The *Olympic* as a troopship is shown in dazzle camouflage during World War I. This scheme was designed to break up the vessel's silhouette and confuse enemy submarines and surface ships as to her true identity. Note also the extra lifeboats. Fourteen are hanging from davits on each side with many more stowed on chocks on the Boat Deck. On the overhead platform on the Poop Deck two 6-inch single mounts are mounted, one to starboard and another to port. (*Courtesy of the Mariners Museum*)

1918 while approaching the Scilly Isles upon concluding a passage from New York with American troops aboard. Around 0345 her lookouts sighted a surfaced *U-103* some 900 yards off the starboard bow. Attempts to sink it with her 6-inch guns proved fruitless, so she rammed the submarine. The force of the collision was keenly felt on the bridge. Damage to the bow was slight with her stem bent eight feet to port and some dented plates. With a frame spacing of 24 inches and a number of non-tight floors between the stem and the collision bulkhead, this was structure that could penetrate the hull of that vintage submarine.

On 12 August 1919, the ship entered a dry dock at Harland & Wolff for an inspection and a postwar refit. Workmen noticed water pouring from an area outside of Boiler Room No. 3. The outer plate of the double hull was found to have a substantial dent and several sprung plates. It was concluded that a dud torpedo must have caused the damage. During a six-month stay at the shipyard, her passenger accommodations were overhauled and her boilers were converted to burning oil instead of coal. The latter change resulted in the reduction of the engine room crew from 650 to 350 men. She sailed for her first postwar transatlantic voyage on 21 July 1920. With the United States restricting immigration, it was decided to convert the third or steerage class into tourist class.

During 1933 the *Olympic* returned to Harland & Wolff for a major overhaul. Many of the rivets in the double bottom in way of the Reciprocating Engine Room had to be replaced. Years of service had taken their toll.

After the war ended the International Mercantile Marine Company began negotiations to sell its British holdings that included the White Star Line. The Royal Mail Steam Packet Company acquired the White Star Line in 1927, but with a world-wide depression, the *Olympic* was assigned to cruising. On 30 December 1933, Cunard and White Star decided to merge with the urging of the British government. The final consummation of the merger took place on 10 May 1934. Five days later the *Olympic* made news. Heading for New York with a small consignment of passengers, the liner encountered a heavy fog south of Nantucket Island. Around 1000 a lookout in the Crow's Nest warned of an object directly ahead. With a speed of 16 knots, Captain Binks ordered a full speed astern to avoid what was to be the Nantucket Lightship. The liner struck the lightship amidships, cutting the little craft in two. The two sections sank almost immediately. Only four of the eleven-man crew survived. Damage to the *Olympic* was minimal as the impact took place in the Forepeak area where the structure was stout.

The *Olympic*, first of the three great White Star liners, terminated her long career on a transatlantic voyage that began on 27 March 1935 and ended in Southampton on 12 April. Her career ended at a ship-breaker's yard in Jarrow in October 1935 after the completion of the *Queen Mary* that made her obsolete. These events were of small consequence to the *Olympic*, however, and of course have none at all for the *Titanic*, lying on the seabed when even the first of them happened. But two earlier incidents had identifiable consequence to the *Titanic*, and one may even have contributed to the evidence of the forensic evaluations of that ship.

Preparation for the Titanic's *Maiden Voyage.* Though somewhat delayed by the sudden necessity to dock the *Olympic*, twice at Harland & Wolff, the transition to ship-in-service began for the *Titanic* in March of 1912. Some of the engineering staff came on board early in the month to familiarize themselves with the propulsion systems, and with the ventilation, water, and electrical systems that served the ship's living quarters. Chief Officer William Murdoch, who would be on watch when ice loomed so close ahead on the night of 14 April, arrived at about the same time. Several deck officers were re-assigned from the White Star liner *Oceanic*, immobilized by a coal strike that had begun in February. The first to report on 20 March was Charles Lightoller, destined to be the *Titanic*'s second officer and highest-ranking survivor. On 27 March junior officers Pitman, Lowe, Boxhall, and Moody reported for duty.

In spite of the rush to completion in the face of delays, minor design changes could still be accomplished in March. None were to have impact on the events of the maiden voyage, but one change increased the ship's register tonnage beyond that of the *Olympic*, thus conferring on the *Titanic* the title of "world's largest ship."[2] This change was merely the enclosure of parts of open areas on Promenade Deck "A" and Bridge Deck "B" by sliding windows that would shield passengers from salt spray (*Olympic* passengers had complained of spray coming over the bow). Enclosure gave the ship a feature that made it visibly distinguishable from the *Olympic*.[3] Enclosure also increased the internal volume that must be measured in determining register tonnage, and thereby made the *Titanic* "world's largest."

The *Titanic* was ready for sea trials as April dawned. The first day of that month brought disappointment however, since a cold and brisk wind roughened the waters of Belfast Lough so that the narrow passage from the river via Victoria Channel seemed too threatening to attempt. The disturbance had passed by the next day, and at 0600 the *Titanic* left the builder's yard, assisted by four tugs. Two tugs forward towed the ship; two aft guided the stern to make sure that she would not swing from the channel. Meanwhile, the boiler room crews were bringing steam to its rated pressure. With boilers hand-fired with coal, such an undertaking always caused a freshet of dense black smoke to erupt from the stacks,[4] impressing the thousands of spectators along shore with the power of the world's largest liner.

At sea, a series of trial runs were made to confirm the expected performance of the propulsion machinery, steering gear, and anchoring outfit. Engine power was increased in steps toward its rated value. The ship was designed for 21 knots with 25 boilers on line at their rated capacity, but no attempt was made on these trials to determine maximum speed. As was customary, the new machinery was to be allowed to "run in" for a period before being pushed to its limits; given that these trials were abbreviated by the delay caused by the *Olympic* propeller replacement, this step could not be taken on the initial trials. Meanwhile, the emergency responses of the ship were found satisfactory in maneuvering trials, and by

The *Titanic* is shown at the outfitting pier. Her boilers and propulsion machinery have been completed. Although her four funnels have been installed, they have not yet been painted the character buff yellow and black top. (*Courtesy of the Ulster Folk Museum*)

a crash stop[5] test that determined stopping distance to be 850 yards from a speed of 20 knots. In all, the trial results satisfied the attending Board of Trade representative, and in consequence the *Titanic* received its certificate for carrying passengers.

The new ship returned to Belfast during the evening of 2 April, and was immediately made ready for departure to Southampton, an event that was scheduled for an arrival at its future homeport coincident with a flood tide. Departure and arrival times were met; the *Titanic* docked at Southampton shortly after the midnight of 4 April, having made the passage at an average speed of 18 knots. Testing of the new ship had continued during the trip, including brief periods when speed was increased to 20 knots. However, Captain Smith did not deign to push speed to its limits just yet.

Coal Fire. Soon after the ship left Belfast on 2 April, the crew of Boiler Room No. 6 noticed the smell of hydrogen sulfide in the aft starboard bunker. Inspection revealed a coal fire smoldering deep within the coal. Such fires occasionally erupted in the bunkers of the coal-burning era, apparently from spontaneous combustion. Careful monitoring of

The *Titanic* is shown after completion of her brief trials and leaving Belfast Lough on 2 April 1912 for Southampton and her fateful maiden voyage across the Atlantic.

coal temperature, sufficient ventilation, and periodic wetting of the coal usually prevented them, but the haste of working up to that first voyage apparently intervened in this case. Fires of this nature were usually difficult to extinguish, and so it proved here: 14 men were kept busy trying to extinguish the *Titanic*'s fire. However, the fire was eventually extinguished, and apparently without consequence to the maiden voyage, though the heat caused by weakening of a boiler-room bulkhead may have affected the happenings that ended it. This possibility is discussed in a later chapter.

Preparation at Southampton. The *Titanic* arrived at Southampton on 4 April; her maiden voyage scheduled for 10 April. The intervening days were a time of frenetic activity, for the ship itself not only had to be made ready, but the enormous complexity of supplies for the passengers and crew had to be taken aboard and stowed in the proper places. The crew itself, especially the stewards department, had to be filled out. Added to the complexity of putting a new ship into service were a thousand tasks to be done in a very short time. That time appeared too short for the completion of outfit, which had not been completed when the ship backed from its builder's pier in Belfast. Some carpets needed to be laid, painting finished, furniture fastened to its foundations, draperies and mirrors hung, and mattresses and bedding put in place.

Easter Sunday 1912, fell on 7 April, putting a temporary halt to the ant-hill activity at the White Star pier in Southampton. The frenzy resumed early on the following morning. It was now the time for frozen and chilled stores, wines, books, tableware, and medical supplies. Five Mail Clerks supervised the loading of mail. The loading of cargo commenced.

These last few days before sailing were also busy days for Thomas Andrews, Chief Designer and Naval Architect for Harland & Wolff, and his staff of eight men. They were busy from 0630 to 2030 in supervising the details of finishing and outfitting, giving special attention to items whose omissions or imperfections might displease passenger and crew. Berth ladders, cabin ventilation, tables, chairs—mundane items such as these were not beneath the notice of the Harland & Wolff people. Andrews himself made copious notes on what he saw, and discussed them with officers, crew, and subcontractors. His efforts were to continue into the coming voyage. It is probably belaboring the obvious to say that he was never to be satisfied, but a last letter to his wife contains the confident remark that the *Titanic* would be a credit to her builders. Beyond satisfaction at the moment, however, was his interest in the third sister, *Britannic*, whose building was just underway at Belfast. All shipbuilders and shipowners will

This is the Gymnasium for first-class passengers located on the Boat Deck of the *Olympic*-class ships. The Gymnasium was equipped with the latest in exercise machines. It would be a space where some of the passengers, who could not access a lifeboat, might have spent their last minutes before the *Titanic* would sink.

The Café Parisienne on “B” Deck was used by younger first-class passengers. It was a popular space to meet other passengers.

surely agree that when sister ships are built in sequence, experience with one always suggests beneficial changes in the following ships. Had Andrews and his notes survived, they would have reminded him of many improvements that could be applied to the *Britannic.*

While attending to the coming advent of the great new ship, White Star also had to contend with a coal miners' strike, an event that might have delayed the *Titanic* had not that ship been given such high priority. As it was, five ships of White Star parent, International Mercantile Marine—*New York, Oceanic, Philadelphia, St. Louis, Majestic*—were immobilized at Southampton by shortage of fuel. Together, their remaining supplies of coal constituted a reserve that could be tapped for their more important sisters.

The strike was settled on 6 April, making unnecessary White Star's announced plan of operating the *Olympic* and *Titanic* at the reduced speed of 20 knots to conserve coal. Normality was restored, but new supplies of coal were not instantly available, making the bunkers of the immobilized liners an invaluable resource. Together those ships supplied 4,427 tons to the *Titanic*, adding to the 1,880 tons, which that ship carried upon arrival from Belfast.

Signing of seamen to complete the *Titanic* crew began on 6 April, the day the coal strike ended. The strike had meant trouble for ship operators from the level of harbor tug to ocean liner, but there was one benefit that befell the *Titanic*: the idling of the five liners then in port had the consequence of swelling the pool of unemployed seafarers close at hand. That, plus the attraction of a new ship, and the high reputation of Captain Smith, made it easy to fill the crew rosters. The names of these men do not resonate in the way that their officers—the Murdochs, the Lightollers, the Boxhalls, the Bells—continue to do, but several were to gain a measure of fame from their testimonies before Senator Smith and Lord Mersey.

The complement of officers had been completed in Belfast, but a change was made in Southampton to bring aboard an officer with experience in handling a ship of the *Olympic* and *Titanic* size. Henry Wilde had served under Captain Smith aboard the *Olympic* since its maiden voyage, and hence had that experience. On the other hand, the initial chief officer, William Murdoch, though former first officer of the *Oceanic*, and highly regarded, lacked it. At Captain Smith's request, Wilde was brought aboard at Southampton to be the *Titanic*'s chief officer. William Murdoch was downgraded to first officer; Charles Lightoller then became second officer, displacing David Blair, who was forced to go ashore because of the lack of any vacancy he could fill. Lower officers remained as they were: Herbert J. Pitman was third, Joseph G. Boxhall was fourth, Harold G. Lowe was fifth, James P. Moody was sixth.

These men, and untold others, were the busiest of humans in those few days between the arrival at Southampton and the sailing on 10 April. In addition to the problems presented by the coal strike, the bunker fire, and the completion of crew and officer staff, were the mundane ones of familiarization with a new ship, its provisioning for the maiden voyage, and preparations for receiving passengers. Among the tasks of familiarization was a lifeboat drill. A drill was conducted, though without that element which would—in the unlikely event of an emergency—so drastically affect the use of lifeboats: the passengers. As events went, the crew was mustered, two boats (numbers 11 and 15) were lowered with about 18 men aboard each, rowed by the crewmen in a brief demonstration of their skill, then re-attached to their falls and hoisted aboard.

Passengers began to arrive soon after the boat drill was completed. Most of them came down from London on special boat trains. The first of these departed Waterloo Station early on the morning of April 10 for the short run to pier-side at Southampton. By noon, passengers and baggage were aboard. The harbor pilot had come aboard at about 1130. Just before noon, three blasts from the *Titanic*'s whistles announced the imminence of her departure. The last of the gangways was hoisted aboard a few minutes later. The order to let go mooring lines came at 1215. The great ship was underway.

Departure. The destination was not yet New York, however. Stops were to be made at Cherbourg and at Queenstown (now known as Cobh) to pick up mail and additional passengers. These two passages were largely uneventful, a seemingly good omen for a new ship on her maiden voyage.

But for believers in omens there were also three of the negative variety. One occurred in the undocking at Southampton. As the ship, assisted by tugs, left its dock, it perforce passed close aboard the docked IMM liners *Oceanic* and *New York*. The "suction" phenomenon that had brought *Olympic* and *Hawke* together acted upon these ships. The attracting force upon the *New York* snapped her lines and drew her toward a collision that might well have delayed *Titanic*'s departure. Quick action by the tugs assisting in the undocking, and a slowing of the moving ship, prevented contact. The lines of *Oceanic* tautened but did not break. Unknown at the time, but discovered later, a sunken barge also was dragged along the bottom by the interaction forces.

These incidents might give support to the notion voiced by the post-sinking critics of the *Titanic* that this ship was "too big." She was indeed too big to be handled as she was on that day in Southampton. However, the successful operation of the many larger ships (as well as the long-lived *Olympic*) built since 1912 indicates that the incident should be charged to the harbor pilot's inadequate training in the handling of ships having the size of the new White Star giants.

But now to speculate: a few days later the *Titanic* at 22 knots passes close aboard—say just a few feet away—a large fixed object. The ship and the "object" would be in water roughly 12,000 feet deep, quite unlike the water scarcely deeper than keel depth in the Southampton dock. But would the ship nonetheless be inexorably forced toward that object? Enough to make a difference between safe clearance and collision? We attempt no answer.

The second event of possibly negative portent was the bunker coal fire. It continued to burn at sea, though under control, and posed no threat to the ship. By judicious application of water, and by using the coal from the bunker in question before any other, the fire was extinguished before the events of 14 and 15 April. But had the fire weakened the watertight bulkhead that bounded this bunker? This was the bulkhead separating Boiler Rooms Nos. 5 and 6. Was its collapse during the sinking a process that led to a rapid flooding of Boiler Room No. 5, and thus hastened the sinking? Would it have held were it not for the consequences

The *Titanic* is being shown leaving the quay at Southampton on 10 April 1912 to begin her fateful maiden voyage. (*Courtesy of the Mariners Museum*)

of the fire? If so, would it have delayed the sinking for a significant time until the *Carpathia* arrived, perhaps?

The third event of possible portent was the disappearance of the binoculars that had been furnished for the Crow's Nest. On the afternoon of 10 April Lookout George Symons searched for them during his first watch there. They were *not* there. The question of their whereabouts fell to Second Officer Lightoller to answer. He determined that binoculars had indeed been available in the Crow's Nest when the ship left Belfast, but as Symons reported, they had disappeared. And they stayed disappeared—only the unaided eyes of the lookouts were available to scan ahead on the night of 14 April.

The story of the missing binoculars came out too late. Second Officer Blair had removed them to his stateroom for safekeeping when the ship reached Southampton. When he left the ship in the shuffle of officers caused by the appointment of Henry Wilde, his cabin remained locked, the binoculars overlooked in his rather sudden departure. But would the binoculars have made a vital difference in the sighting of the ice? A case for such a conclusion is easy to make, but it would not be universally accepted. As we shall note in a later chapter, the 1914 SOLAS convention—acting in full knowledge of the 14 April 1912 events—recommended that lookouts *not* be given binoculars. The apparent reason is the principle that lookouts in the Crow's Nest are supposed to maintain a wide range of vision to give early warning of distant objects. Identification of the object is then the province of a binocular-equipped watch officer on the bridge.

Notes for Chapter 11

1. Most of the events related in this chapter constitute a background for those that occurred after the ship and the ice began to appear on the same page, so to speak. Being background only, they are described rather sketchily. Because they are familiar, and moreover constitute no challenge to the *Titanic* canon, sources are generally not cited.
2. World's largest ship. Even had the *Titanic* not sunk, this title would have been lost to a later ship. At the time of this writing, the world's largest passenger ship would be the *Queen Mary 2* to start service in 2005. The tanker *Globtik Tokyo* at a displacement of approximately 600,000 tons is the world's largest ship.
3. After the *Titanic* tragedy, makers of postcards relabeled pictures of the *Olympic* at sea with the name of the lost ship. The ship can be identified as the *Olympic* by noting the absence of the windows.
4. We should remember that only three of the four stacks served the ship's boilers. Any picture that shows heavy smoke from all four is evidence of artistic license.
5. A crash stop test (a standard feature of new ship's sea trials) consists of sudden reversal of engines, with time and distance to stop being measured.

CHAPTER 12

Titanic: *The Event*[1]

Captain Edward Smith retired to his cabin for the night, unmindful of an important wireless messages from the steamer *Mesaba* of a dangerous icefield on the ink-dark sea directly in the path of his ship. First Officer Murdoch frantically ordered starboard helm in a desperate effort to avoid collision with an iceburg. The crew of the *Californian* does nothing, even as they wonder about a ship that fires rockets into midnight's skies. As the ship tragedy begins to reach its climax, Ida Strauss refuses to go where her husband cannot go. A funnel falls into a sea crowded with swimmers in freezing water, while propellers slowly rise from the sea. Captain Rostron of the *Carpathia* drives his ship beyond its limits in an attempt at succor. Later in the day, a vast throng of the curious and the fearful presses upon the New York offices of White Star, begging for news. Here is the story that has been retold in seemingly every way possible, and especially following the 1997 *Titanic* movie.

The preceding chapter provided only very sketchy accounts of the events from her arrival in Southampton and her departure from Queenstown (now Cobh). However, it has been suggested that the ship was not traveling with technical perfection. Two accidents to sister *Olympic* had delayed her maiden sailing, leaving time for icebergs to drift farther south in the North Atlantic. A fire in one coal bunker may have weakened a main transverse bulkhead whose collapse could hasten the sinking. Binoculars that could have enhanced the lookouts' ability to detect the iceberg were not in the Crow's Nest. A last minute change in officers so that Henry Wilde could be Chief Officer may have been the reason that the binoculars were missing—they were in the locked stateroom of the officer who had to be left ashore to make room for Wilde. A near-collision in harbor with IMM liners *New York* and *Oceanic* gave a hint that seamanship had not quite kept pace with the increase in ship size exemplified by the *Titanic*.

During the passage from Southampton to Cherbourg, Lookout George Symons did not find binoculars in the Crow's Nest. After the *Titanic* was clear of the Nab Lightship he went to the officers' mess room and asked Officer Lightoller for the binoculars. Lightoller went into Officer Murdoch's stateroom and came out to say, "Symons, there are none." With that Symons went and told the other lookouts, Fleet, Lee, and Evans. The rest of the trip to Cherbourg was uneventful with the ship arriving around 1830. Passengers and mail were picked up here and the ship departed around 2000 for her passage to Queenstown. Early in the morning of 11 April the rudder and steering gear were checked. Father Browne, who rode the ship to Queenstown and provided the only photographs of the ship during this portion of the voyage that survived, took pictures of a zig-zagging wake. Arriving in Queenstown, the ship discharged passengers and mail and picked up her last consignment of passengers and mail.

The ship sailed at 1330 from Queenstown on the passage to New York during the afternoon of 11 April. The immigration officer at this Irish port had issued a certificate showing a total complement of 2,208 passengers and crew. As a result of the sinking, 706 persons were saved (one died before reaching New York), and the commonly accepted number of those lost is 1,523. The difference between this number and the 1,502 that would agree with the certificate is supposedly an indication of stowaways and a few passengers who were inadvertently not counted.

It should be noted that the *Titanic* carried only about half of her rated passenger load. Was this a matter of forebodings? Not at all. The peak period for transatlantic travel in 1912 was midsummer. Mid-April travel was distinctly off-peak, a fortunate circumstance indeed, since every additional passenger aboard would almost certainly have meant one more casualty. Those who were aboard, especially those in first class, gave strong representation to the higher levels of American society; some names would resonate with the many fascinated by the *Titanic*. On board were such notables as Ben Guggenheim, a wealthy industrialist who was said to have donned his evening clothes, then lit a cigar to "die like a gentleman." Ida Strauss, wife of Isador Strauss, owner of Macy's in New York, who refused to enter a lifeboat when "women and children first" kept her husband out was returning from an extended vacation in Europe. Major Archibold Butt, diplomat and military aide to President Taft, was returning from a visit to Europe. John Jacob Astor and his recent bride, Madeline, were returning from their honeymoon in Europe. W. T. Stead, renowned English journalist was taking a trip to America to give a lecture. George Widener, Philadelphia street railway magnate, with his wife and son, was returning from a business trip. Margaret (Molly) Brown, the socialite who was soon to become "The Unsinkable Molly Brown," had been on an expedition in Egypt. These names constitute a sample of prominent people aboard the newest luxury liner of her time. We may introduce others as they appear in the story.

The opulence of the new ship, its novelty as the world's largest, the cachet of being on its maiden voyage . . . some or all such factors must have appealed to the affluent people. So, too, must a remarkable new feature—the wireless—a useful novelty that must have had an impact like that of the Internet in the 1990s. Of course, it was not a true novelty on this particular voyage, since it was first installed on the German passenger liner, *Kaiser Wilhelm der Grosse,* in 1900. Her use a few years later in saving 1600 passengers and crew from the White Star liner *Republic,* after collision with the Italian Line *Florida*, demonstrated her immense value at sea, and made it essential for all ocean-going passenger liners. More recently, its use by passengers for communicating with the shore had become commonplace.[2]

The *Olympic*-class outfit consisted of the sending and receiving equipment, a five-kilowatt motor generator that produced high voltage

This is one of the last photographs taken of the *Titanic.* She is shown arriving at Queenstown on 11 April to pick up her last consignment of passengers before beginning her voyage to the open sea and her doom. (*Courtesy of the Mariners Museum*)

from the ship's 100-volt system, and twin-T antennas hung over the 600-foot span between foremast and mainmast. The ship provided housing: a sleeping compartment for the two operators, a second room for the receiver, and a third for the transmitter and the motor generator.

In 1912 three companies competed for the privilege of providing wireless service: the Marconi Company of wireless pioneer Guglielmo Marconi, the German Telefunken, and the American DeForest. These companies did not sell their equipment for shipboard installation, but leased it and its operator(s) to the shipowners. Commercial traffic was passed only between sets of the same company, but messages at sea concerning safety and navigation passed freely among ships regardless of company affiliation.The two radio officers, Jack Phillips and Harold Bride, were employees of the Marconi Company and were listed as second-class passengers.

Of course, the wireless would play an essential part in rescue of the 706 survivors of the *Titanic*, and with a bit of luck might have led to the rescue of all those who were lost. Unfortunately, it did not. As we shall see, the blame for that failure might lie with the crowding of the airwaves by traffic to and from the *Titanic*'s affluent passengers.

While the *Titanic* was taking on passengers and mail in Queenstown, Chief Engineer Joseph Bell and Bruce Ismay, Chairman of the White Star Line, met in the latter's stateroom on "B" Deck to discuss several important aspect of the coming voyage. A vital point of discussion was the status of the ship's fuel supply. Supplies were low in Southampton when *Titanic* arrived there on April 3 in consequence of the coal strike that had idled many ships. As it turned out, of course, the new ship was to be adequately fueled when its departure time arrived, but not to the extent that coal could be lavishly expended.

Mr. Ismay then spoke to the Chief Engineer about the voyage, telling him that it would not be possible for the ship to arrive in New York on Tuesday, April 16. In consequence, there was no purpose in pushing the ship at full speed. The time for arrival at Ambrose Lightship was, therefore, set at 0500 on Wednesday morning, a time more consistent with the desire to arrive at the White Star piers in New York in mid-morning. A gala reception had also been planned for the new ship as she made her way to her berth on the Hudson River. However, the two men also discussed the feasibility of running the ship at full speed during the voyage ahead. Mr. Ismay told Bell that there might be an opportunity of driving her at full speed for a few hours on Monday 15 April, or Tuesday 16 April if the weather was unsuitable the day before. Consistent with this possibility, and from evidence presented at the post-accident hearings, all 24 double-ended boilers were in operation by nightfall of 14 April, but the furnaces for the five single-ended boilers in Boiler Room No. 1 were not fired.

The decisions of the Chairman and the Chief were later amended to include a full-speed run of a few hours. That speed run would take place on Monday afternoon, 15 April, to confirm that the ship could sustain its rated power for an indefinite time. This test never took place, of course, for by that time the remains of the Great Ship lay among the creatures of the abyss, two miles down.

Among the many canards laid against the memory of the men who managed the *Titanic* is the belief that her high speed on the night of 14 April was part of an attempt at the North Atlantic speed record. There can be no truth to that supposition—the ship's maximum speed was known to be at least 3 knots less than that of *Lusitania*, the record-holder at the time.

The Delayed Voyage Begins. As Ireland dropped below the eastern horizon on the evening of Thursday 11 April, three days and nights remained before the kiss of the ice queen was to immortalize the ship and its people. Meanwhile the passengers enjoyed the comfort and luxury of a great hotel that in a miracle of 20th-century technology was speeding along a well-known track between Europe and America, and with scant need to care for perils of the sea. No wind that blew dismayed the crew, nor troubled the captain's mind . . . nor those of the passengers. Even in the spartan accommodations of third class, the words of Dr. Johnson that spoke of prison and the threat of drowning were the stuff of humor rather than wisdom. All aboard—save, perhaps, the sweating men of the firing aisles—had reason to be happy. Of course, the many works of fiction set aboard the *Titanic* inevitably illuminate the smolderings of interpersonal conflict waiting to be resolved when the shortage of lifeboats separates those who may live from those who must die. The mirror of fiction surely shows real life with at least partial accuracy, and hence we must accept that even the comfort and luxury of first class did not guarantee happiness. But the White Star Line had done its part.

The vernal equinox had banished official winter from the North Atlantic three weeks before the *Titanic* departed Southampton on 10 April. Winter's legacy lingered, however, and would for several more months. Icebergs do not melt quickly, and they can drift far south before they finally do. From their standpoint, the winter just passed had been a good one, since it was milder than normal, as was also the case with several preceding winters. Paradoxically, mild weather is good for iceberg culture because it increases the "calving" of bergs from the icecaps of the far northern islands. The ice floe of 1912, 10 miles wide and 300 miles long, was one of the worst of the 20th century. (However, its magnitude was much less than that of 1986—the worst ice year of that century.)

Sunday, 14 April 1912—the day of agony and apotheosis for the Great Ship—dawned clear and cool, the sea calm with slight swells. The record ice of that year was about 350 miles away at dawn. Some wireless warnings on icebergs had reached the *Titanic*, and Fourth Officer Boxhall had plotted them on his chart as he kept a record of the ship's course. Only a chill to the early morning air made the morning less than perfect for the few passengers arising early to pace the promenades.

A little-known event took place on 11 April 1912. The eastbound steamer, *Niagara*, had a brushing encounter with an iceberg about ten miles from where the *Titanic* would meet her fate. The *Niagara* had a steel hull and was 8,590 gross tons, but was capable of a speed of 15 knots. The damage was confined to two places below the waterline, but pumps kept her afloat. She was able to make port for repairs. Two other ships, the *Manchester Inventor* and the *Corsican* also collided with ice in 1912, but survived. From 1874 to 1912 some 70 ships had encounters with icebergs and ten of these sank. The most famous iceberg collision before *Titanic*'s time was the steamer *Arizona* that struck a very large iceberg head on 7 November 1879 at a speed of 15.5 knots. There were no known deaths, but there were some injuries to some of the 509 passengers. Her bow was pushed back about 20 feet, but no damage was done to the collision bulkhead. She backed all the way to St. Johns for repairs where 200 tons of ice was wedged into the hull.

Ice Warnings. At 0710 Captain Barr of the Cunard liner *Caronia* alerted the *Titanic* to ice danger ahead, reporting that westbound steamers were reporting icebergs, growlers (small bergs), and field ice at 420 north from 490 through 510 west. But this early message appeared to have little urgency in it since it also carried congratulations to Captain Smith on his new command. After studying this transmission, Smith ordered this message posted in the chart room. He then retired to the first-class Dining Room.

Another ice warning came via wireless at 1342. The liner *Baltic* reported moderate but variable winds and clear weather, and that she had received a report during the night from the westbound German tanker *Deutchsland*. The tanker was short of coal and had lost steering control. All of that was little more than seafarers' gossip having no consequence to the *Titanic*, but the ice warning was included. The *Baltic* had received from the Greek steamer *Athinai* an account of passing icebergs and a large quantity of field ice at 41051’ N, 49052’ W. Captain Smith acknowledged receipt of the *Baltic*’s report at 1355.

The captain showed this report to Bruce Ismay, and the two men discussed its implications for the scheduled arrival at New York. Part of the lore of the *Titanic* is that Ismay, Managing Director of White Star, and hence Smith’s employer, ordered (or persuaded) the latter to maintain speed in spite of the apparent hazard. Ismay, who survived to testify—and to endure criticism on several counts—denied the allegations. It does seem likely the two men discussed the full-speed run that had been set for Monday afternoon.

At lunch after his discussions with Captain Smith, Ismay found that word of the ice somewhere ahead had reached the passengers. As he ate, he was quizzed by his tablemates about the ship’s speed and the possibility of docking on time in spite of dangers inherent in ice. Perhaps to demonstrate his in-the-know status, he showed the *Baltic* telegram. One of the passengers, Mrs. Arthur Ryerson, later remembered that she had asked “I suppose you are going to slow her down?” and that the managing director had replied “Oh, no. On the contrary, we are going to let her run faster and get out of it.”

Ismay’s alleged remark echoes the commonly held doctrine of the era’s ship captains: in poor visibility the probability of collision is minimized by steaming as fast as possible. However, if his captain had read the contemporary edition of *Modern Seamanship*,[3] the following passage might have given him pause:

> A reason frequently given for advocating high speed in a fog is that, the fog belt being of definite width, the danger of collision will be reduced by getting across it as quickly as possible. This is like saying that if one is called on a dark night to cross a public square in which people are moving about in all directions, it will be safer to run across at full speed than to walk slowly.

“Fog belt” could as well be written as “ice belt.” Unfortunately, “let her run faster” so dovetailed with the imperatives of schedules that any doubt the captain may have had about the go-fast doctrine must have been overridden. This veteran skipper of White Star ships, so aware of those imperatives, probably needed no go-fast urgings from Bruce Ismay.

What about the contrary words taken from the well-regarded book? The shipping-line executive, for all that he may have enjoyed his status as an authority in the eyes of *Titanic* passengers, could not have been expected to read a book on seamanship. What has this to do with Captain Smith? His high reputation was based on a distinguished career, and not on his book knowledge. He charmed the ladies as well, but he was not the studious type. Like that other renowned seaman, Joseph Conrad (whose words about the *Titanic* event are quoted later), he may well have scorned the wisdom of doctrine that was not based on his own experience.

At 1345 the German liner *Amerika* informed the United States Hydrographic Office that she had passed two large icebergs at 41°27'N, 50°08'W. Although the message was not addressed to the *Titanic*, it was received and retransmitted by that ship's wireless operators. In spite of its obvious implications for all ships in the western North Atlantic, operators Bride and Phillips treated it as a private message. They did not notify the navigating personnel on the Bridge. It should be noted that these two men were employees of the Marconi firm, not the White Star Line.

At 1452 a wireless message from Captain Krol of the liner *Noordam* added to the congratulations on his new command coming to Edward Smith, and reported westerly winds, fair weather, no fog. In addition, it added to the tide of incoming ice warnings: Krol reported major ice in the area 42°24' to 42°45' N, 49°50' to 50°20' W. The warning prompted a second course change that would take the *Titanic* farther south of the shipping lanes that were normal for April; after discussion with Chief Officer Henry Wilde, Smith opted for south, 860 west (266°, magnetic). But the schedule, the near-sacred schedule? It could still be met if speed was not reduced. It remained unchanged at 22.5 knots.[4]

At 1800, Second Officer Lightoller took the bridge as watch officer, relieving Henry Wilde. Captain Smith left the bridge to get ready for a party to be given by the Henry Wideners, prominent first-class passengers. About this time the temperature on deck began a steep decline. Only the hardy remained outside, because within another hour the air temperature had fallen to 33°F, a low temperature that was exacerbated by the strong wind caused by the forward speed of the ship into calm air.

Captain Smith's night order book, in an order written at 1720, specified a course of south, 62° west (242° magnetic). For reasons unclear, the ship's heading was changed to south, 710 west (251° magnetic) at about the time that the forward external lights were being covered to avoid blinding the watchstanders in the Wheelhouse.

Thirty minutes after coming on watch at 1800, Lightoller obtained a stellar fix with his sextant, and gave his data to Fourth Officer Boxhall for a computation of position. Captain Smith, when he returned to the Wheelhouse after the dinner party given by the Wideners, had the information entered on the ship's chart.

By 2040 the seawater temperature had fallen to 31°F, prompting Second Officer Lightoller, then on watch, to request that Fourth Officer Boxhall have the Ship's Carpenter[5] check the fresh water tanks for possible freezing. These tanks were fitted with steam heaters to prevent that troublesome event; the Carpenter made sure that the heaters were regulated to act before freezing could begin.

Meanwhile, the star-crossed captain of the Leyland liner *Californian*, Stanley Lord, had appeared on the fringes of the *Titanic* drama. His ship was confronted by a large expanse of field ice. Sensing the danger, he sent a message to the *Antillian*: "Six thirty P. M., apparent ship's time, latitude 42° N, longitude 49°9' W. Three large icebergs five miles to the south of us. Regards, Lord." Harold Bride on the *Titanic* overheard this transmission. He sent a copy to the Bridge, but he could not remember when testifying later to whom he had delivered the message.

After the Wideners' dinner had concluded, the party moved to the Palm Court, where Captain Smith smoked a cigar and listened to the orchestra play a selection from the "Tales of Hoffman." At 2055, he thanked his host for his cordiality and bade his goodnight, and within a few minutes was back on the Bridge, where Second Officer Lightoller was still on watch. There he found that the ambiance of the North Atlantic, so favorable since Ireland sank below the horizon on Thursday, had taken an ominous turn, for all that neither man recognized the signs as such. The air and sea were cold, much colder than when the sun had set, colder because ice was about. The dark ocean in unquiet slumber lay—in slumber, but menacing, for its stillness precluded the warning sound or sight of wave against ice. The moon that might have negated these evil circumstances still hung below the horizon. But those warnings by wireless, those dot-dash rumbles of doom, like something from the tales of Tolkien, should have connected. Someone on that bridge, the captain or his officers, should have dared to say "We've been told of ice, repeatedly. Now it's near, it's deadly, and we're blinded to its presence by darkness."

Perhaps an officer did have such thoughts, but cowed by the reputed wisdom of his captain, thought it insolent to speak. Whistle blowing as an acceptable line of conduct was not heard of until late in the 20th century, and never become acceptable on a ship's bridge. This was still Balaklava-at-sea, with only one man permitted to reason why. Perhaps that one man, the captain, saw the converging signs of possible doom, but was fatally handicapped by that doctrine so presciently rebutted by Knight[6]—that when endangered by poor visibility one lessened danger by speeding through the affected area. Perhaps it was the imperatives of meeting schedule, reinforced by his just-past conference with Bruce Ismay on that topic, and the decision it had generated. Perhaps it was a belief that his ship was invulnerable. *That* is highly unlikely; Captain Smith may have been misled by those unwise bruitings of unsinkability, but he surely knew that a collision with a large mass of ice would inflict severe damage on his ship, and on his reputation as a prudent shipmaster.

Around 2100 the course was still south 71° west, according to testimony by Quartermaster Hitchens, the helmsman at that time. Captain Smith came to the Bridge at 2105 and remarked that it was very cold. Second Officer Charles Lightoller, the watch officer on duty, responded, "Yes, it is very cold, Sir. In fact it is only one degree above freezing. I have sent word down to the Carpenter and rung up the engine room and told them that it is freezing or will be during the night." Their conversation continued as follows:

Captain Smith: There is not much wind.
Lightoller: No, it is a flat calm, as a matter of fact.
Captain Smith: A flat calm?
Lightoller: Yes, quite flat; there is no wind. It is rather a pity that there is no breeze whilst we were going through the ice region.

Captain Smith knew the implications of Lightoller's statement. A breeze would mean waves that would break at the base of an iceberg. The foaming white of a breaking wave against the waterline ice of the iceberg made it easier for the lookouts to detect its presence. As Captain Smith's eyes adjusted to the darkness, the conversation turned to the weather.

Captain Smith: The sky seems quite clear.
Lightoller: Yes, it is perfectly clear.

It was a beautiful night, with not a cloud in the sky. The crescent of a waning moon would not rise until several hours later. In fact, there was to be an eclipse of the sun to take place on 17 April. The sea was apparently smooth, and there was no wind. The stars could be seen rising and setting with absolute distinctness at the horizon line. With conditions so clear and starlight reflecting off the water surface, it was also difficult in determining the horizon line. Overhead, the sky was also quite clear. Such conditions are favorable for the phenomenon of "looming" (in other venues known as a "mirage") in which objects just below the horizon are visible. Looming might well have allowed an early warning of an iceberg, but not if haze from field ice clouded the horizon.

The conversation turned to the problems of sighting ice under these conditions.

Lightoller: In any case there will be a certain amount of reflected light from the bergs.
Captain Smith: Oh, yes, there will be a certain amount of reflected light.
Lightoller: Even though the blue side of the berg might be towards us, probably the outline, the white outline, would give us sufficient warning. Therefore, we should be able to see it at a good distance.

Lightoller was making the point that not all icebergs are white, but that even so, starlight reflecting from the ice should form an outline that could be seen.

During this conversation, the ship was not in the ice region, but the wireless warnings had shown these two officers that ice was not far away. They also knew from many years on the North Atlantic that ice could be encountered, even though it had *not* been predicted. Precautions had to be taken. Younger officers had to be told of what to expect in the presence of ice.

Captain Smith remained on the Bridge for 20 to 30 minutes, but during that time he did not discuss the ship's speed with Second Officer Lightoller. The latter, who survived, was to testify that the ship was making about 21 knots. The helmsman, Hitchens, maintained in his testimony that it was 22.5 knots. The Captain spent a few minutes telling Lightoller about the party he had just attended, and then left the Bridge. As he departed, he instructed Lightoller, "If the conditions become at all doubtful let me know at once; I will be just inside. If there is the slightest degree of haze, the slightest haze to any degree noticeable, immediately notify me." Captain Smith retired to his quarters.

Haze—which some mariners call "smoke"—could indeed signal the presence of an iceberg otherwise unseen. On the other hand, large bergs frequently generate a little low-lying haze, which can make their visual detection more difficult.

As soon as Captain Smith retired, Lightoller called the Crow's Nest and told the two lookouts, George Symons and Archie Jewell, who were at their post on the foremast to keep a sharp eye for the slightest presence of haze and ice. They were relieved at 2200. Just before he went off duty, Lightoller again called the Crow's Nest to repeat his admonitions to the relieving lookouts, Frederick Fleet and Reginald Lee. In both instances he warned the men that the absence of waves breaking at the waterline of a berg would make their task more difficult.

The Bridge watch also changed at 2200. First Officer Murdoch took over the Bridge, selected by chance to be the one on watch an hour and forty minutes later when the panicked alarm of ice came from the Crow's Nest. Lightoller left him in charge after briefing him on his conversations with Captain Smith earlier, and then retired after a short tour of the ship.

As the night wore toward its climax, Thomas Andrews was busy in cabin A-36, studying his plans and notes. The stubborn coal fire had been extinguished the previous day, but its consequences were a worry, and he had made a note to have the bulkhead between Boiler Rooms Nos. 5 and 6 checked for fire damage. He had noticed in his own inspection that some stiffeners on that bulkhead were slightly out of line, but that circumstance was not a source of worry. In a matter far more trivial than possible damage to ship's structure, he had noticed that hat racks were fastened to the stateroom bulkheads by four screws. Wouldn't three of these be just as effective? Surely, and though the cost of the screws themselves was of scant consequence, saving of the labor of drilling and installation would amount to a worthwhile saving on the third sister, *Britannic*. Stateroom décor could have a different impact, that of passenger satisfaction. It was an intangible thing in itself, but one that could have a tangible impact on a passenger's "next time" choice of ship. Andrews asked, and found that yes, the color of the pebble dashing on the private decks was not well liked. He made note to have it changed during the *Titanic*'s first return to its builder's yard. Overall, however, he appears to have been highly pleased with the ship. A survivor reported that earlier in the day he had stated "The *Titanic* is as nearly perfect as human brains can make her to be," echoing sentiments that he had confided to his wife in a letter sent on the day of departure.

Ice warnings still dominated the inter-ship wireless traffic. At 2150 a signal was received in the Marconi Room from the *Mesaba*:

> In latitude 42 N to latitude 41.25 N and longitude 49 W to longitude 50.30 W saw much heavy pack ice and a great number of large icebergs. Also field ice. Weather good, clear.

The *Titanic*'s position and course at 2150 would take the ship through the area cited by the *Mesaba*. The importance of the *Mesaba* message justified its immediate delivery to the watch officer on the bridge. That officer, either Lightoller or Murdoch (who were both present for the 2200 change of watch) presumably would have made the necessary plot on the chart, recognized the implications of this message, and probably summoned the captain. Captain Smith, surely, would then have given in to the necessity of abandoning the schedule in favor of slowing the ship or making a further course change. But none of that happened—the message was not delivered to the Bridge.

The message was never delivered. The prime reason was the priority that passengers were given in wireless use. In that, the operators Harold Bride and Jack Phillips were only following the rules set by their employer, the Marconi Company. They were allowed to use their own discretion in assigning priority to messages, for all messages relating to navigation were supposed to have first priority. However, the numbers describing location of ice meant nothing to Jack Phillips, then on duty. Wireless traffic was heavy from passengers' messages being sent up by the Purser, and Phillips may not have been at his best because of exhausting hours spent with the repair of the wireless set the night before. Harold Bride, his assistant, was no help since he was off duty, asleep. The vital message went into a pigeonhole for delivery when circumstances were less taxing.

A long day was nearing its end, but Sunday is still a Sabbath until midnight strikes. The Reverend Ernest Carter, beginning around 2030, held a hymn service in the second-class Dining Room. About 100 people gathered around a piano to sing hymns. "For Those in Peril on the Sea" had long been *de rigeur* at divine services on English-speaking ships, but it was almost overlooked on this ship that had moved seafaring beyond hazard. Just at the end, however, someone asked that this tradition be honored. Reverend Carter acquiesced, but with the remark "In this ship we are not at peril, but others may be. Let us sing this hymn for them." Afterward, in thanking the Purser for use of the room, he remarked that though it was the first time that hymns had been sung there, he did trust and pray that it would not be the last time.

Messages emphasizing the word "ice" still continued. The one at 2230 did not depend on the cooperation of *Titanic*'s Marconi people, since it came directly to the bridge by signal lamp from a ship passing abeam. The ship was the Furness Withy freighter *Rappahannock*, outward bound from Halifax. The signal read "Have just passed through heavy field ice and several icebergs." It also mentioned rudder damage caused by ice. The *Titanic* acknowledged receipt, with thanks. The other ship had the last word, which was simply "Goodnight." First Officer Murdoch, then on watch, took no significant action because of this additional—and what should have been redundant—message.

Shortly before 2255, the Leyland liner *Californian* and her captain, Stanley Lord, soon to share the role of villain with Bruce Ismay, entered the arena. Earlier that evening, Lord had to put his engines full astern to avoid hitting some field ice. The ship stopped, swinging onto a northeast heading, and with no clear path of escape apparent, was left to drift with the ice for the night. At 2310 Captain Lord ordered his wireless operator, Cyril Evans, to inform other ships of the *Californian*'s position, and of her circumstances of being stopped amid ice floes. The lights of an approaching steamer, thought to be the *Titanic*, indicated a possible hazard, or a ship that might herself be unknowingly at risk. Evans put on his earphones and sent "Say, old man, we stopped and are surrounded by ice." At the receiving end, the strength of a message originating so close at hand interrupted Jack Phillips in the process of sending a number of passengers' messages to Cape Race.[7] Not at his best from overwork and lack of rest, an angry Phillips instantly responded "Shut up! Shut up! I am busy. I am working Cape Race."

Undaunted, Evans continued to listen in hopes of getting through, and heard Phillips telling Cape Race, "Sorry. Please repeat. Jammed," though the *Californian* wireless set could not pick up the reply. Evans hung on for 25 minutes, listening, until at 2325 he shut down his set and went to bed.

The fate of the *Titanic* now lay with the eyesight, alertness, and quickness of response of the two lookouts in her Crow's Nest on the foremast.

Iceberg Ahead. The two lookouts, Frederick Fleet and Reginald Lee, were posted in the Crow's Nest, 95 feet above the waterline on the foremast, about 120 feet aft of the bow. They had relieved the earlier watch at 2200, when they had been advised by Jewell and Symons, as well as Officer Lightoller, of the necessity of keeping a sharp eye for icebergs and growlers. So calm was the sea that the whiteness of foam from waves breaking against ice would not help them, but the new men were impressed with the importance of any sighting. The retiring watch had also spoken of a strange, clammy smell of ice about, all the more reason for vigilance.

Vigilant they were, so far as we know. There was no witness to say otherwise. Their position was not a comfortable one, since they were in the open air, exposed to the cold wind, a wind brewed from still air by the forward motion of the ship. Warm clothing was a help, but it obviously could not cover the eyes. The canvas screen on the aft side of the Crow's Nest shielded its occupants from the ship's lights, but did essentially nothing to help them endure the wind in their faces. Whose eyes do not water when they stare into a wind of 20-plus knots and freezing temperature? What watering eye can pick up a faint object, white though it may be, out of total darkness? An unprotected eye, it had to be. A screen of glass would have helped under the conditions of that night, but in general, the likelihood of vision blocked by spray or rain on glass[8] meant that the eyes of the lookout must be exposed.

Whatever the difficulties, whatever the delay in recognizing hazards that watering eyes might have caused, Frederick Fleet testified that he had seen a haze to starboard about 2300, and that it had troubled him. Turning to Lee, he mentioned that he wished for binoculars[9,10] so he could determine what might lie under the haze. In spite of his concern, he did not convey it to First Officer Murdoch on the Bridge. The lookouts had not been told that Captain Smith regarded haze, a sure sign of ice, as so threatening that he was to be awakened if it were sighted.

Just before 2340, Reginald Lee spotted a slight mist stretching across the horizon, while Frederick Fleet saw a mist occupying about 45 degrees of the horizon. Both lookouts strained to see if anything lay beyond the mist until Fleet finally became aware of an object, a dark mass with pointed top that obscured the stars near the horizon that looked like the size of two tables. The dark object lay direcly in the path of the *Titanic*. Lee remarked to Fleet that he did not like the look of this black mass, and well he might not, for it was the shadow at the end of the valley of death for 1523 innocents. Hopeless... but if he had recognized it a few seconds earlier? If he had grabbed the bell rope and telephone at the first suspicion, could events have been different?[11]

Fleet, after seeing the black object high above the water, reached immediately for the rope to ring the bell in the Crow's Nest three times. He then went to the starboard side of the Crow's Nest to pick up the telephone to ring the telephone on the Bridge. Officer Murdoch picked up the phone and the conversation went like this:

Murdoch: What do you see?
Fleet: Iceberg right ahead.
Murdoch: Thank you.

Fleet and Lee continued to keep their eyes ahead and saw that the ship was slowly beginning to turn to port, but the distance between the ship and iceberg was quickly closing. At first, the berg seemed to be small, but quickly grew in perceived size as the ship sped toward it.

Personnel on duty on the Bridge consisted of First Officer William Murdoch, the watch officer, standing on the starboard bridge wing, Sixth Officer James Moody within the Wheelhouse, Quartermaster Robert Hitchens at the wheel, and Quartermaster Alfred Olliver, standby to Hitchens, and acting as messenger. Fourth Officer Joseph Boxhall was approaching the bridge from the starboard side. Moody answered the phone, and repeated Fleet's warning verbatim. After a quick glance ahead, Murdoch ran in from the starboard wing and yelled at the helmsman "Hard a'starboard"[12] with the intention of turning the ship to port. Quartermaster Hitchens responded, and quickly had the wheel hard over. Almost simultaneously, Murdoch reached the engine telegraphs to signal STOP, followed quickly by FULL ASTERN.[13] He pushed the button that activated a 10-second alarm that watertight doors were closing, then pulled the lever that put that closure into effect. Murdoch had done all that he possibly could. He may have done more than he should. The same *Modern Seamanship* that decried high speed in poor visibility also offered this advice:

> A vessel turning away from another vessel to avoid collision should always continue at full speed, as the effort involved in this course is an attempt to cross the other vessel's bow. To turn away and slow is the surest way of bringing about collision.

The iceberg was simply a vessel with speed of less than a knot. Even if Murdoch did not accept advice from books on how to handle his ship in an emergency, he should have known the *Titanic*'s characteristics well enough to realize that an attempt to stop would produce virtually no effect in the distance available. Perhaps he fell victim to panic. (The technical aspects of stopping and turning the ship are discussed in Chapter 28.)

Once he had done what either calm analysis or panic demanded, it was only left for him to order Olliver to note the time, and Moody to make suitable entry in the log.

Turning to avoid was doubtless the best measure available, but ships are not automobiles, nor are they small motorboats. Unlike these comparatively minute objects, their inertia does not allow them to respond instantly to helm inputs. The iceberg is thought to have been about 30 seconds ahead when the alarm reached the Bridge. For far too many of those seconds, the *Titanic* continued without change of course, and without slackening her speed. She did, too late, begin to turn, but at the moment of impact the *Titanic* had turned only about 22.5 degrees[14] with speed not significantly reduced.

In the Crow's Nest, Fleet and Lee instinctively clutched the railing as they felt their bodies being thrust to starboard as the ship began to turn. The turn was encouraging, but they sensed that it was far too slow—the black mass of the iceberg was rapidly closing the ship's bow. Fleet, worrying about the consequences to their perch when the inevitable contact occurred suggested that Lee leave. After a few seconds of discussion, Lee started down the ladder, but apparently realizing that he couldn't reach the deck before the iceberg came, scrambled back to face the inevitable with his mate.

Notes for Chapter 12

1. The ultimate sources for the events related in this chapter are the testimonies recorded at the post-*Titanic* hearings of the U.S. Senate (Senator William Alden Smith) and the British Board of Trade (Lord Mersey). These testimonies have been organized into coherent stories by many authors, and in that way retold many times. Particular authors that we acknowledge as helpful secondary sources are Walter Lord, 1995, *A Night to Remember*, Henry Holt, New York; Charles Eaton and John Haas, 1986, *Titanic, Triumph and Tragedy*, W. W. Norton, New York; and Geoffrey Marcus, 1969, *The Maiden Voyage*, Viking Press, New York.
2. The early "wireless" electromagnetic wave was not modulated by a superimposed audio-frequency signal, but communicated in coded interruptions of a continuous wave—the familiar dots and dashes. Furthermore, it was not confined to specific frequencies (channels) as was universal practice after voice radio was introduced, thus vital messages concerning navigational hazards had to compete for air time with the prattle of passengers calling home.
3. Austin M. Knight, 1912, *Modern Seamanship*, fifth edition.
4. The ship speed of 22.5 knots was taken from the testimony of Helmsman Robert Hitchens at the British Inquiry.
5. The position of Ship's Carpenter was a carryover from the days when repairs to wooden ships while at sea were often required. In the days when steel construction left little to saw and hammer, the position was more that of a general handyman. Tradition preserved the ancient title, however.
6. Knight. See Note 3.
7. Cape Race lies at the southeastern tip of Newfoundland. In the early days of wireless telegraphy its station served the vital function of relay point for messages between ships in the Atlantic and the North American continent.
8. In later years the problem of vision in the face of flying water (liquid or otherwise) was solved by the "clear view screen," which is a circular panel of glass rotated at high speed by an unobtrusive electric motor. The rotational speed is so high that nothing sticks to the glass. If the glass is otherwise clean, one sees into the murk as if the glass were not there.
9. Why no binoculars? Recall that their disappearance in Southampton has been recounted in the preceding chapter.
10. Would the binoculars have made a vital difference in the sighting of the ice? A case for such a conclusion is easy to make, but it would not be universally accepted. Recall the mention in the preceding chapter that the 1914 SOLAS Convention recommended that lookouts *not* be given binoculars.
11. The memory of the minute or less when they held the fate of the *Titanic* in their hands was to haunt both men. Although they were never taxed with any blame, they must have felt that quicker recognition and reporting might have saved the ship. Lee eventually turned to alcohol, Fleet to a psychologist.
12. Paradoxically, starboard helm turned the ship to port. This is the way it is with a tiller, as anyone with experience in small sailboats will recognize. The wheel had begun to replace the tiller at least 300 years before 1912, but maritime tradition can easily outlast such a brief interval. However, by the mid-20th century ships were being steered like automobiles: turn the wheel (push the top of it, that is) in the way the ship is to turn.

13. Some authors have expressed the belief that this order was SLOW ASTERN. This possibility is discussed in the following chapter.
14. The 22.5-degree turn estimate comes from a post-accident maneuvering trial of the *Olympic* in which time and speed estimates from the *Titanic* were used.

CHAPTER 13

Collision

Lookouts Frederick Fleet and Reginald Lee, beset by the uncertainties of haze and shadowy shapes lying before a ship careening too fast through eye-watering 27°F cold of a North Atlantic night, had made the best of their circumstances. They had struggled to keep warm from a cold wind from the ship's forward motion. They had been told to be alert for icebergs by the preceding watch. Now they had seen the danger a little over a half-mile ahead and had spoken the doom-invoking word "iceberg" into the telephone connecting their lookout position to the Bridge. Roughly half a minute was left for them to flee to some spot less exposed than the Crow's Nest. Fleet wanted Lee to go below and Lee did start down the narrow passage in the mast structure. However, he came back up to watch this sinister object they had warned the Bridge about appear to take a path that would intersect the *Titanic*'s hull at a point forward of the Crow's Nest according to their testimonies before the American and British inquiries. Mercifully, their post proved to be safe from the immediate consequences of the collision, and both men also survived the ultimate consequences. Fleet testified before both the British and American inquiries while Lee testified only at the British one.

Fleet[1] did testify to his feelings at a more practical level. He described the iceberg as first appearing in size "like two tables put together"; and as it neared the ship he estimated its height to be no higher than the ship's forecastle, or some 50–60 feet above the water. As he watched, the bow made contact about 20–50 feet aft of the stem. A brushing encounter of ice against hull continued as the ship steamed forward with its speed not noticeably reduced. Fleet maintained that the ship was moving throughout the contact and did not come to a stop until the iceberg had disappeared astern. This testimony is consistent with that of other witnesses. On the other hand, their testimony that the ship started to turn while Fleet was completing his warning to the Bridge is inconsistent with other testimony, and with the known maneuvering characteristics of the ship. As the iceberg disappeared from their view, Fleet turned to Lee and said, "Well, I thought it was a narrow shave."

On the Bridge, First Officer Murdoch, fated to be the man upon whom immediate responsibility had fallen, reacted as we have seen in the previous chapter. He took the call from the lookouts and shouted the iceberg warning to Sixth Officer James Moody. Murdoch, in a panic, shouted the helm order for an emergency turn to port. He rang an ASTERN order on the engine room telegraphs (there were four, two were for emergencies). He pushed the switch that sounded the watertight door alarm. After holding the alarm switch down for 10 seconds, he pulled the lever that initiated closure of the doors. At that point, he had done all that was humanly possible—he could then only stare in dismay at the approaching iceberg, by then probably no more than 10 seconds away. More ominously was the underwater mass of the iceberg, hidden by the seas and darkness.

The ship was beginning to turn. Lookouts Fleet and Lee testified to that during the inquiries. The startled engineers must have been hastening to respond—but an adequate turn to avoid a collision was impossible, and the ship could not stop in time. A jar and then a grinding noise as ship and iceberg made intermittent contacts along 300 feet of hull signaled failure of everything Murdoch had tried to do.

A small degree of uncertainty surrounds the engine order initiated by Murdoch. ASTERN, it surely was, but was it FULL ASTERN? Logical thought says that it was, and the 1997 movie *Titanic* shows it thus, and also has Chief Engineer Bell shouting to his men, "FULL ASTERN" after a quick glance at the telegraph dial. The movie was in error on Chief Engineer Bell. He was in his quarters when the emergency began. The only first-hand testimonies are those of Thomas Dillon, a trimmer who had been assigned to cleaning duty in the reciprocating engine rooms, and Frederick Scott, a greaser who was standing just inside the watertight door between the Reciprocating Engine and Turbine Rooms. Dillon reported seeing one of the telegraphs there set at SLOW ASTERN. However, in his testimony he estimates the time of that observation to be about two minutes after he had felt the shock of hull against ice.

Frederick Scott was standing at the watertight door on the starboard side of the Turbine Room, when he heard a noise that he thought emanated from the Reciprocating Engine Room. He heard all four engine telegraphs ring, including the emergency ones. He did not say that they rang in unison, nor was he quizzed on that point. He remembered that the first order was STOP. Two greasers on duty in the Reciprocating Engine Room rang back confirming the order. He was uncertain of the orders after this, but did see SLOW ASTERN on the engine telegraphs in the Turbine Room. He testified at the British Inquiry that he estimated that the ship went astern for about five minutes. By that time another STOP came down and a SLOW AHEAD order followed. By the time that this order came down all the engines and the turbine were not operating. By this time, we believe that Chief Engineer Bell and all his engineering staff had gathered in the Reciprocating Engine Room to discuss the situation and alternatives.

We speculate that Murdoch *may* have given the telegraph that illogical setting—but not with intention, but in unthinking haste. There were four engine telegraphs on the Bridge and Murdoch would have had to move quickly to produce the orders that Scott heard in the Turbine Room. On the other hand, there are several firm reasons for rejecting the value of the Dillon and Scott testimonies, and accept the conventional doctrine that Murdoch's order was FULL ASTERN. Any SLOW ASTERN order came later, and as part of the effort to stop the ship after the collision.

We discuss the procedures necessary to reverse the *Titanic*'s propeller thrust in a note.[2] Commentary on the wisdom of Murdoch's futile attempt to stop the ship appeared in the preceding chapter.

Captain Smith did not need to be called. He was resting in his cabin, perhaps asleep, but the jar awakened him. Startled, he rushed to the Bridge to confront his watch officer.

Captain Smith: What have we struck?
First Officer Murdoch: An iceberg, Sir. I hard-a-starboarded and reversed the engines, and I was going to hard-a-port round it but she was too close. I could not do any more. I have closed the watertight doors.[3]
Captain Smith: Have you rung the warning bell?
First Officer Murdoch: Yes.

The captain and officers Murdoch and Boxhall (who had just entered the wheelhouse after the collision occurred) then went to the starboard bridge wing to catch a glimpse of the iceberg as it receded astern. Boxhall testified that beyond the starboard quarter aft they saw a black mass receding into the distance. It appeared to him to be low-lying, and not to reach high above the water—more like a low-lying growler than a true berg. However, he also testified that his eyes were not accustomed to the darkness as he had just come out of the light so that he was unsure just what he had seen. Boxhall also testified that he saw the engine telegraphs set on FULL ASTERN. This is why most writers—though not Dillon and Scott—believed the order was FULL ASTERN.

Events on the Docking Bridge. Quartermaster Alfred Rowe was on watch on the docking bridge near the stern, and from there had his own close encounter with the iceberg as it drifted by. He was to testify that the mass of ice was much larger than Boxhall's estimate. Rowe's observation on the iceberg's size and shape is more in line with Fleet's and Lee's testimonies that it was somewhat equivalent to two tables at the estimated 1400-foot distance when first observed.

Rowe's initial realization of something amiss came from a slight jar. He first checked his watch, noting the time to be 2340, and then looked forward along the starboard side. His first hint of ice came from a shower of ice crystals that flashed the colors of the spectrum as they were caught briefly in the ship's lights. He then saw a "mass of ice" that he first thought might be a windjammer with sails set, then realized that the object was indeed ice, a berg that he estimated to rear 100 feet above the sea. He stared in awe as it slid silently away, to be lost in the darkness astern. Rowe remained at his post oblivious to the drama being played out on the Bridge or in other parts of the ship.

Activities of Fourth Officer Joseph Boxhall. Officer Boxhall had the 2000 to midnight watch and was working on the ship's position, based on stellar observations, in the Chart Room. Sometime after 2300, he left there and decided to go out on deck. Exiting the officer's quarters on the starboard side, he heard the three bells from the Crow's Nest. Walking forward toward the Bridge, only a minute later, he heard Officer Murdoch give the order, "Hard a Starboard" followed by the ring of the engine telegraphs. Then came the shock of impact as Boxhall began to head into the wheelhouse. When he entered, he noticed that the two engine telegraphs were set on FULL ASTERN. However, Boxhall was not on the Bridge when the engine telegraph was rung and did not know of the sequence of events in the telegraph orders. None of those who survived from the engine room remember a FULL ASTERN. Also, he saw their position when entering the Bridge after the impact.[8] Turning around, he saw Captain Smith alongside him. He also saw Mr. Murdoch closing the watertight doors by pulling the lever.

Events on the Bridge. Returning inside the wheelhouse, Captain Smith moved the engine telegraph to SLOW AHEAD, but his quick look at the commutator revealed that the ship had a starboard list of five degrees. His immediate reaction was a whispering "My God"! This statement is the first indication that Captain Smith was aware that his ship may have sustained serious underwater damage. The starboard list resulted from the quick flooding of Cargo Holds Nos. 2 and 3, flooding that was initially blocked by the Firemen's Passage and Pipe Tunnel that ran along the centerline of the ship between the Tank Top to the Orlop (or "H") Deck. The Firemen's Passage and Pipe Tunnel extended from the forward bulkhead of Hold No. 2 to the forward bulkhead of Boiler Room No. 5. When the flooding waters eventually reached "H" Deck, the flooding water poured over the obstructing tunnel to the port side of the two cargo holds. In consequence, the list decreased toward zero degrees over a period of an hour.

Turning to Officer Boxhall, Captain Smith asked his fourth officer to go below and inspect the forward compartments for damage from the impact. He then asked Quartermaster Olliver to find the Ship's Carpenter and have him sound all the tanks forward. Quartermaster Olliver left the Bridge and went down to "Scotland Road" (officers referred to this as "Park Lane"), a large passageway on the port side of the ship that extended almost its entire length on "E" Deck. There he found the Ship's Carpenter, John Maxwell, in the process of completing the sounding of the tanks as the Captain had ordered. When Olliver asked him to sound the tanks, the reply was "I am doing it." His duty now completed, Olliver returned to the Bridge around 2350 to report that the tanks were being sounded.

The Note. After Boxhall left to make his first inspection of compartments forward, Captain Smith dispatched Quartermaster Olliver to deliver a message to Chief Engineer Bell in the Engine Room, and to wait for a reply. Olliver left the Bridge with the message and proceeded at a great haste to the Reciprocating Engine Room. The message was on a folded piece of paper which—unfortunately for history—Olliver did not read. Considering the situation, the note said either of two things. One would be to reduce the steam pressure in the boilers or an inquiry on the flooding status of the boiler rooms and what measures to control any flooding that should be taken by the chief engineer.

Chief Officer Henry Wilde. Once Boxhall and Olliver had left the Bridge, Chief Officer Henry Wilde approached the Captain and asked, "Captain, is the damage serious?" Captain Smith responded, "Certainly . . . I am afraid it is more than serious." This reply is the first indication in which Captain Smith is known to admit concern that the ship was in mortal danger. Probably he also was aware that if, in fact, the ship should sink, there would not be enough lifeboats for everyone aboard. Captain Smith, now looking directly at Officer Wilde, ordered him to "go below and inspect the forward compartments for flooding." This is a further indication of Captain Smith's concern of the flooding that could produce a significant starboard list.

Around this time Captain Smith must have moved the engine telegraphs to the STOP position. This was a prudent precautionary measure until he received answers from those he had dispatched to various portions of the ship to determine the magnitude of the flooding. By midnight, the *Titanic* was adrift, without headway, and moving southward in the prevailing ocean current.

Boxhall's First Inspection. Around 2345, Officer Boxhall went right down below in the areas of Cargo Holds Nos. 2 and 3, in the lowest steerage, as far as he could possibly go without entering the cargo portion of the ship. He went down a staircase under the port side of the forecastle head which took him down into "D" Deck, and then he walked aft along "D" Deck to just underneath the Bridge. He went down the staircase there on the port side, and then came down to "E" Deck near the "E" Deck watertight doors, the working alleyway (Scotland Run), and then crossed over to the starboard side of "E" Deck to go down another accommodation staircase onto "F" Deck. He came up the same way as he went down. He found no indication of damage on the inside of the ship. The port holes that he could see in the steerage quarters, when he visited down there, to the best of his memory were closed. That was in the forward part of the ship, between the forecastle head and the Bridge. When he came up onto "C" Deck, he saw some ice there on the deck (the Forward Steerage promenade). He took a small piece of ice (about the size of a small basin) out of a man's hand as he was heading toward the passenger accommodation. Boxhall took the piece of ice and walked across the deck to the starboard side. He discovered some ice on the Well ("C") Deck covering a space just a few feet from the bulwark. Completing his inspection, he returned to the Bridge and reported to Captain Smith and First Officer Murdoch that he did not see any damage.

Boxhall's Probable Report to Captain Smith. This is the essence of the report that Boxhall probably gave to Captain Smith, based on his testimony at the two inquiries:

> Boxhall reported that he had inspected the steerage-class areas on "G" Deck in the vicinity of Cargo Holds Nos. 2 and 3, but observed no damage. He also inspected the portholes there and to the best of his knowledge they had been closed. This was in the forward part of the ship between the forecastle head and the Bridge. He also inspected all the decks in the area (Cargo Holds Nos. 2 and 3) where he thought contact had been made. He went to the shell in the area below the gangway doors on the starboard and port side of "E" Deck. Boxhall went forward, down to third-class accommodation on "F" Deck, and walked to the area where he thought she had struck ice. He went down the staircase under port side of the forecastle head to "D" Deck and walked along that deck to an area underneath the Bridge. He observed no damage to bulkheads C and D (see Table 8-2 on page 104) on "D" Deck. The watertight doors in the bulkheads on the port side of "F" Deck were open at the time of his inspection. Boxhall went right out to the ship's side via athwartships alleyways (Starboard Gangway Door on "E" Deck), but found no damage. He went back to "E" Deck where he found the door at forward end of starboard alleyway on "E" Deck to be closed. (He described it in the British Inquiry as being an iron door but not watertight.) Boxhall then went up to "C" Deck in the area of the Well Deck, but saw no damage. However, he found ice deposits there, covering a space of 3 to 4 feet from the bulwark along the Well Deck. "Small stuff," he reported it.

Several minutes after giving his report, a skeptical Captain Smith approached Boxhall and told him, "Go down and find the Carpenter and have him sound the ship in the area of the iceberg impact." This is a second indication that Captain Smith was concerned about potential flooding in the lower parts of the *Titanic*. The time was around 2350. On his way down from the Bridge, Boxhall encountered Carpenter John Maxwell on the ladder leading from the Bridge down to "A" Deck. Maxwell wanted to know where the Captain was and Boxhall told him he was on the Bridge. Boxhall told Maxwell that the Captain wanted him to sound the starboard forward side of the ship. Maxwell told Boxhall, "The ship is making water fast." Maxwell, passing Boxhall, continued up toward the Bridge to make his report to the captain on what he had found.

Officer Boxhall thought he would go down to investigate where this water was coming in. As he continued his descent below, he met a Mail Clerk, John Smith. An excited Smith asked where the Captain was. Boxhall responded, "He is on the Bridge." Smith then said, "The Mail Hold is filling rapidly. Boxhall told him, "Well, you go and report it to the Captain and I will go down and see." Boxhall decided to go down to the

Mail Room, going down the same way as he did when he visited the third-class accommodation previously. He went down as far as "B" Deck and then went to the starboard alleyway on "E" Deck, but the watertight door at the end of this corridor was closed. Unable to go forward, he went into Scotland Run and then headed for the Mail Room.

When Boxhall went to the mail hold on "F" Deck, he went down as far as the sorting room and found the Mail Clerks down there working. They were taking letters out of the racks. However, he was not sure whether they were putting the mail into bags. Through the open door to the sorting room he also noticed that water in the mail hold had risen to 2 feet from the deck he was standing on. The water was rising rapidly up the ladder there. He could see this through the opening in the staircase that led down to the lower place. Bags of mail were floating about on top of this water. He also could hear the water coming into the ship. Boxhall stopped any further inspections. He realized that there was no use trying to get the bags out so he decided to go back again to the Bridge. On the way, he met the Second Steward, George Dodd, in the saloon companionway on his way up to the Bridge. Dodd asked Boxhall about sending men below to help recover the mails. Boxhall told Dodd, "You had better wait till I go to the Bridge and find out what we can do." Arriving on the Bridge after an inspection that took a little less than 15 minutes, Boxhall reported to the Captain on what he had seen in the Mail Room and the request of Steward Dodd. All that the Captain said was, "all right." Then Captain Smith gave the order to clear the lifeboats. Boxhall went around the decks and helped to take the canvas covers off the lifeboats, except for the sea boats and assisting generally around the decks.

The Reciprocating Engine Room. Olliver, the bearer of a written message from Captain Smith, found the Chief Engineer in conversation with a group of his engineers. He also noticed that the two reciprocating engines were stopped. Approaching Chief Engineer Joseph Bell, he handed him the message from Captain Smith. Bell first walked toward Olliver and then away to continue his conversation with his engineers. Olliver stood alone waiting for some action from Bell for two or three minutes. While waiting for an answer, Olliver noticed that engineering officers were busy conversing, but he could not hear what they were saying. He saw an engineer leave the Reciprocating Engine Room and enter Boiler Room No. 1. The watertight door between these spaces was open the entire time that Olliver was in this space. He also noticed that the lights in the aft Stokehold in that Boiler Room were not lit but those in the Engine Room were. Seeing Olliver still standing there, Bell approached him.

Chief Engineer Bell: What do you want?

Quartermaster Olliver: I am waiting for an answer to the message I brought you.

Chief Engineer Bell: (Writes a brief note on the paper and folds it and hands it back to Olliver). All right, take this paper back to the Captain and tell him that I will get it done as soon as possible. Return this to the Captain.

The contents of that note are not known, but can be surmised that Captain Smith wanted to make sure that his chief engineer was aware of the ship's peril, of the necessity of keeping it afloat for as long as possible, and to keep the lights burning. Reviewing the situation from a naval architect's viewpoint, it is very possible that Captain Smith wanted Chief Engineer Bell to make an assessment of the flooding and how many boilers were affected. Captain Smith also may have alerted Bell to his potential visit there with Thomas Andrews. Smith also knew that the boilers in Boiler Rooms Nos. 2, 3, and 5 supplied steam to the emergency and ship service dynamos. Alternate means were a cross-connection with main steam lines in the Reciprocating Engine Room. The dynamos could take steam from Boiler Rooms Nos. 1, 2, and 4. They were crucial in keeping the electric power on for any evacuation. Captain Smith also probably told Bell that he was coming to the Reciprocating Engine Room for a discussion of the ship's condition and if there were measures that could be taken to save her.

Scene on the Bridge. Around 2350, about 10 minutes after the collision, Captain Smith summoned naval architect Thomas Andrews, the managing director of the design department at Harland & Wolff. Here was an unusual resource for a ship in peril, a naval architect, who perhaps better than anyone else, knew the characteristics of the ship. Such a person, as Andrews was to demonstrate, was able to make calculations that could predict how the events might follow.

The Ship's Carpenter, John Maxwell, arrived on the Bridge out of breath from running and ascending ladders. Maxwell approached Captain Smith in conversation with Thomas Andrews and said, "Captain, she's making water, and lots of it. Cargo Holds Nos. 1, 2, and 3 are flooding quickly. The Mail Room is also filling with water." Captain Smith thanked Maxwell for his report. No sooner than Maxwell had completed his report on the soundings of tanks, an excited Mail Clerk John Smith arrived on the Bridge just as Captain Smith was about to leave.

Mail Clerk John Smith: Captain! The Mail Hold is full of water!
Captain Smith: The Carpenter has just reported this to me. I have also alerted Bell that we would like to come and talk with him about the situation. What are you doing with the mail?
John Smith: I am salvaging what mail I can from the Mail Hold and am having the others moved to "F" Deck, but the water is still rising.
Captain Smith: Find the Purser. He will give you some instructions.

After this exchange, Mail Clerk Smith went to see Chief Purser McElroy who then went with Smith to assess the flooding situation in the Mail Room.

Naval Architect Thomas Andrews, who had been alone in Stateroom A-36, had felt a slight jolt, but apparently was unaware of its significance. Now, summoned to the Bridge, he arrived there about 2355. Although their conversation was not observed by anyone who survived, it may have gone as follows. Approaching Captain Smith, Andrews probably said, "Captain, what has happened?" The captain in all probability explained to Andrews the situation as far as he knew it.

Captain Smith: Mr. Andrews, the forward part of the ship has struck an iceberg. The Ship's Carpenter has sounded the compartments forward and has reported that Cargo Holds Nos. 1, 2, and 3 are filling rapidly. One of the Mail Clerks has just told me that the Mail Hold is flooded. The ship has a five-degree list to starboard. I have directed Fourth Officer Boxhall and Chief Officer Wilde to inspect the forward compartments. I also have sent a message to Chief Engineer Bell to give me a report on the conditions in the engine and boiler rooms.
Thomas Andrews: The five-degree list and the amount of flooding are troubling. Has the list been increasing? How much is the trim forward? To assess the situation, we must determine the compartments flooding and the rate of that flooding. I agree that we should find out from Bell if there is any flooding in the engine room and boiler rooms. That could be very serious.
Captain Smith: I am waiting for a report from two of my officers who have gone forward to inspect for damage and flooding. Let us see then what they and Bell have to say. I have also alerted Bell that we would like to come and talk with him about the situation. What are your opinions, Mr. Andrews?
Thomas Andrews: Captain, I cannot assess the situation until we

see Mr. Bell and have an indication of the extent and rate of flooding below. If all of the compartments forward and Boiler Room No. 6 are flooding, we will have a serious problem to contend with. We can then look at the drawings and make a better assessment and make some quick calculations. When I know what compartments are flooding and the rate of that flooding, then I can make a better assessment.

Quartermaster Olliver: (Returning from delivering the note to the Reciprocating Engine Room): Captain, the Chief Engineer says that he will get it done as soon as possible. (Olliver is later ordered by Chief Officer Henry Wilde to assist in uncovering the lifeboats.)

Captain Smith: (After reading the note handed to him by Olliver) Mr. Andrews, Boiler Room No. 6 has been abandoned.

Thomas Andrews: The situation in Boiler Room No. 6 is indeed critical. That is where we put out a coal fire yesterday (Saturday morning, April 13). The heat from the fire could have weakened the after bulkhead of Boiler Room No. 6. Captain, Sir, I believe we should meet with Mr. Bell at once and see how we can contain the flooding there.

Between 0005–0010, Chief Officer Henry Wilde returned from his inspection of the Forepeak Area. He probably told Captain Smith, "I have inspected the Forepeak Area and the peak tank has flooded, but the spaces over it are dry. The storekeeper inspected all those spaces for me." Captain Smith, with some degree of resignation told Wilde that he should begin uncovering the lifeboats.

Thomas Andrews, seeing that the Captain was preoccupied with damage reports coming to him on the Bridge, went alone to meet with Chief Engineer Joseph Bell. The reports of flooding really concerned him. As he left the Bridge, Thomas Andrews probably said, "I hope Bell has a plan to deal with this." Thomas Andrews then proceeded to the Grand Staircase on the Boat Deck and walked down to "D" Deck where he went through the first-class Dining Saloon to a stairs that led down to "E" Deck. From there he turned left and entered the Reciprocating Engine Room.[4]

Before Mr. Andrews left the Bridge, Captain Smith told him that he was going to the Wireless Room to give the ship's position to the radio operators. After that he would join him down in the Reciprocating Engine Room. However, before he could leave Officer Boxhall returned from his second inspection.

Officer Boxhall: Captain! The Mail Hold is almost completely flooded. The water level already is about 2 feet below "G" Deck! Steward Dodd wants to know what should be done with the mail.

Captain Smith: All right, I just received the reports on this from one of the Mail Clerks and the Carpenter. You should help uncover the boats. Go rouse Lightoller and Pitman!

After this exchange, Captain Smith left for the Marconi Room with the din of steam roaring from the escape pipes that were fitted to the three forward stacks.

The Marconi (wireless) Room. Captain Smith entered the Marconi Room and told radio operator Harold Bride that the ship struck an iceberg and he was having the ship inspected to determine the extent of the damage. He also instructed him to be prepared to send a call for assistance, but to not send it until he told him to. Radioman Jack Phillips came into the room, and looking at Radioman Harold Bride, asked Captain Smith, "Should I send a distress call? The regular international call for help? Just that?" Captain Smith told Phillips to send the regular international call for assistance and he gave Phillips the ship's position. The two radio operators were not aware of the seriousness of the situation and joked about sending a CQD as the Captain left the space. However, Phillips took over the transmissions and sent a message immediately.

Captain Smith left the Marconi Room and proceeded to the Grand Staircase following the path of Thomas Andrews down to "D" Deck and through the first-class Dining Saloon about four minutes afterward. Nightwatchman James Johnson[4] saw and followed him to the door of the Reciprocating Engine Room.

In the meantime, Phillips sent out several CQDs.

0015: (approximate time) *La Provence* received *Titanic* distress signals. "CQD *Titanic* in 41–46N 50–24 W"

0015: *Mount Temple* heard *Titanic* sending a CQD. Says require assistance. Gives position. Cannot hear me. Advise my Captain his position 41.46 N 50.24 W

0015: Cape Race hears *Titanic* giving position on CQD. 41.44 N. 50.24 W.

0018: *Ypiranga* hears CQD from *Titanic*. *Titanic* gives CQD here. Position 41.4 N, 50.24 W. Require assistance (calls about 10 times).

The *Frankfurt* was the first ship to answer the calls for help, followed by the *Carpathia* a few minutes later. The *Carpathia*'s response to the CQD was:

0025: CQD call received from *Titanic* by *Carpathia*. *Titanic* responded, "Come at once. We have struck a berg. It's a CQD O(ld) .M(an) Position 41.46 N, 50.24 W."

The Conference in the Reciprocating Engine Room. Captain Smith arrived in the Reciprocating Engine Room around 0010. Meanwhile, Thomas Andrews had been able to inform Chief Engineer Bell of the flooding in the forward cargo holds, since he had been present on the Bridge when the Ship's Carpenter and the Mail Clerk made their reports. The following is a reconstructed conversation between Smith and Andrews, based on facts and events as it is known from testimony given at the two inquiries. There were no witnesses to these conversations that survived the sinking.

Thomas Andrews: Captain, in addition to the flooding forward, Mr. Bell has just informed me that Boiler Room No. 6 is flooded and abandoned. The Firemen's Passage is also flooded. The pumps in Boiler Room No. 5 can control the current flooding in that space. As a precaution, the men there are drawing the fires in the furnaces. There is also minor flooding in Boiler Room No. 4, but that is also being controlled by the pumps there. It is my belief that we will need to control the flooding in Boiler Room No. 5 for as long as possible. Once that space floods, the ship will be surely lost, Captain. The ship is not designed to take that length of flooding.

Captain Smith: The situation forward is worse than we thought. Officer Boxhall just informed me that the water level in the Mail Room is two feet below "G" Deck and Mr. Wilde reported that the Forepeak Tank has flooded, but the spaces over it are still dry.

Chief Engineer Joseph Bell: Captain, as we lose boilers, we lose steam pressure to the dynamos and pumps. I have also begun to lower the steam pressure to lessen any chance of a boiler explosion. Also, we no longer need a great amount of steam pressure with the engines on STOP. Comment: With propulsion engines stopped, the problem was not losing steam pressure, but too much pressure. Pressure for generator engines and pumps probably did not begin to fail until fires were out in all boiler rooms, late in the sinking process.

Chief Electrician Peter Sloan: My men will shut off the engine and boiler room fans. This will help extinguish the fires in those boilers you want to take off line. But we must keep adequate steam pressure for the main and emergency dynamos.

Chief Engineer Bell: Captain, we have raised all the watertight doors aft of Boiler Room No. 4. This was done so we could easily get to all the pumps and valves in each space. We intend to rig a suction hose to the after pumps to help empty spaces that are flooding forward.

Thomas Andrews: Do not raise the watertight door between Boiler Rooms Nos. 4 and 5. Because of the coal fire in aft stokehold of Boiler Room No. 6, we had planned to inspect that bulkhead further when the ship docks in New York. With the flooding of Boiler Room No. 6, water pressure can start a leak at any time in the area of that fire. The pumps in Boiler Room No. 5 are pumping at capacity; they cannot remove any additional water. Captain, you should not move the ship. Any forward motion of the ship will increase the pressure on the main transverse bulkhead between Boiler Rooms Nos. 5 and 6. This would prevent any further flooding into Boiler Room No. 5 which must be kept dry or else the ship will sink.

Captain Smith: Chief, see to it.

Chief Engineer Joseph Bell: The turbine has been already put on standby. We will now make sure that the reciprocating engines will stay on standby.

Captain Smith, Mr. Andrews, and Chief Engineer Bell understood very well that not maintaining internal integrity was a calculated risk in trading off suction hoses for active intervention against upflooding against the loss of internal integrity. However frantic times call for desperate men to reach bold decisions.

Captain Smith: Mr. Andrews! Would filling some of the aft ballast tanks help counter the bow's sinking?

Thomas Andrews: No, Sir. Adding ballast in those aft tanks will only hasten the sinking by putting extra stress on the hull and increasing the draft. No, do not do that.

Captain Smith: (*turning to Thomas Andrews*) How long do you think the ship has?

Thomas Andrews: (*talking to Captain Smith*) I believe it has another hour or more. Boiler Room No. 5 is the important space. I will have to make some calculations to be sure. Before doing that, I will need to determine the rate of flooding in the forward compartments. I strongly urge you, Captain, to begin the evacuation of the passengers as quickly as you can. I am sure Mr. Bell will do what is necessary to keep as many compartments dry as he can.

Chief Engineer Bell: (talking to Chief Electrician Sloan in a separate conversation) When will you have the lights back on in the Boiler Rooms?

Chief Electrician Sloan: (answering Chief Engineer Bell) As soon as we can isolate the circuit to Boiler Room No. 6. I have my men working on it.

The futility of the effort to contain the flooding is evident from the total pumping capacity that Chief Engineer Bell had available. The total pumping capacity of the ship was 1,700 tons/hour, but the total flooding of the ship in the first hour amounted to 25,000 tons.

Chief Electrician Peter Sloan: I will check on the emergency dynamos (generators) on "D" Deck.

Captain Smith: Chief, as we begin the evacuation of passengers, send whatever men you can spare up to the Boat Deck to man the lifeboats.

Thomas Andrews: Captain, you realize that there are only seats in the lifeboats for half of those on board.

Captain Smith (with a dejected tone of voice): Yes, I realize that and it will have to be women and children first. I will also release the watertight doors to make it easier for the engine crew to rig their pumps.

Chief Engineer Joseph Bell: I have some of my engineers containing the flooding in Boiler Rooms Nos. 4 and 5. I will do my best, Sir, but with the amount of flooding occurring forward, I do not think we have the pumping capacity to save her.

The conference adjourned with Captain Smith going forward with Thomas Andrews and then returning to the Bridge. Thomas Andrews made an inspection of spaces forward, Chief Electrician Sloan went to "D" Deck, and Chief Engineer Bell directed engineers in setting up the pumping activities and lowering of the steam pressure.

Chief Engineer Joseph Bell called Second Engineer Hesketh in Boiler Room No. 5 soon after the conference concluded around 0015–0020. Hesketh was instructed to extinguish the fires in the furnaces and lower the steam pressure in the boilers there. He was also told of the importance of keeping this space from flooding. Chief Engineer Joseph Bell told Hesketh that he considered Boiler Room No. 5 as the critical space. Acting on these instructions, Hesketh ordered leading fireman Frederick Barrett to draw the fires in its furnaces and Engineers Shepherd and Harvey to monitor the pumps. The personnel in Boiler Room No. 4 were also instructed to extinguish the fires in their boiler furnaces. The pumps in both spaces were kept running. This contained the water flowing in from the forward coal bunker on the starboard side in Boiler Room No. 5. Heavy suction hoses connected to pumps in Boiler Rooms Nos. 2 and 3 assisted the pump in Boiler Room No. 4. Therefore, the small amount of water that was collecting there from leaking rivets in its bilges was contained by this pumping setup. The watertight door between Boiler Rooms Nos. 4 and 5 was kept closed. These measures, Chief Engineer Bell felt, would provide the first line of defense. The watertight door between Boiler Rooms Nos. 3 and 4 was raised enough to allow a suction hose from a bilge pump[1] in the Reciprocating Engine Room to clear the door opening. It was necessary to use this pump to assist the bilge and ballast pump already working to keep the water level down below the grating level in Boiler Room No. 4. Meanwhile, the stokers in Boiler Rooms Nos. 4 and 5 were busy trying to extinguish the coal fires in the furnaces. The steam pressure had been reduced in Boiler Rooms Nos. 2 and 3, while the five boilers in Boiler Room No. 1 had not been in use during the voyage.

Chief Engineer Joseph Bell had at his disposal five ballast and bilge pumps, each capable of discharging 250 tons of water per hour, and three bilge pumps, each of 150 tons per hour capacity. There was one ash ejector in each of the large boiler compartments and pumps to work the ash ejectors and to circulate or feed the boilers as required. The circulation or feed pumps could be connected to the bilges, except in the case of three of the boiler rooms, where three ballast and bilge pumps were placed. The pumps in each case had direct bilge suctions as well as a connection to the main bilge pipe, so that each boiler room might be independent. The remaining auxiliary pumps were placed in the reciprocating and turbine engine rooms. Two ballast pumps were placed in the reciprocating engine room, with large suctions from the bilges direct and from the bilge main. One bilge pump was placed in the turbine room and one of the hot salt-water pumps had a connection from the bilge main pipe for use in emergency. A 10-inch main ballast pipe was carried fore and aft through the ship with separate connections to each tank, and with filling pipes from the sea connected at intervals for trimming purposes. The five ballast pumps were arranged to draw from this pipe. A double line of bilge main pipe was fitted forward of Boiler Room No. 5 and aft of Boiler Room No. 1. See the sketch of the bilge and ballast system in Chapter 10.

Captain Smith returned to the Bridge after walking forward on "E" Deck to look at the flooding there and talk briefly with the Mail Clerks. On his way up to the Bridge, Captain Smith stopped on "C" Deck to briefly talk to Chief Steward Andrew Latimer and the Chief Purser Hugh

McElroy outside of the Purser's Office. He told them that evacuation of the passengers and crew should commence with women and children first. Once back on the Bridge, Captain Smith instructed Officer Boxhall to give the updated ship's position to the wireless operators, telling him, "I want you to go over the 7:30 (P.M.) stellar positions and determine our position for the wireless people. Figure the ship speed as 22 knots." (Boxhall also testified that he guessed at that speed.)

Before leaving for the Marconi Room, Captain Smith spoke to Second Officer Charles Lightoller and Sixth Officer James Moody. He wanted them to go to the Boat Deck and have the crew ready the lifeboats for the passengers to board. Captain Smith then went to the Marconi Room around 0023. Just as he was ready to leave, Quartermaster Robert Hichens was relieved at the helm by Quartermaster John Perkis. One of the officers in the Bridge instructed Hichens to assist in loading the lifeboats.

Marconi Room and the Distress Message. Captain Smith explained to the radio operators that the ship was badly damaged and he now wanted them to send a distress call detailing the damage. The following messages were sent and received over the next two hours by the *Titanic* and other ships responding to her calls of distress:

0025: Cape Race hears M.G.Y. [*Titanic*] give corrected position 41.46 N, 50.14 W, calling him, no answer.

0026: M.G.Y. [*Titanic*] says, "CQD Here corrected position 41.46 N, 50.14 W. Require immediate assistance. We have collision with iceberg. Sinking. Can hear nothing for noise of steam. Sent about 15 to 20 times to 'Ypiranga.' "

0027: *Titanic* sent following: "I require assistance immediately. Struck by iceberg in 41.46 N, 50.14 W."

0030: *Titanic* gives her position to *Frankfurt*, and says, "Tell your Captain to come to our help. We are on the ice."

0030: *Caronia* sent C.Q. message to M.B.C. *Baltic* and CQD: M.G.Y. [*Titanic*] struck iceberg, require immediate assistance.

0020: *Mount Temple* hears M.G.Y. [*Titanic*] still calling CQD Our Captain reverses ship. We are about 50 miles off.

0026: D.K.F. [*Prinz Friedrich Wilhelm*] calls M.G.Y. [*Titanic*] and gives position at 12 A.M. 39.47 N, 50.10 W. M.G.Y. [*Titanic*] responds: "Are you coming to our assistance?" D.F.T. [*Frankfurt*] responds: "What is the matter with u? M.G.Y. [*Titanic*] "We have collision with iceberg. Sinking. Please tell Captain to come." D.F.T. [*Frankfurt*] responds, "O.K. will tell?"

0038: *Mount Temple* hears *Frankfurt* give M.G.Y. [*Titanic*] his position 39.47 N, 52.10W.

0045: *Titanic* calls *Olympic* S.O.S.

0050: *Titanic* calls CQD and says, "I require immediate assistance. Position 41.46 N, 50.14 W." Received by "*Celtic*."

0053: *Caronia* to M.B.C. [*Baltic*] and SOS, M.G.Y. [*Titanic*] CQD in. 41.46 N, 50.14 W. Wants immediate assistance."

0100: M.G.Y. gives distress signal. D.D.C. [SS *Cincinnati*] replies. M.G.Y.'s position 41.46 N, 50.14 W. Assistance from D.D.C. not necessary as M.K.C. shortly afterwards answers distress call.

0100: *Titanic* replies to *Olympic* and gives her position as 41.40 N, 50.14 W, and says, "We have struck an iceberg."

0102: *Titanic* calls *Asian* and says, "Want immediate assistance." *Asian* answered at once and received *Titanic*'s position as 41.46 N, 50.14 W, which the radio operator immediately takes to the bridge. Captain instructs the radio operator to have *Titanic*'s position repeated.

0102: *Virginian* calls *Titanic* but gets no response. Cape Race tells *Virginian* to report to his Captain the *Titanic* has struck iceberg and requires immediate assistance.

0110: *Titanic* to M.K.C. [*Olympic*], "We are in collision with berg. Sinking head down. 41.46 N, 50.14 W. Come soon as possible."

0110: *Titanic* to M.K.C. [*Olympic*], Captain says, "Get your boats ready. What is your position?"

0115: *Baltic* to *Caronia*, "Please tell *Titanic*. We are making towards her."

0120: *Virginian* hears M.C.E. (Cape Race) inform M.G.Y. [*Titanic*] "that we are going to his assistance. Our position 170 miles N of *Titanic*."

0125: *Caronia* tells *Titanic*, "*Baltic* coming to your assistance."

0125: *Olympic* sends position to *Titanic* 4:24 GMT 40.52 N, 61.18 W. "Are you steering southerly to meet us?" *Titanic* replies, "We are putting the women off in the boats."

0125: *Titanic* and *Olympic* work together.

0127: M.G.Y. [*Titanic*] says, "We are putting the women off in the boats."

0130: *Titanic* tells *Olympic*,"We are putting passengers off in small boats."

0135: *Olympic* asks *Titanic* what weather he had. *Titanic* replies, "Clear and calm."

0135: *Baltic* hears *Titanic* say "Engine room getting flooded."

0135: *Mount Temple* hears D.F.T. [*Frankfurt*] ask, "Are there any boats around you already?" No reply.
0137: *Baltic* tells *Titanic*, "We are rushing to you."
0140: *Olympic* to *Titanic*, "Am lighting up all possible boilers as fast as can."
0140: Cape Race says to *Virginian*, "Please tell your Captain this: 'The *Olympic* is making all speed for *Titanic*, but his [*Olympic*'s] position is 40.32 N, 61.18 W. You are much nearer to *Titanic*. The *Titanic* is already putting women off in the boats, and he says the weather there is calm and clear. The *Olympic* is the only ship we have heard say, "Going to the assistance of the *Titanic*." The others must be a long way from the *Titanic*.' "
0145: Last signals heard from *Titanic* by *Carpathia*, "Engine-room full up to boilers."
0145: *Mount Temple* hears D.F.T. [*Frankfurt*] calling M.G.Y. (*Titanic*). No reply.
0147: *Caronia* hears M.G.Y. [*Titanic*] though signals unreadable still.
0148: *Asian* heard *Titanic* call SOS "*Asian*" answers *Titanic*, but receives no answer.
0150: *Caronia* hears *Frankfurt* working to *Titanic*. *Frankfurt* according to position, 172 miles from M.G.Y. [*Titanic*] at time first SOS sent out.
0155: Cape Race says to *Virginian,* "we have not heard *Titanic* for about half an hour. His power may be gone."
0200: *Virginian* hears *Titanic* calling very faintly, her power is being greatly reduced.
0210: *Virginian* hears 2 v's signaled faintly in spark similar to *Titanic*'s probably adjusting spark.
0217: *Virginian* hears *Titanic* call CQD, but unable to read her. *Titanic*'s signals end very abruptly as power suddenly switched off. His spark rather blurred or ragged. Called M.G.Y. [*Titanic*] and suggested he should try emergency set, but heard no response.
0220: *Olympic*, his signs strong, asked him if he had heard anything about M.G.Y. [*Titanic*] he says, "No. Keeping strict watch, but heard nothing more from M.G.Y. [*Titanic*]. No reply from him."
0220: This was the official time the *Titanic* foundered in 41.46 N, 50.14 W, as given by the *Carpathia* in message to the *Olympic*; about 2:20 A.M.

After rousing Officers Lightoller and Lowe, Boxhall went to the Chart Room to work out the ship's position in accordance with the stellar positions taken at 1930 that he had been working on since coming onto his watch at 2000. He had been clearing boats for a little while, and then went in and completed his work. He then went to the Bridge and gave the position report to Captain Smith, who told him, "Take it to the Marconi Room." Boxhall went to the Marconi Room to see the radio operators Phillips and Bride. There was too much noise from the steam escaping, so Boxhall simply wrote down the position on a piece of paper—Forty-one, forty-six, north; fifty, fourteen west—and left it on a table there. Phillips spotted it. He was in the midst of a call and listening. Boxhall did not interrupt him. As it turned out, this plot was 14 miles to the northwest of where the ship was actually located.

After the order had been given to clear the boats, Smith had an occasion to ask Officer Boxhall something about what the men were doing. This conversation occurred after the Captain had returned to the Bridge from the Reciprocating Engine Room. Captain Smith asked him how the radio operators were doing.

Boxhall: Yes, they are carrying on all right. Is it really serious?
Captain Smith: Mr. Andrews tells me he gives her from an hour to an hour and a half.

After this Boxhall went around the decks assisting to prepare the lifeboats for lowering and send distress signals. He remained around the Bridge most of the time, sending off distress signals and tried to signal a ship that he thought was ahead of the *Titanic.* Quartermaster Rowe would assist him in these efforts.

Quartermaster Rowe remained on the after bridge to await any orders on the telephone. At 1225 he spotted a lifeboat being lowered. Not realizing the perilous circumstances developing forward, he telephoned the Bridge to let them know about the lifeboat. He was questioned, "Are you the third officer?" He replied that he was the quartermaster, stationed on the after bridge. His interlocutor was surely aware that orders had been given to begin the evacuation. Rowe was instructed to bring the detonators for the rocket distress signals to the Bridge. He quickly complied, handing the detonators to Fourth Officer Boxhall. Together the two men would fire off the distress rockets for the next 45 minutes. The significance of this incident is its evidence of insufficient communication from the bridge to the crew regarding the *Titanic*'s condition at this time, over a half hour after the collision.

Evacuation. At 0015, the order was given to the officers to load women and children first. This was a Victorian response and also a prudent mariner's answer to a difficult situation, the presumed lack of ability to reduce the flooding rate, and, in general, a prudent decision given the lack of sufficient lifeboat capacity for all those aboard. Once Captain Smith had been advised of the enormity of the situation, he realized that the lack of lifeboats was a disastrous flaw. He instructed his officers to maintain calm.

Andrew's Inspection Forward. Thomas Andrews went along "E" Deck to determine the flooding forward after leaving the Engine Room between 0015–0020. He went past the Racket Court on "E" Deck that by this time was filling rapidly with water. The lights in that space were still on but were below the water level. This gave an eerie look to the scene. Andrews probably stared at a reference point to assess how fast the water level there was rising. Knowing the width of the wood paneling there, he took out his watch and calculated the time it took for the water to rise over that distance. Using this time and computing the volume of water, he could determine the rate of flooding. Using this flooding rate, he could then roughly project how much time it would take to sink the ship. Descending down to "F" Deck, he noted that the Mail Hold had already flooded. Having seen the flooding situation in the area of Cargo Hold No. 3, he recognized that the situation was indeed as precarious as he had heard from Captain Smith. He headed for his stateroom on "A" Deck to look at the floodable length plan and make a few calculations to determine where the flooding level might be. While walking toward the staircase on "E" Deck, Andrews encountered Stewardess Annie Robinson. She had already got the blankets and the life belts out of the rooms, which were unoccupied at the foot of the staircase.

Mr. Andrews (*stops and looks at Robinson*): Put your life belt on and walk about and let the passengers see you.
Stewardess Annie Robinson: It looks rather mean.
Thomas Andrews: No, put it on. (*Pauses, then adds*) Well, if you value your life, put your belt on.

Passing the Pursers Office on his way to his stateroom A-36, Andrews spoke briefly to Chief Steward Latimer and Chief Purser McElroy before going back to his quarters. Andrews also passed Albert Dick with whom he had eaten dinner earlier in the evening under more pleasant circumstances (Albert Dick and his wife, Vera, were newlyweds). Andrews continued up the stairs without making any comments, but the look on his face was grim. Once in his stateroom he made a few calculations and went over the floodable length plan. He returned to the Bridge with his notes between 0020–0025. A grim-faced Thomas Andrews walked up to Captain Smith and this probable interchange took place.

Captain Smith (*Hesitating before asking the question, upon looking at the serious face of the ship's architect*): How long do we have?
Thomas Andrews (*jotting a few figures on a scrap of paper*): An hour and a half. Possibly two. Not much longer.

The flooding was more serious than Andrews had realized after his first visit to the Bridge before midnight. Once he had seen the flooding in the Racquet Court and conversed with the Chief Engineer, he realized that the flooding was far worse than he had thought. Looking at the plans that he had available, the ship could stay afloat with the first four compartments flooded, but five were already flooding and there was danger that the sixth, Boiler Room No. 5 could succumb at any moment. Once this occurred, the ship would surely sink. Captain Smith was stunned by this assessment, however, he had presence of mind to say to his officers, "Put the women and children in the boats and lower away!"

On the way down from the Bridge, Thomas Andrews encountered Steward Henry Etches on "B" Deck. Andrews asked Etches, if he had wakened all the passengers in his area of responsibility. "No," was the response. A passenger, Mr. Harrison, approached as Etches went to the quarters of the Carter family. Harrison told Etches, "They are already up. I have just come out of my cabin." Andrews then asked Etches to come with him to "C" Deck and they went down the pantry staircase together. On the way down the staircase, Thomas Andrews told Etches to make sure that all the passengers open their doors and to tell them that the life belts were on top of the wardrobes in racks. "Make sure to assist these passengers in every way to get their life belts on," Etches was told. As they walked along "C" Deck they came upon a large group of ladies outside the Purser's Office. Purser McElroy was telling these ladies to do what he was instructing them to do. Etches remembered Andrews telling these ladies, "Go back to your rooms and do not be frightened. However, as a precaution, you should put on your lifebelts." Mr. Andrews then continued, "That is exactly what I have been trying to get them to do." After this statement, Andrews left Etches and headed down to "D" Deck.

Chief Electrician Peter Sloan went back to the Electrical Workshop on "E" Deck to have Thomas Ranger begin the effort of shutting down the 45 fans by starting with those of Boiler Room No. 6 first. Sloan then left to go to "D" Deck to supervise the effort in checking the emergency generators there. Working his way up to the Boat Deck, Thomas Ranger completed his duty and went out onto the Boat Deck where he would enter a lifeboat and be rescued.

Damage Control. After Captain Smith returned to the Bridge, but before the arrival of Thomas Andrews, he ordered that all the watertight doors be opened aft, starting with the bulkhead door between Boiler Rooms Nos. 3 and 4. This measure allowed Chief Engineer Joseph Bell to control the flooding of Boiler Room No. 4 with 3- or 5-inch hoses rigged to 150- and 250-ton pumps from spaces aft. Boiler Room No. 4 had a single 150-ton pump. This effort controlled the flooding there until Boiler Room No. 5 flooded. Water on "E" Deck then found its way into this space and spilled down into this boiler room through non-tight openings in the deck. The increased hydrostatic head from a greater bow trim increased the flow of water through failed rivets. This began around 0115.

There were no wooden timbers aboard to reinforce weakened bulkheads or to shore up those that would be subjected to hydrostatic loads beyond their design limits. Unlike naval ships, there were no collision mats or sail cloth to lower over the openings created by the collision, nor were there divers aboard to place them accurately in position. Such measures would have slowed the flooding, but damage control of this nature was only used in naval ships. The crew of the *Titanic* was not trained for such procedures and, even if such a measure was available, many were new to the ship and unfamiliar with any equipment or its location that would have been provided. Also, many of her crew did not serve in the Royal Navy where rudimentary damage control was practiced as a routine procedure.

Bruce Ismay. The chairman of the White Star Line[7] was awakened by the impact in his cabin that was just aft of the Grand Staircase on "B" Deck. Wondering what it could be, he left his quarters to ask a steward what had happened. Unable to get a satisfactory answer, Ismay returned to his cabin and put on a coat over his pajamas and in his slippers struck out for the Bridge. The chairman of the White Star Line made his way to the Bridge around 0025 to confer with Captain Smith. Arriving on the Bridge, he asked Captain Smith, "What happened?" Captain Smith replied, "We have struck ice." A concerned Ismay then asked, "Is the damage serious?" Captain Smith answered, "I am afraid it is." Ismay then decided to return to his quarters on "B" Deck and on his way he met Chief Engineer Bell at the top of the stairs to "A" Deck. Bell was on his way to the Bridge to determine what Andrews had found and also to tell Captain Smith what measures he had instituted since their meeting. Ismay asked Chief Engineer Bell, "Do you think the ship is seriously damaged?" Bell answered, "Mr. Ismay, I believe she is sir, but I hope (or think) the pumps can control the flooding." A stunned Bruce Ismay could not recall where he went next. The seriousness of the damage was now apparent from the tone of Smith's and Bell's responses to his questions. Ismay now realized that the ship could sink. Therefore, he did go up to the Boat Deck and assisted the officers there in loading the lifeboats.

Situation in the Propulsion Plant. With the reciprocating engines and the turbine stopped and drawing no steam, the thermal inertia of the boilers caused the steam pressure to rise until the safety valves lifted. The boiler draft dampers were also closed to retard combustion. (They were likely to have been at least partially closed on the engine STOP order.) Shoveling of coal into the furnaces was also halted. However, steam production still continued because of the great mass of burning coal still in the furnaces. To diminish the steam pressure, where there was time, the burning coals were raked from the furnaces. Without such action, steam pressure would continue to rise. The escape valves were intentionally opened to decrease this pressure before seawater reached the boiler units. Steam line stop valves were closed to isolate the abandoned boilers from still functioning systems. As the flooding worked its way aft through additional boiler rooms, this process was repeated. Boiler room crews were sent to "E" Deck as they became superfluous and it is from there that the survivors of the engine department came. The sound of escaping steam from escape pipe outlets high on the first three stacks was remarked by many that survived this tragedy. Second Officer Lightoller said of the escaping steam, "it sounded like the noise of one thousand steam locomotives." Radio Operator Harold Bride recounted that Jack Phillips found it so difficult to concentrate with such a cacophony that Captain Smith sent a request to Chief Engineer Bell that he diminish the flow of steam. Even after the steam flow through the stacks had been diminished, there was the noise of air escaping from the forward expansion joint that was located just forward of the Marconi Room. As the bow kept settling deeper from the flooding in the forward part of the ship, stresses began to build around this joint, causing it to gradually open. In addition, as the flooding water began to move aft into spaces below the expansion joint, air was being forced through this space in the superstructure causing a whistling type of noise that made it hard for Phillips to hear incoming radio messages on a radio set that was also losing power.

Aftermath of the Iceberg Impact. With the initial sighting of an object only 1,500 feet ahead and a ship moving at 22.5 knots, neither a hard-a-starboard, nor its combination with a hard-a-port "fishtail" maneuver, not even an emergency reversal of the engines, could have prevented the fatal impact of steel and ice. With the rudder hard over to port, there was a small initial transfer of the ship to starboard before the rotation generated enough forces on the hull to begin the port turn. The testimonies of the surviving crew and passenger survivors differ on the severity of the impact. Whereas the crew in the bow area likened the impact to a slight jar, some of the passengers noted stronger jolts. From the Crow's Nest, Fleet and Lee noted that there was not much noise other than the sound of ice falling onto the Forecastle and Well Decks. Other survivors noted the large amounts of ice scattered over these decks. A scraping noise and slight jolt were among the descriptions of the encounter by others. Second Officer Lightoller recounted the noise as being like an anchor chain being paid out. In the forward boiler room, the collision sounded like thunder.

Third Officer Herbert Pitman was off watch when the collision occurred. He heard what he thought was the anchor chain running out, like Officer Lightoller, but with a slight vibration. He went out on deck, but saw no one and, therefore, returned to his stateroom. A few minutes later, Fourth Officer Boxhall came into his stateroom and informed Pitman that the Mail Room was filling with water because the ship hit an iceberg. Alarmed, Pitman now dressed, put on his overcoat and decided to make his own inspection. On the way out on the Boat Deck, he met Sixth Officer James Moody who told him that there was ice on the Forward Well Deck. Pitman's curiosity was now really aroused. He saw a little ice on the Well Deck and proceeding to the Forecastle Deck he met a group of firemen carrying their duffle bags.

Pitman: What is the matter, mates?
Firemen: Plenty, Sir, water is coming into our place from somewhere.
Pitman: That is strange.

Pitman went to the hatch of Cargo Hold No. 1 and looked down. He could see water overflow the hatch coaming on the deck below. Recognizing the danger, he proceeded to the Boat Deck.

What about the iceberg? This question is not one of mockery, since knowledge of damage to the iceberg would provide a better understanding of where the damage to the hull could have occurred. The day after the sinking, there was a photograph in the vicinity of the sinking taken of an iceberg from the *Carpathia* that had a red smudge at her waterline. Red paint, a ship's anti-fouling paint, covering its hull below the waterline, and finding its way to the surface of an iceberg only by means of a collision. It was reasonably reported that this was the one that the *Titanic* had struck. A careful examination of this photograph shows a freshly made cavity in the iceberg well above its waterline as well as above the red paint believed to be from the *Titanic*. It was indeed the iceberg that the ship had struck. We now postulate that a small spar from that iceberg probably was dislodged during the ship-iceberg collision, leaving that cavity and causing the ice to topple onto the forward Well Deck.

Below the level of the sea where eyes could not see, the ship probably cut away or broke off a portion of the iceberg. The damage to the riveted seams that was discovered in 1996 is indicative of this. The openings in the Forepeak and Cargo Hold No. 1 are a strake higher than those discovered aft from Cargo Hold No. 2 to Boiler Room No. 5. This helps explain why those spaces filled more rapidly. They were over a longer length and deeper underwater. Fleet and Lee made mention that the iceberg started to make contact with the ship around the bulkhead between Cargo Holds Nos. 1 and 2 as ice fell on the Forecastle Deck aft of there. Forward of this, it was a grazing encounter with the iceberg.

Survivor Reports. Many of the crew off duty in the forward part of the hull were immediately aware of the happening and rushed out to investigate. Able Seaman Frederick Clench was one who survived to tell what it was like. Awakened from a sound sleep by the crunching and jarring noises, he dressed quickly and rushed out to the forward Well Deck where he spotted ice covering the starboard side of that deck. He then went to the alleyway under the Forecastle Deck to put on his shoes. While there, one of the crew asked him did he hear any rush of water. "No," he responded but 10 minutes later was told to look under the hatchway which was covered with a tarpaulin which was "bellied out as if there was a lot of wind under it and I heard the rush of water coming through." Although he was aware that the ship was taking on water, he was not concerned that she would sink. The water was flooding compartments below and trapped the air, then forced it out.

Able Seaman Frank Oliver Evans had a similar experience as Able Seaman Clench. He was reading a book in the Seamen's Mess when he felt a slight jar. He did not take much notice of this until one of his fellow seamen came into the space with a large lump of ice. Letting it fall to the deck, he remarked, "Look what I found on the Fore Well Deck." Evans wanted to see this for himself and started to go topside when he met Sixth Officer Moody, who instructed Evans to find the Carpenter and sound all tanks and report this to the Bridge. Obeying the command, Evans found the Bos'un instead. The Bos'un told Evans that the ship had struck an iceberg; that the Ship's Carpenter had completed the tank soundings and gone topside. Looking down at the forward hatch, Evans noticed that the tarpaulin over it was bulging upward from the force of air escaping from below. The Bos'un instructed him to return to the Able Seamen's Bunk Space and rouse everyone there.

Seaman John Buley was in the "C" Deck Crews Mess ready to go on watch at midnight. He felt a slight jar and a scrapping noise along the hull. Putting on his overcoat and curious to find out what it was, he started up the access stairs to the Forecastle Deck when he met two firemen who were off watch. "They told me that she had struck ice. One of our chaps went and got a hand full of ice and took it down below to "E" Deck. They turned in again."

Seaman Frank Osman was in the Seamen's Mess with Evans when he also heard some noise. He went into the forward Well Deck where he discovered some ice on the starboard side. Going below, he also noticed that the ship was listing slightly to starboard. From his Royal Navy experience, he recognized that something was wrong and reported to the Boat Deck.

Seaman Samuel Hemming was asleep in his bunk when he was awakened by an impact. Opening the porthole on the portside, he looked out to see what it was, but was unable to see anything, probably due to nightblindness. Turning to his roommate he said that there was no ship in sight, so it must be ice. Leaving his space, he proceeded to the forecastle head where he heard a loud hissing sound. Looking for its source, he went down several decks to the Orlop Deck to a Forepeak storeroom over the Forepeak Tank. He did not find any trace of flooding water; therefore, he determined it must be air escaping through the air escape pipe from that tank. Returning to "E" Deck, he met Chief Officer Wilde who was in this area of the ship to determine where water might be coming into the ship. The lamp trimmer, Mr. Hemming was also present. Hemming had come there to determine where the hissing noise was coming from. He determined that it was air escaping from the air escape for the Forepeak Tank. Around this time of discovery, the chief officer, Mr. Wilde, put his head around the hawse pipe and said:

Wilde: What is that, Hemming?
Hemming: The air is escaping from the Forepeak Tank. She is making water in the Forepeak Tank, but the storeroom is quite dry.
Wilde: (*responding*) All right.

Wilde asked if there was any water in the Forepeak spaces, and the storekeeper went into those spaces over the Forepeak Tank, and found there was no water there. However, he discovered that the Forepeak Tank was full. The chief officer then left and went to the Bridge to give his report on what he had seen and been told. Chief Officer Henry Wilde gave his report to Captain Smith.

Bos'un Mate Albert Haines went down to look at Cargo Hold No. 1. He noticed that the tarpaulin was "bellying up," or rising above the hatch, showing that the water was coming in from air being ejected under the tarpaulin. He left there immediately for the Bridge to report this to the Chief Officer.

Arriving on the Bridge, Bos'un Mate Haines explained to Chief Officer Wilde that Cargo Hold No. 1 was filling quickly. Chief Officer Henry Wilde instructed Haines to go below and get all Bos'un Mates to the lifeboats and uncover them. "Get them ready to be lowered. Tell the men that Mr. Andrews gives the ship about one hour to live. However, they are not to tell the passengers, otherwise there will be a panic."

Lookout Reginald Lee left the Crow's Nest around midnight to go below to the Crew Mess on "C" Deck. His quarters were on "E" Deck and below that there were the firemen's quarters on "D" Deck. He could hear water coming into "D" Deck. This would be around 0005. Lee went over to the hatch coaming to Cargo Hold No.1 on "C" Deck and looked down below under the tarpaulin. He could see the water flooding in rather quickly from the starboard side at the level of "G" Deck through a door there. It also made a discernible noise, somewhat like a brook flowing by only louder. He also noticed that the firemen and trimmers who had the 12–4 watch were coming up with their belongings. They left their belongings on the Forecastle Deck before reporting for duty. Around 0020–0030 Lee heard the Bos'un Mate Haines call for all off-duty seamen to report to the Boat Deck. Lee went to the starboard side of that deck to uncover the lifeboats.

Shortly after midnight Hemming returned to his bunk on "E" Deck. Satisfied that there was no danger, Hemming climbed into his bunk and tried to go back to sleep. However, he was roused again by another shipmate who informed those in their bunks, "If I were you, I would turn out, you fellows. She is making water one-two-three, and the Racket Court is full of water." Just after this man left, the Bos'un came into the Able Seamen's Bunk Room just after midnight and shouted, "Turn out, you fellows. You haven't half an hour to live. That word is from Mr. Andrews. Keep it to yourself and let no one know."

Mr. and Mrs. George A. Harder, returning from their honeymoon, were awake in their cabin on "E" Deck. They heard a dull thump and then the ship seemed to shake a little, followed by a "sort of rumbling, scraping noise" on the starboard side of the ship. Mr. Harder sprang out of bed and went to a porthole to see what had happened. He caught a glance of a wall of ice gliding by. This sighting would indicate along with the officers and crew in open deck positions that the ship had indeed struck an iceberg.

Passenger Russell[5] had just entered her starboard cabin (A-11) on the Boat Deck, one deck below the Bridge, to retire for the night. She noticed a very slight jar, then a second quick one, that seemed closer and a little stronger. The third jar came as a shock, strong enough to make her grab her bedpost. She also felt that the ship had come to a stop. Turning toward the starboard window, she observed a white shape, like a mountain, gliding by. Curious to determine what the object was, she donned her fur coat and ran onto the deck, where she learned that the ship had hit an iceberg. Looking aft, she was able to see something drifting astern and soon it was out of sight. Looking down to the second-class deck, she observed some stokers walking across an ice-covered deck and could hear it crunch under their boots.

In Edith (nee Brown) Haisman's biography, she remembered feeling three distinct blows, verifying Edith Russell's testimony. Such testimony is important to the analysis of the bow damage because the Haisman and Russell recollections confirm the impact and momentum theory that we think was associated with the iceberg-ship encounter. The severity of these responses would explain the damage in Cargo Holds Nos. 2 and 3 and Boiler Room No. 6. Such an impact would be sufficient to snap off the heads of wrought iron rivets. It also provides a measure of the severity of the ship response. When she collided with the iceberg, the *Titanic*, being the lighter mass and having the greater velocity of the two bodies, rebounded. The ship's forward momentum, coupled with its increasing beam, brought it into contact with the iceberg for several encounters, each a little stronger than the first. These combined factors helped account for the intermittent scraping sounds. Inside the ship nearer to the encounter, it probably sounded more like a sledgehammer banging and dragging along an empty steel drum. This is why, as will be seen later, the damage to the hull is isolated by compartments and not continuous from the point of encounter until the point of departure between the two bodies. It also appears that the iceberg may have had a small projecting ridge or spar that was massive and hard enough to cause riveted seams to part and the wrought iron rivets to fail. It is important to note that at main watertight bulkheads or web frames, the likelihood of punctures was greater. Most of the damage is hidden by sediments, but author William Garzke actually spotted some of the damage from the encounter at the sediment line from images sent back by *Magellan 725* from an area outside of Boiler Room No. 6.[6] The plate separation was very narrow and there were a few missing rivets in the upper rivet row (the lower one being covered by sediment), indicating the area of entry for the flooding that ensued was very small and isolated. Also, it is possible during the encounter, the ship may have sliced into and sheared off part of this ridge or spar of the iceberg.

The plate separation was indeed tiny and was further exacerbated when the bow portion of the wreck plowed into the sediments upon contact with the seabed. In that process some 30,000 tons of mud was pushed aside or along before the bow wreck came to a halt. Therefore, the seam might have parted further as the rivets had already been weakened or failed by the iceberg impact. The seam that was spotted was only 10.5 feet above the keel and with the ship's draft of 33 feet, there was a substantial head of water that forced water through the tiny opening. The separation was only about 0.5 inches initially, but there were also some failed rivets that fell out of their holes creating an additional area through which water could pass. To an observer within the space, the scene would look like a sheet of water with jets of water coming through some of the 1⅛-inch failed rivet holes and the separated seam. Also, there is the small possibility that the plates may have sustained some small fractures due to the brittleness of the steel at a temperature of 28–31°F. The opening appears to be greater at the forward end of Boiler Room No. 6 than that at the aft end. This is verified from the testimony of Frederick Barrett and George Beauchamp who were in the space at the time of impact. Barrett saw a seam open and water pour in two feet above the floor plate on which he was standing. He made a quick exit into Boiler Room No. 5 with Engineer James Hesketh. Beauchamp, on the other hand, was in the forward stokehold, shutting the dampers. In a matter of minutes, water was up to his waist. He ended his activity of extinguishing the fires in the furnaces and made his way up an emergency escape ladder.

The amount of ice on the Forward Well Deck indicates that the ship-iceberg collision was significant enough to jar a spar loose from the iceberg and the debris to fall onto the ship's starboard side. Assistant Cook

John Collins described that the starboard side of that deck was packed with ice that extended almost to the vessel's centerline. This was verified by the ship's barber, A. H. Weikman, who left the barber shop aft and made his way forward on "E" Deck to his quarters forward on "C" Deck. On his way, he saw a baggage man who informed him that the baggage room had filled with water. He also came upon Thomas Andrews trying to get all the steerage passengers to the lifeboat stations. The Second Steward Dodd then ordered Weikman topside to man the lifeboats.

Further confirmation on the amount of ice comes from Major Peuchen, an amateur yachtsman, who made several visits to the forward end of "A" Deck to observe ice on deck, shortly after the collision with the iceberg. When he saw the scene at first, he noted that the soft or shell ice had fallen some 4 to 4.5 feet inside the railing. Upon his return with Mr. Charles Hays, he saw steerage passengers were kicking ice chunks around on deck like they were having a game of soccer. At this moment, Peuchen detected a slight trim by the bow and noted this to Hays, adding that the sea was calm and the ship was no longer moving. Expressing concern, Hays responded by saying, "Oh! I don't know; you cannot sink this boat. No matter what we have struck, this boat is good for 8 to 10 hours." Statements like these confirm the attitude of some passengers as to the danger confronting the *Titanic* at this moment in her sinking process.

The significance of the ice on the Forward Well Deck was finally determined during an examination of the starboard shell below that deck in the August 1996 and 1998 Expeditions. The side of the ship was torn open at a seam above the waterline and the starboard gangway door was missing, its frame being partially torn away. It is suspected that the force of the iceberg collision at this point in the ship was the greatest, enough to dislodge ice from the iceberg and cause ice to tumble onto the deck of the ship. The ship's beam almost reaches a maximum at this point and as the two bodies, ship and ice, interacted with one another, the bumping and jarring noises heard by the survivors above reached a maximum at this point. It was enough to weaken the riveted seams around the area of the gangway door, but not enough to cause the seam to part in the manner that is seen on the bow wreck today. That would come later when the ship hit the seabed below. As the ship slid along the seabed, it plowed up a massive amount of sediments and came to a sudden crunching halt that caused some of the entrained water in the hull to be expelled out of the ship through that weakened area. Three rivets were spotted below this hole and failed plate, giving evidence that it was more of the ship's jarring impact with the seabed than the iceberg itself.

The collision did not bring immediate concern of the ship sinking;

however, the symptoms of flooding in the forward compartments quickly manifested themselves. The five-degree starboard list was the first indicator. The presence of ice on the Well Deck was only symptomatic of a strong contact with the ice mass. The Captain, perhaps sensing danger based on the list, was not convinced of Boxhall's first report and feared that his ship may have been severely damaged. He then ordered his fourth officer to have the Ship's Carpenter sound all tanks to confirm his suspicions on the magnitude of damage. It is interesting to note that second-class passenger Lawrence Beesely had some difficulty in climbing or descending stairs around this time, confirming the fact that the list and trim, within 30 minutes after the collision, was severe enough to make walking up and down stairs more difficult.

Every professional aboard must have known that "unsinkability" did allow for some flooding. They must also have known that there were severe limits to that allowance. The designers had provided a watertight subdivision that would allow the ship to float if any two adjacent compartments flooded. In fact, the first four compartments could flood and the ship would barely survive because the flooding water would be just below the margin line at the forward bulkhead of Boiler Room No. 6. In actuality, the ship was built to the following rule for ships over 450 feet of the Board of Trade's Bulkhead Committee:

> Vessels may be considered able to float in moderate weather with any two adjoining compartments in free communication with the sea if fitted with efficient transverse watertight bulkheads so spaced that when two such compartments are laid open to the sea the uppermost watertight deck to which all the bulkheads extend and which we will call the bulkhead deck is not brought nearer to the water surface that would be indicated by a line drawn round the side at a distance amidships of 3/100 of the depth at side at that place below the bulkhead deck and gradually approaching it at the ends where it may be 3/200 of the depth below it. This we will call the margin of safety line. With longitudinal bulkheads in conjunction with transverse the vessel should not, under the above conditions, list so as to bring the deck on either side below the water level.

This was the criterion used by Harland & Wolff to design the *Olympic* and *Titanic*. From the calculations of Edward Wilding, it was shown that the Bulkhead Deck was never less than 2' 7" above the waterline with two-compartment damage and this exceeded the requirements of the Bulkhead Committee. A ship should be able to survive the flooding of any two adjoining compartments—that was the rule. Open too many compartments to the sea and the ship would founder.

It is not often that a captain has a naval architect aboard to consult in an emergency such as this. Captain Smith at once sought the advice of Thomas Andrews, foremost among the Harland & Wolff naval architects. Among the many plans in Cabin A-36 was the floodable length curve that demonstrated how the *Titanic* conformed to the rules above and how she exceeded the rule in the forward portion of the ship. Whichever compartments were flooding, Andrews could look at his curve and determine if the ship would remain afloat with those compartments open to the sea.

Andrews and Smith went below to look for evidence of seawater. They found far too much. Cargo Holds Nos. 2 and 3 were flooding at an alarming rate. The testimony of Seaman John Buley before the American inquiry gives us an idea of the inflow of the flooding water in Cargo Hold No. 2. He noted that you could hear it coming into the ship and that the force of that water coming in caused a tarpaulin over a hatchway to bulge up. Finally, the escaping air from that hold caused that cover to come off. Strangely enough the water in Cargo Hold No. 2 had an eerie green glow—the flooding had submerged the lights, but had not yet put them out. The Mail Room over Cargo Hold No. 3 had completely flooded and water had almost filled the Squash Court, which meant that all of the spaces below "G" Deck, including Cargo Hold No. 3 and the reserve coal bunker were flooded. With the two holds filling with water and flooding reported forward of these, the ship could barely survive such flooding as the floodable length curve showed. However, two more compartments, Boiler Rooms Nos. 5 and 6 were also reporting flooding. An inspection of that space, however, showed that the pumps there were containing the flooding. Boiler Room No. 4 reported some minor flooding but at such a low rate that the bilge pumps were easily holding that flooding in check. Although there was no immediate danger of the flooding in Boiler Room No. 5, Andrews was concerned with the flooding in Boiler Room No. 6 because it could be enough to cause the ship to sink.

The floodable length curve will not predict the amount of time that will be required for a ship to sink, since this calculation is not time-dependent since the rate of flooding is an unknown. Time, however, is important, because it will determine how much time is needed for the transfer of passenger and crew into lifeboats, and for rescue ships to arrive. During the initial stages of flooding in the *Titanic*, the sea poured into Boiler Room No. 6 at such a rate—about 100 tons per minute—that the personnel there fled before taking the preventive measure of securing their boilers. In Boiler Room No. 5, however, flooding was being controlled by pumps in that space. At some point, there would be downflooding of this space through non-tight openings once water began to cover "E" Deck in that area. Water would also find its way aft through passageways, pipe openings, accesses to spaces below, cable penetrations, non-secured drains in lavatories and bathrooms, and ventilation trunks. There was also the danger of non-structural bulkheads collapsing from pressures that they were not designed to resist. Most prominent of these were the forward nontight coal bunker on the starboard side of the ship in Boiler Room No. 5. The inevitable would happen, but if time could be bought, a wireless message indicating that the ship was sinking would bring rescuers. To that end Chief Engineer Joseph Bell organized his engineering crew to activate all pumps that served the bilge and ballast systems, and to open the bilge suction valves in the affected spaces.

There had been a coal fire in the after coal bunker on the starboard side of Boiler Room No. 6. The coal in the adjoining bunker in Boiler Room No. 6 had also been removed as a precautionary measure. Andrews had inspected the main transverse bulkhead that had been heated by fire, but had found no hazards. At the time of his inspection, he was not thinking of a water head on a weakened section of bulkhead. A failure here and the sea, would in effect, have a short cut through the ship's inadequate subdivision. As it was to happen, about 80 minutes would pass before there was any bulkhead failure in this area.

In the meantime, Thomas Andrews returned to Cabin A-36 where he made some quick computations to determine the time of survival taking into account the rate of flooding he had seen in those compartments that he and Smith had inspected. He returned to the bridge to summarize his findings and to announce his final pronouncement—two more hours afloat at the most.

Smith had already told Phillips and Bride to commence sending distress calls. With the damage being extensive and time short, he needed to let the ships know of his position. The last known position had been recorded at 1930, based upon celestial sightings by Second Officer Lightoller. Boxhall corrected this position by dead reckoning, making the present position to be 50° 14' West, and 41° 46' North. After approval by Captain Smith, Boxhall took the estimated position to Phillips and Bride with instructions on what was to be told of their plight.

Notes for Chapter 13

1. In an interview with Leslie Reade some 50 years after the tragedy, Frederick Fleet revealed that Reginald Lee was haunted by the incident and died from excessive drinking. Fleet had trouble sleeping and eventually sought medical help. [Reade, Leslie, *The Ship That Stood Still*, W. W. Norton & Company, New York, 1993.]

2. Recall that the propulsion machinery was arranged in a series-compound manner. Steam expanded through a pair of reciprocating engines, one driving each of the wing propellers, then exhausting its steam into a single turbine that drove the center propeller. The ultimate destination of the steam was the condensers, one on either side of the turbine. Given the technology of the time, this was an arrangement that was commendable from several standpoints and one that served the *Olympic* through her long career. However, it was awkward when going astern because the turbine could not be reversed. The expedient was for the engineers to switch the reciprocating engine exhaust from the turbines to the condensers. The exhaust change-over valves were of the butterfly type, requiring no more than a 90-degree rotation for the change. A steam cylinder effected the change, so that the only human action required was the change of a small steam valve from the turbine to the condenser position. Once a crewman's hand had reached this valve, only 10 seconds were required for the changeover to be completed. Operation of the reciprocating engines was not affected.
3. The report of Third Officer Joseph Boxhall is based on his testimony at the British and American Inquiries.
4. Testimony of James Johnson, Nightwatchman, taken from the British Inquiry.
5. Edith Russell gave an account of her experiences aboard the *Titanic* in the May 1964 issue of the Ladies Home Journal.
6. Author William Garzke has examined several video tapes taken of the iceberg damage outside of Boiler Room No. 6. He has determined that the floodable area of Boiler Room No. 6 was 2.62 square feet. Bedford and Hackett have estimated that the floodable area was 2.80 square feet in their analysis using computer techniques. The sonar scan by Paul Matthias during the 1996 *Titanic* Expedition, the analysis of Bedford and Hackett, and the observations of author William Garzke prove that the damage caused to the ship by the collision was small—11.91 square feet or the size of two city sidewalk squares, but deep enough to cause uncontrollable flooding.
7. Ismay testified to Senator Smith, "I presume the impact awakened me."
8. The evidence of Scott and Dillon indicates that there was an intermediate stop request and that they saw SLOW ASTERN on the engine telegraph. Whether the ship was at FULL ASTERN or SLOW ASTERN is immaterial because the *Titanic* was going to collide with the iceberg in any event. In *Triumph and Tragedy*, page 45, Charles Haas and Jack Eaton reported that during her April trial the *Titanic* came to a full stop from a speed of 20.5 knots in less than half a mile when the engines were reversed.

CHAPTER 14

Evacuation

Ships have carried passengers across the seas since the day when humans first fitted together a few planks or animal hides as means of crossing open water. The Celts to Britain, the Ainu to Japan, the Polynesians to Hawaii . . . all of these were "passengers" of a sort. On the other hand, the "passenger ship," i.e., a ship designed and built for the purpose of carrying people who would board voluntarily, would pay for their passage, and would expect to be fed and berthed while aboard, was a quite recent concept in 1912. The early Cunard transatlantic paddle steamers and the Australia-bound immigrant ships of the early White Star Line might justifiably be called the first passenger ships. Only three quarters of a century, say, separated the *Titanic* from these first-of-its-kind examples. Just as there were survivors from that ship still living three quarters of a century after it sank, the *Titanic* passenger roster could have included individuals that had crossed the Atlantic on the first-ever true passenger ship.

Seagoing from the first cockleshell sailer into the early days of Cunard and White Star had been a dangerous venture, one in which fire, stranding, overwhelming storm, even human enemies, were ever-waiting hazards beyond the horizon. The "lifeboat" that would save everyone—or anyone—aboard if the gamble of going to sea were lost was still to be conceived. Only when the growth in size and power of ships (and perhaps of *Pax Britannica* along the major shipping lanes) made transoceanic travel comparatively safe, and yielded deck space that could be devoted to small boat carriage, did this concept take hold and become a feature of ships.

As progress in ship technology continued from the 19th century into the 20th, grim statistics such as deaths per passenger-mile would surely show continuing decline, but loss of passenger ships with consequent fatalities continued. However, by 1912 incidents in which time and circumstances permitted all aboard to escape to the safe haven of lifeboats were rare. Rapid sinking, severe list as part of the incident, fire racing through wooden structure, capsizing, sea state too hazardous for small boats: one or several of these conditions usually accompanied a life-threatening occurrence. Small wonder, then, that the British Board of Trade[1] did not think of specifying space in small boats for everyone carried by a ship. Rules for safety equipment were, and are, usually based on that best measure of commercial ship size, gross tonnage. Without experience showing the merits of a boat seat for everyone, boat capacity for this new notion of saving people by putting them off into small boats was based on that parameter.

Armed with that invincible weapon, hindsight, many have condemned the doddering decision-makers in the Board of Trade who were reluctant to accept the technical progress going on around them. However, some justification can be found for their failure to update some of their outdated regulations. It was believed with the large numbers of ships plying the oceans that the safest place would be on a large ship rather than a small lifeboat. The lifeboat would simply be a conveyance for the passenger or crew to another ship. The theory seemed sound, but what if no ship could get there in time? In the first decade of the 20th century the goal of unsinkability still seemed attainable to many professionals. Given those many hazards that could make taking to the boats a grave hazard in itself, unsinkability seemed to be a more certain route to safety for passengers. Indeed, this belief continued in professional circles even *after* the *Titanic* experience. This was certainly true in the case of the *Lusitania*. As we shall see in a later chapter, some writers on maritime technical matters looked upon that ship's longitudinal subdivision as a protection against sinking might indeed bring with it the elusive quality "unsinkability." They were to be disabused, of course, but until the foundering of the *Lusitania*, providing unsinkability could be reasonably argued as the best guarantee of safety for the sea-going human.

The Board of Trade does deserve criticism for not extending its rules for lifeboats past the 10,000 ton limit they thought adequate in 1891, but it should be awarded some credit just for *having* rules requiring boats, even though the 16 for *Titanic* were far from enough.

Lifeboats thus might be called an expedient unlikely to answer in the exigencies of many kinds of maritime emergencies. Here on the fated maiden voyage was a rare instance in which the unlikely occurred: a ship with a large number of people aboard was sinking slowly with a heavy bow trim, but without significant list, into a calm sea. Conditions were ideal for use of the lifeboats. If the task of organizing and carrying out an evacuation could be put in motion and carried out in the approximately two hours available, then some passengers and crew would live to reach New York. If there were enough boat capacity, then *all* of the 2,228 persons might live. Otherwise, those left behind in freezing water would surely die in the unbearably cold North Atlantic.

The boat capacity, however, was far from adequate, as anyone who has heard the *Titanic* story surely knows. There were 14 lifeboats of 65-person capacity. There were two cutters or emergency boats of 40-person capacity. There were four collapsible lifeboats[2] of 47-person capacity. These last had canvas sides that had to be raised and then secured before lowering into the water. The total lifeboat capacity was 1,178 persons. The four "collapsibles" were lashed down on deck and located near the boat falls for the two emergency boats. The standard lifeboats were "under davits," meaning that they were mounted on deck, resting in support chocks, and lashed down, but attached to the falls of the lowering gear.[3] The davits that launched the cutters were supposed to be used a second and a third time to launch the collapsibles. They were successfully used to launch the first two collapsibles. Because the remaining two boats were stowed on the officers' quarters deckhouse, they had to be lowered to the Boat Deck before being attached to the davits for the emergency boats. With this complication slowing the launching process, the march of events intervened. The boats floated free as the ship sank under them.

Captain Smith may not have had that exact 1,178-person lifeboat capacity total in his head, but he surely must have been aware that only a fraction of the people in his care would find space in lifeboats. If not, he was certainly reminded by naval architect Thomas Andrews. The only mercy was that the ship carried enough lifebelts for everyone, but that circumstance was most truly a cold comfort, since anyone whose only survival aid was a lifebelt would surely succumb within a few minutes to hypothermia in the frigid 28–31° Fahrenheit water.

For a time after the collision Captain Smith was probably confident of a rescue ship's timely arrival. His wireless was sure to reach many ships, including, most likely, that one whose light was visible only a few miles away. The shortfall in boat space was probably only a nascent worry in the beginning. It was to grow into a devastating realization when the closest ship to respond to wireless call for assistance (the *Carpathia*) was more than three hours away, and that nearby ship could not be roused by any means. He nonetheless ordered the boats under davits uncovered 25 minutes after the collision (i.e., at 0005). The order to swing them out (swung over the side and lowered to a position from which they could be boarded) came 20 minutes later. At 0030 the order to place women and children in boats was given, and at about 0045 the lowering of the first boat, number 7, began.

At the time of the first of these orders, confidence in the timely arrival of a rescue ship must have remained high, but an evacuation by boat was nonetheless in order. Getting 2,228 persons (even had there been space for that many), most of them either frightened and unused to the procedures of evacuation, into boats, then lowering those boats slowly into the sea, was going to take time, a long time, regardless of the number of boats or the number of rescue ships hovering nearby. The putative rescuers would surely send their own boats, but getting passengers (in particular) from a deck of *Titanic* to a small boat standing by alongside would also be difficult, and require time. Best to get the evacuation started, regardless. There was barely enough time to lower 18 boats before the ship began her final death throes.

The uncertainty and disbelief among the passengers that the ship would sink, and the lack of practice by the crew, were factors that caused 467 lifeboat spaces to be unfilled. Now, critics have taxed the survivors for their failure to bring their boats in to a point where they could rescue those unfortunates who floundered in the water after the ship sank beneath them. The fear of suction as the ship went down, and the fear of desperate swimmers swamping their boats, have been cited as justification for inaction.

The low number of people in the first lifeboats launched can also be attributed to the desire to have these boats take some of the passengers from the gangway doors located on "E" Deck, forward and aft. Six Bos'uns were dispatched by Officer Lightoller to open the forward ones, but these men were never heard from again. They were, perhaps, victims of the flooding ongoing in the bow. Some of the partially filled lifeboats were summoned to return, but the occupants were worried that they would be swamped by those wanting rescue or the suction forces of the sinking ship.

Every boat had assigned crew, but the weekly lifeboat drill (normally on Sunday at 1100) had not been held. This omission left many crewmen unaware of their lifeboat assignments, and at best, they were slow to report to their stations. Some of the crew had not looked at their lifeboat-station list, while a few had difficulty finding their way to the lifeboats. The situation among the deck officers was much better. Boat loading was supervised by the senior officers; Lord Mersey was to conclude "The evidence satisfies me that the officers did their work very well and without any thought for themselves."

The term "total confusion" finds good application in describing the early responses of the passengers. As has been seen, many passengers felt or heard something that turned out to be the scraping-by of an iceberg, and that those who went on deck found that, indeed, their ship had hit that dark shape disappearing astern. Those venturing forward found that it had delivered that scatter of shattered ice in the forward well deck. Some, like Major Peuchen, told of being almost immediately aware from a slight inclination of the deck to starboard that some kind of trouble had come from that iceberg. But would the ship sink? Was it necessary to get into lifeboats? In the absence of central communication system, that news took time to get around, and met resistance even from those who knew what was happening.

The steerage passengers quartered low and forward were the first to be affected enough to take action. Water began seeping into their accommodations early. As it began to rise, many of the male passengers housed forward sought safety aft, taking their possessions (perhaps their only possessions in this world) with them along "Scotland Road," the passage that led to the steerage quarters aft. This same passage gave access to the boiler rooms, and had become by this time the gathering place for the blackened stokers rousted from Boiler Rooms Nos. 5 and 6. They milled about in Scotland Road, possibly in as much uncertainty and fear as these men who brushed by them.

First class, isolated in its lofty accommodations from the peril maturing far below, was a more difficult sell, especially since the hour was far past the bedtimes of most passengers. It fell to the cabin stewards to rouse them and convince them to go to boat stations, but even they needed convincing. Thomas Andrews, designer, naval architect, now troubleshooter with the bad news from his floodable length information, was an important resource in the struggle to get the word around, and convincingly. He apparently did not pound on stateroom doors himself, but persuaded some of the reluctant stewards and stewardesses to do that very thing. Steward Henry Etches testified that he was on "B" Deck in the process of waking passengers, without great forcefulness, but knocking because he had been told to do so. He encountered Thomas Andrews, who with no authority but his recognized stature, told him to have the passengers open their doors to confirm that they were indeed awake, to have them take their lifebelts from the rack in their wardrobe lockers, and then to make sure that they put on these seeming inconveniences.[4]

Etches went to work, but had his difficulties. On "C" Deck, several encounters went this way:

> I found cabin door (C-78) shut, and I banged with both hands on the door loudly and a voice answered, "What is it?" Then a lady's voice said, "Tell me what the trouble is." I said, "It is necessary that you open the door, and I will explain everything, but please put the lifebelts on or bring them into the corridor." They said, "I want to know what is the matter." I said, "Kindly open the door," and I still kept banging. I passed along, and I found one cabin empty, and then I came to another cabin and a lady and gentleman stood at the door. They were swinging a lifebelt in their hands.

Did that first couple finally emerge, girded for the sea? The exploration of the wreck and a study of the large piece of shell plating taken from the seabed in August 1998 indicates that stateroom C-78 was in the area of the hull fracture. The shell-plating fragment offers no clues.

The steward who encountered Helen Bishop and her husband on "A" Deck had a different experience. The couple were awakened by the collision, dressed, and went up to "A" Deck. There they met a steward who told them that the ship had struck a little piece of ice, but there was nothing to worry about. "Return to your cabin," he told them. They complied, but soon another passenger, John Jacob Astor, was at their door to tell them to go above. They again believed the news they were offered, and were soon being lowered in Lifeboat 7, the first to go, at 0045. The order "women and children first" had been given, but Mr. Bishop was allowed to board because not enough of these priority-holders were in sight to fill the boat. During this time in the evacuation, people were very reluctant to board these little crafts and be lowered onto a dark and forbidding sea.

Third-class Steward John Hart was asleep in the aft quarters on "F" Deck when the collision occurred. Awakened by his fellow stewards, he was informed there was an accident. Not thinking it was serious, he went back to sleep only to be awakened by the chief third-class steward, James Kieran. He told Hart to go to his rooms in his aft accommodations for third class and wake the single women and nine married couples with children (58 persons), see that they find their lifebelts, and make sure that they are wearing those lifebelts. Kieran also said that there would be further instructions and that Hart should stay with his people. These instructions were given to eight other third-class stewards who had rooms assigned to them. Hart went to "E" Deck with his group of passengers. While waiting for instructions there, Hart placed some lifebelts in Scotland Run as there were a large number of men with baggage heading aft and going down to the third-class Dining Saloon on "F" Deck. Kieran came by with his assistant to see how the evacuation was going.

Around 0030 the word came down to bring women and children up to the Boat Deck. Hart said that those that were willing to go to the Boat Deck were shown the way. All the third-class stewards that had passengers waiting to go to the Boat Deck also heard these instructions. Hart took his entourage to the aft Well Deck ("C") and the Boat Deck via "B" Deck. Hart testified at the British inquiry that there were barriers that in ordinary times were closed, but they were open during the emergency. There were some of his people that did not want to go and stayed behind on "E" Deck. Others proceeded to the Boat Deck, found it rather cold, saw the lifeboats being lowered away, and thought themselves more secure on the ship. Consequently they returned to their quarters. Several said they preferred to remain on the ship than be tossed around on the water like a cockleshell. The remainder of his people entered Lifeboat 8.

Hart went back to "E" Deck. He had difficulty going down the stairwells, as there was a large number of men exiting from below. However he found another group of 25 third-class passengers (22 women, three children and one man holding a baby). He led them up to the Boat Deck to Lifeboat 15, which was in the process of loading. The port boats had already been launched. All of the second group as well as Hart entered the lifeboat.

That women-and-children-first dictum was not an expedient invented on that strange Monday morning, but it always evokes images of the *Titanic*, and probably because that disaster was one of the few in maritime history where sexes and ages could be carefully sorted. On the port side, Second Officer Lightoller commanded the Boat Deck, and he was strict in that regard. A total of 349 women and children were saved by portside boats while only five men escaped Lightoller's vigilance to join them. On the starboard side, First Officer Murdoch commanded, but he was more lenient in that he allowed men to go when no women were in sight to fill his boats. A total of 355 women and children were joined by 36 male passengers.

In general, the evacuation of passengers began and continued for a while as a leisurely and panic-free affair because of the difficulty of convincing those being saved that the ship was sinking. Many seemed to have believed that it was a mere precaution, and that they would be picked up again as soon as the ship had been checked. Eva Hart, a seven-year-old survivor, recalled many years later the sound of footsteps on the boat deck. Early on, they impressed her as being those of people in no hurry at all, but later they suggested a chaotic milling of people, worried and maybe even fearful as they began to recognize the inevitable would happen. Near the end, however, officers loading the lifeboats were armed with pistols to make sure that no male passenger tried to board lifeboats already crammed with women and children.

In the end, only 706 persons found places in the boats, (six of whom died of exposure before being rescued), leaving 472 unused spaces in the boats. In part, this dreadful discrepancy arose during that time of disbelief when the officers in charge struggled to find those willing to risk entering the lifeboats. Some of it may be charged to the alleged fear of at least one officer, Second Officer Charles Lightoller, that the boats were not strong enough to support their rated loads while suspended in air. He was questioned about this point at both the Smith and Mersey hearings, and managed to deflect serious criticism. A damning point, however, is that the lifeboats had been found safe for their rated capacity by a test at Harland & Wolff, and Lightoller should have known that. We have further discussion of this point in Chapter 27.

Distress rockets were an expedient that should alert any ship within visual range; the first was fired at 0045 by Fourth Officer Boxhall and Quartermaster Rowe from the starboard side of Boat Deck. In all, eight rockets were fired at five-minute intervals. These rockets were standard distress signals that produced a brilliant white explosion at about 800 feet above the sea. Since the rockets carried no message other than "distress," Boxhall sent the message "Come quick, we are sinking" by means of a signal lamp. Testimony before Senator Smith was to reveal that the rockets were seen by the *Californian*, but that their import was not recognized, nor were the coded light flashes from the signal lamp seen. Captain Smith ordered the evacuation, and it was he that proclaimed the rule of women and children first, but overall he seems to have left management of the evacuation in the hands of his officers. In the later stages of the evacuation he retreated to the Bridge to contemplate the unspoken demand that he go down with the ship. Or did he want to retreat from the unwilling company he would have after all boats were gone? He was a respected mariner, a favorite of passengers who traveled regularly on White Star ships, and a favorite of his employer—he was to become commodore upon his retirement at the end of this maiden voyage—but none of that outweighed his inability to handle the ultimate responsibility that now faced him—possible death in that frigid water.

Wireless calls for help began at 0015, but those signals never reached the ship—presumably the *Californian,* whose lights may have been visible from the sinking *Titanic*. Or were they the lights and masts of the *Sampson*, who had been illegally catching seals in the vicinity of the tragedy in hours preceding the accident? That question was raised in 1912, but could never be answered.

Near the end of the first hour following the collision, the captain must have become aware that rescue could not come in time. The ships that acknowledged receipt were all too far away, and that one that appeared so near made no reply nor showed any indication of approaching. After that apparent realization, Captain Smith kept to the wheelhouse, save for an occasional appearance at the door to check on the progress of the evacuation. Instead of being the exemplar of strength and comfort that might have been expected, he appeared unwilling to face either crew or passengers. When a woman from steerage, confused and uncertain, asked him what she should do, he had no answer. He turned and walked away.

The leader of the ship's musicians, Wallace Hartley, seems to have handled the situation better than his captain, and his eight musicians stuck with him in good spirit so far as we know. It is believed, though without confirming testimony, that the entire group (who usually split up to play for the several dining rooms) assembled in the first-class Lounge around 0015 to play selections from the White Star song book. The book included hymns, ragtime, waltzes, and selections from operettas. Judicious selections might well have a favorable effect on the moods of passengers: diverting their thoughts from the troubles that confronted them, calming them in the face of danger, or offering solace as the inevitable approached. Whether this was a command performance or Hartley's inspired contribution is not known.

The advancing flood caused Hartley to move his performers several times. From the Lounge they were forced to retreat to the Boat Deck foyer in the Grand Staircase. They later moved to the deck outside. Flooding in this area around 0210, about the time the collapsible boats were being chivvied off their chocks, forced everyone in the vicinity to retreat aft, and that event was probably the end of the music. Survivors who believed they had heard the final selection differ as to whether it was "Nearer My God to Thee," or the lesser-known hymn "Autumn." Hartley is known to have said in an interview before his last voyage that he would choose the former for his own funeral. We may regard this comment as strong evidence concerning what he would order when he could see no succor, no hope, but only the solace that his favored hymn might convey.

Captain Smith and Hartley's musicians, each apparently moved by the imperatives of duty and the unspoken demands for self-sacrifice, did not participate more than peripherally in the evacuation. Other professionals, moved by the same imperatives, distanced themselves from it entirely, even though contributing in important ways to the success of the evacuation. Chief Engineer Bell and the several hundred of his department who died with him belong in this category. So do the deck officers who directed the evacuation, but stayed with the ship. So does Thomas Andrews, who did all he could to get passengers to the boats, but never approached a boat himself.

The passengers could be characterized in a general way as a herd of sheep: some were herded into boats, and lived; others were barred from the boats, and died. When their experiences are examined under sharper focus, however, stories of self-sacrifice also emerge: the decision by Ida Strauss to refuse a place in a boat to die with her husband, the decision of Ben Guggenheim to dress in formal attire and "die like a gentleman". . . these stories are such well-known elements of the *Titanic* story that we shall not repeat them here. There must have been many other victims of the tragedy that behaved with self-sacrificial intent, but who did not attract the notice of a survivor who could tell of it.

Appendix 4 contains a chronology of the boat launchings, with times, number of persons taken, and remarks on incidents of interest associated with particular boats.

Notes for Chapter 14

1. The American equivalent to the Board of Trade was the federal Steamboat Inspection Service, long since merged into the U. S. Coast Guard. It, too, had no rule requiring boat space for every person aboard.
2. Collapsible lifeboats don't sound like something to brave the North Atlantic in, but they continued to be acceptable under the stringent lifeboat rules put into effect post-*Titanic*. The term comes from construction that allowed the upper level of their sides to be collapsed to reduce the stowage space required.
3. The more technical aspects of the *Titanic*'s lifeboat outfit are discussed in Chapter 26.
4. Lifebelts may themselves be hazards if not worn properly, and hence latter-day boat drills include instruction in this matter. Not having had a boat drill, many passengers may not even been aware of this equipment—unless, of course, it took up too much room in those lockers.

CHAPTER 15

Damage to the Hull

The *Titanic* ground her hull against the flank of an iceberg in a collision that was sensed in the first-class passenger quarters as a faint jarring, but one that more ominously sounded like thunder in Boiler Room No. 6. Regardless of how the humans aboard felt or heard it, the consequence of this encounter was an immediate inpouring of the sea into the first five watertight compartments. In those five damaged compartments, the flooding was at a rate beyond the capacity of pumps to counteract. A sixth compartment (Boiler Room No. 5) was flooding at a rate that the bilge pumping system was able to control. A seventh compartment (Boiler Room No. 4) experienced a minor seepage.

As the ship's designer Thomas Andrews knew, and 2,227 other souls were soon to discover, five forward compartments that filled to the level of the sea would eventually send the ship to the bottom.

There was little chance for the crew or the eight men from Harland & Wolff to inspect the damage before the ship sank since the parted riveted seams and torn plates (if there were any) were submerged long before anyone could inspect them from the inside of the ship. The only post-accident accounting that could be given in 1912 was provided by naval architect Edward Wilding of Harland & Wolff at the British Inquiry. He testified that the damage extended over a length of about 300 feet and had an area of inflow of 12 square feet. However, the extent of that damage was not known until 1996 when Paul Matthias made a sonar survey of the hull submerged in sediment. In Boiler Rooms Nos. 5 and 6, the impact of ice occurred almost under the eyes of firemen at their labors. One of them, Leading Fireman Frederick Barrett, who survived, gave testimony that suggested failure of riveted seams rather than

This is the "Big Piece" that is evidence of the break up of the *Titanic*. It is a section of the ship just forward of the second expansion joint on the starboard side between "B" and "D" Decks and yielded much information on the quality of the steel plate and rivets used in the construction of the *Titanic*. This 22-ton section of the ship was raised in August 1998 as part of an expedition sponsored by RMS *Titanic*, Inc. and the Discovery Channel. (*Courtesy of RMS Titanic, Inc.*)

holes being punched or slashed through the three-quarter to one-inch hull plates. There was no sign of ice floating on the water within Boiler Room No. 6. The observations of Barrett were prescient, indeed, but the "classical" belief held that the damage was a continuous gash, formed by the slicing action of a sharp edge of the iceberg. This belief influenced Joseph Conrad, in his comments on the Mersey hearings, to compare the *Titanic*'s hull to a biscuit tin that had been ripped open. Although not a verbatim quotation, his words were soon being repeated as "sliced open like a biscuit tin," words eagerly seized upon by critics of the *Titanic*'s design. The concept of a gash may not have had its origin in Conrad's intemperate analysis (quoted in Chapter 19), but his writings surely popularized the concept. Whatever its origin, the classical "gash" view prevailed until 1993.

The area of the putative gash was fairly well estimated soon after the sinking by Edward Wilding of Harland & Wolff. He analyzed survivor testimony concerning depths of flooding at certain times at particular places, and then calculated from that and the early rate of sinking how fast the water was entering. At the British Inquiry and later during the liability hearings held in New York, 1915, he testified that he had calculated the total area of the opening to be only 12 square feet, an area no more than two city sidewalk squares or that of an ordinary closet door. However, he did not intend his statement to imply a single gash of a uniform 0.5-inch width over 300 feet, since he knew that the six compartments holed had flooded at distinctly different rates. Wilding's conclusion was reinforced eighty some years later in an analysis aided by computer simulation of the sinking, made by naval architects John Bedford[1] and Chris Hackett of Harland & Wolff in 1995–96. They treated the openings as orifices with a flow coefficient of 0.485, and in so doing, found the area that would account for estimates of flooding was an average of 12.6 square feet.

That the area of initial opening was far from being uniformly distributed was attested by surviving stokers who witnessed the initial flooding in the affected boiler rooms. In Boiler Room No. 6, the sea poured in at a rate far beyond the capacity of the bilge pumping system to counter. In Boiler Room No. 5, water entered at a much slower rate into an empty forward starboard coal bunker. Leading Fireman Barrett likened the inflow to the outpouring from a three-inch fire hose. Although the door giving access into this bunker from the boiler room was not watertight, closing it slowed the inflow to the point where the ship's bilge pumping system could keep the water level below the floor plates. Even after some minutes of confusion over whether Boiler Room No. 5 should be evacuated immediately, Barrett testified that he and other firemen were recruited to enter that space and draw the fires in its furnaces. Boiler Room No. 4 experienced a much slower rate of flooding through leaking steel rivets, a rate low enough for easy management by the bilge pumping system. Subsequent study of the construction plans has shown that the rivets in the area of Boiler Room No. 4 were steel and not wrought iron as were those from Boiler Room No. 6 and forward. Furthermore, the longitudinal seams in Boiler Room No. 4 were joined by three rows of 1.125-inch rivets, whereas they were double riveted forward of Boiler Room No. 5. The joint efficiency for the triple riveted seam was substantially higher than that of the double riveted row. For the plate and rivet sizes used in that portion of the *Titanic*'s hull, the efficiency of a double row of wrought iron rivets, spaced up to 6 inches apart, is estimated at no more than 27 percent, while a triple row of steel rivets at the same spacing would have an efficiency of almost 53 percent. These efficiency values compare the shear ultimate strength of the rivets with the tensile ultimate strength of the plates, taking into account the loss of plate area due to the presence of the rivet holes. The rivet efficiency increases slightly over time due to rusting in the mating riveted plates, which augments the rivets' resistance with frictional forces between the plate surfaces. However, the *Titanic* was new and it is likely that there was very little to no rust buildup.

Trimmer Thomas Patrick Dillon[2] went to Boiler Room No. 4 from the Reciprocating Engine Room shortly after the collision. His duty was to assist the engineers to open the watertight doors so that they could attend the valves and pumps in Boiler Rooms Nos. 1–4. He testified at the British Inquiry that he saw water coming into Boiler Room No. 4 at its forward end and beneath the floor plates there. He noted that it was coming in at a low rate, and that the only indication of flooding there was that the surface of the floor plates was damp.

Fireman George Cavell[3] was stationed in the forward coal bunker of Boiler Room No. 4. When the ship hit the iceberg, coal slid down and partially buried him. However, he was able to work himself free and escape from the bunker. As he entered the forward stokehold of the boiler room, the lights went out. His superior ordered him to bring flashlights from the Reciprocating Engine Room. Because watertight doors had been closed, he perforce had to climb to Scotland Run on "E" Deck to reach his destination. The lights were on in the passage and a number of third-class passengers were moving aft to escape the flooding forward. It was from them that he learned that the ship had struck an iceberg.

Cavell found the flashlights and returned to Boiler Room No. 4, where the lights were back on again. Orders were given to draw the fires in that room. As the men began that task, water gradually started to seep above the floor plates. By the time the men had cleaned the coal from the furnaces, a foot of water had accumulated above the stokehold plates. Cavell left the space for the relative safety of Scotland Run, but decided to go back. Halfway down the escape ladder, he decided it was time to go to the Boat Deck.

The difference in the testimony of Cavell and Dillon is difficult to reconcile unless Dillon was not involved in the extinguishing of the fires in the furnaces in Boiler Room No. 4. However, the flooding there was indeed gradual.

The uncontrollable flooding of Boiler Room No. 6 and the four compartments forward of it made Thomas Andrews realize that the ship was going to sink. However, if flooding in the other compartments could be stalled until the water from Boiler Room No. 6 found its way aft, the inevitable foundering would be delayed. Andrews could not have known as he made his analysis how near or how far potential rescue ships might be, but delay was obviously worth every effort that would extend the time these distant rescue ships would need to reach the *Titanic*. He spoke of this to Captain Smith, who passed the advice to Chief Engineer Bell, who then put his engineers to do the work of containment.

Initially, the inflow of water approached 400 tons per minute. This rate slowed to about 50 tons per minute about 45 minutes later when incoming water had to find its way aft along passageways, down stairwells, and through nontight openings for cables, pipes, and ventilation trunks. It also found its way up through submerged plumbing drains with no check valves that were fitted above the margin line. However, when the structure of the non-tight bunker bulkhead in the forward starboard corner of Boiler Room No. 5 failed or the nontight coal bunker doors were forced open by the rising hydrostatic pressure, the rate increased dramatically. By 0200 it had reached a rate of around 700 tons per minute.

Able Seaman John Poingdestre[4] was attending the lifeboats on the Boat Deck when he decided to go to his quarters near No. 2 Hatch on the port side of "E" Deck. The time was approximately 0030. While he was there, a wooden joiner bulkhead, separating the third-class and crew quarters collapsed under hydrostatic pressure. The water rushed into the Seamen's quarters. The source of this water was almost certainly one or all of the several lavatories located in this part of the ship. The drains for these facilities discharged overboard above the margin line, and as appropriate for that location, were not fitted with check valves to prevent backflow. At the time, the water level was well above "E" Deck in this vicinity (near Hatch No. 2). Poingdestre fled, and survived to testify, but by the time he could reach the stairs to "D" Deck, the water had risen to his waist level.

Further insight into hull damage began with visits to the wreck that followed Dr. Robert Ballard's pioneering exploration in 1985. The principal discovery of his first visit was that the ship had broken in two. The 1998 *Titanic* Expedition determined that the hull broke into four sections during its plunge to the bottom. The bow portion had plowed into the ocean-bed sediment to such an extent that the area of ice damage was covered, but in September 1993 Arnie Geller of RMS *Titanic*, Inc. told author William Garzke that a portion of a failed riveted seam in way of Boiler Rooms Nos. 5 and 6 was exposed. Failed rivets and a parted seam would give much credence to the testimony of Leading Fireman Frederick Barrett to the American and British Inquiries that he had seen water flowing in from a seam about two feet below the level of the floor plates in Boiler Room No. 6. In August 1998, author William Garzke was at a television monitor tracking the pictures being taken by the Remote Operating Vehicle *Magellan 725* that was positioned along the starboard shell plating that covered Boiler Room No. 5. Suddenly, he noticed a shadow where one should not be. *Magellan 725* was brought closer, allowing viewers to determine that the cause was a shell seam that was opened only a fraction of an inch with some rivets missing in the upper row of the double riveted seam. That opening is just where Leading Fireman Barrett had testified it would be at both the American and British Inquiries—two feet above the floor plates in Boiler Room No. 6.

In a general way, the bow portion has a humped appearance consistent with its coming to an abrupt halt as it encountered the seabed. Within the mounds of sediment displaced by the encounter, and now piled against the hull, there is evidence that the bilge keel on the starboard side was ripped away. The bow is slightly bent to starboard, also evidence of a terminal plowing into the seabed. On the port side, near the transverse bulkhead between Boiler Rooms Nos. 3 and 4, there is a 10-foot bend in the hull plate from a few feet above the mudline and below. Damage of these kinds, almost certainly occurring in situ rather than at the surface, may mask some of the damage caused by the iceberg, and hence becomes a factor to be accounted for in any attempt to analyze the ice damage.

Was the flooding aided by brittle fracture? Charpy V-notch tests done on samples of the hull steel in 1992 at CANMET in Ottawa, Canada and the University of Missouri, Rolla in 1996 indicated that the steel at ice temperatures was more brittle than at higher temperatures. Separate tests were made at the National Institute of Standards and Technology in Gaithersburg, Maryland in 1998 under the auspices of Dr. Timothy Foecke. Taking a piece of steel from a large piece of the ship's shell plating that came from an area between "B" and "D" Decks and a few feet away from the second expansion joint on the starboard side (referred to as the "Big Piece"), Dr. Foecke put the specimen under a slow bending process at ice temperatures. The steel bent and did not exhibit any tendency for brittle fracture. This situation confirms the behavior of some plates on the bow wreck, such as the one cited previously where a fold in the plating along a 10-foot length took place without any indication of brittle failure. The steel's fracture mode depends on the temperature and the speed of the crack. For this material, when the crack was growing slowly, it was quite ductile. When the crack moved at very high speeds (like when the ship broke in two), it probably was mostly brittle. During the breakup, when things were happening very fast in an uncontrolled fashion, there most likely was a very complex combination of ductile and brittle behavior.

In preparation for the 1996 *Titanic* expedition, that of IFREMER-RMS *Titanic*, Inc., the submersible *Nautile* was modified under the direction of Paul Matthias to carry a side-scan sonar profiler. This device was able to penetrate the sediment to produce images of the damaged area. The resulting sonar scan showed that the iceberg damage was only 110 feet in length but it was spread over a length of ship that was not survivable by the *Titanic*. The flooding openings were concentrated in six small slits, later traced to hull plate seams, whose total area was between 12 and 13 square feet. Naval Architect Edward Wilding's estimates made seven decades earlier, and from quite indirect evidence, were thereby substantiated. John Bedford wrote author William Garzke that he was pleased with the findings of Mr. Matthias because it confirmed his and Wilding's analyses for the initial flooding opening.

Specifics of Damage Shown by Sonar. The ship made contact with the ice for about 11 seconds that produced six distinct openings in the hull.[5] All these openings follow riveted seams.

The profiler revealed comparatively little damage in the Forepeak area; that which did occur was just below the tank top near the top of a riveted seam. Compartments above this tank remained dry for more than an hour after the collision. Because the hull was narrow in this extreme forward area, and because the ship was turning to port, contact must have been a glancing brush against the ice. However, the hull widens rapidly aft of the Forepeak Tank, making subsequent blows more direct and more severe during the iceberg and ship encounter. Two separate areas of damage, 4 and 5 feet in length, appear along what appears to be a riveted seam, just below the waterline, in way of Cargo Hold No. 1. After this initial encounter, pieces of the iceberg are likely to have been sheared away by these two contacts; a supposition based on the much lower damage caused by the next contact.

The fourth opening appears in way of the transverse bulkhead between Cargo Holds 1 and 2 at a level of from 10 to 14 feet above the keel. The damage covers an area about 15 feet long at the midspan of a plate. There was also damage to the inboard bulkhead of the access to the Firemen's Passage, 3.5 feet inboard from the outer skin of the ship. It seems unlikely that the ice could penetrate that distance into the hull; it seems more likely that impact forces were transmitted from the shell through transverse structure to an inner longitudinal bulkhead that bounded the Firemen's Passage. Rivets failed in that bulkhead and a seam was opened slightly. Stoker Charles Hendrickson testified that water entered the bottom of the passage through this bulkhead. Edward Wilding believed otherwise; he testified that he believed that ice had penetrated the steel and caused this flooding. However, we do not subscribe to that theory. The thickness of steel plate and the frailty of the iceberg makes it unlikely that ice punched (or sliced) holes through the steel. Given the relative properties of ice, hull steel, and the wrought iron rivets, the most likely failure at the shell would be that of the brittle wrought iron rivets. Failure of the rivets in turn allowed the seams to open, but the damage 3.5 feet inboard must have come from forces transmitted through intervening structure.

There was a slight recoil from the contact in the vicinity of Cargo Holds Nos. 1 and 2. The ship and ice may have been separated for a moment, but the ship's forward motion brought the widening portion of the hull back into contact for a far more serious blow at a depth of approximately 13 feet above the keel. A principal point of impact was along Cargo Hold No. 3, a space whose rapid flooding had led Edward Wilding, and later Bedford and Hackett, to label it as the locale of most serious damage. (Their conclusions were based largely on testimony indicating more rapid flooding here than elsewhere.) The sonar images show a damaged area of about 33 feet in length. Plate samples retrieved from the seabed have missing rivets, and rivets missing their heads. One plate shows a bearing-type failure where a rivet had been pulled through the plate edge. Such failures were also seen in the "Big Piece" of hull that was recovered from the seabed in August 1998.

Momentum exchange between ice and hull must have reduced ship speed by the time Cargo Hold No.3 had passed by the iceberg, but subsequent contact in way of Boiler Rooms Nos. 5 and 6 caused further serious damage. The ship may have rebounded slightly from the contact at Cargo Hold No. 3, but made hard contact again in way of those first two boiler rooms at a distance of about 23 feet below the waterline. The sonar images show a 45-foot length of scraping contact and damage there. Its most serious aspect is the parting of a riveted seam along Boiler Room No. 6 and into the forward corner of the coal bunker in Boiler Room No. 5.

Leading Fireman Barrett looked into the coal bunker for Boiler Room No. 5. He maintained that the damage extended for about 2 feet aft within the forward bulkhead of the bunker, just aft of the main transverse bulkhead. Its area is about 0.4 square inches, allowing ingress of water that witnesses described as similar in magnitude to water flowing from a fire hose. The extent of damage to these two boiler rooms would determine whether the ship would sink or survive. Had there been no flooding of Boiler Room No. 6, the ship would have remained afloat, for the *Olympic*-class ships were designed to survive flooding of the first four compartments.

The hull in way of Boiler Room No. 4 was lightly damaged by a brief parting contact with the ice. As the ship pivoted around the iceberg, the contact was made in the region in way of this boiler room where mild steel rivets were employed. These rivets are believed to have offered greater resistance to the forces of impact with the iceberg than did the wrought iron rivets in the area of collision from the Forepeak to Boiler Room No 5. However, the damage in the area of Boiler Room No. 4 was of near-trivial magnitude, since the bilge pumping system was able to keep the water level below the floor plates until hydrostatic pressure created by the plunging bow increased the flow, and later progressive flooding from forward made its way into this space.

There is also a large dent in the hull near the transverse bulkhead separating Boiler Rooms Nos. 5 and 6. The dent appears to be damage sustained later as the hull decelerated against the seabed.

Testimony by surviving stokers that water poured into Boiler Room No. 6 from a seam about two feet above the stokehold floor plates (about 10.5 feet above the keel) is consistent with the sonar images—and an observation by author William Garzke during the 1998 RMS *Titanic*, Inc./IFREMER Expedition. Ten minutes after flooding began, Leading Fireman Frederick Barrett and Engineer Jonathan Shepherd returned to this space to find water at a level about 8 feet above the inner bottom. This level at that time would require a flooding rate of about 200 tons per minute, a rate consistent with the forensic observations of the damage, and with the 2.85 square foot flooding area estimated for that room by Bedford and Hackett.[6]

The foregoing descriptions have the hull damage occurring in six distinct areas. In each of those six, the damage appears to have occurred along a riveted seam. Although one might expect brittle fracture of plating, especially since the hull was in contact with very cold water, the sonar imaging shows what appears to be separation of riveted seams. This is the case even in the area of the most severe damage, that in the vicinity of Cargo Hold No.3. Further inspection of the shell expansion plan shows a riveted seam at the point where Leading Fireman Frederick Barrett, on duty in Boiler Room No. 6, spoke of seeing the initial inrush of water.

The flooding through the damaged area in way of Cargo Hold No.3

was very rapid, as evidenced by the rapid rise of water in this area. For example, the Racquet Court on "G" (Lower) Deck was half filled with water by 0010. The Mail Room on the same deck was filled to within two feet below the level of "G" Deck by the time Fourth Officer Boxhall made his second inspection a few minutes before midnight. The two compartments forward of Cargo Hold No. 3 probably received some overflow from this compartment before being filled by flooding from their own lesser damage.

Several survivors, for example Edith Russell and Edith Haisman, testified that they felt the ice encounter as a series of jolts, and they remembered that the third one was the strongest. We can readily surmise that the strongest was the blow that opened the hull in way of Cargo Hold No. 3 or Boiler Room No. 6.

Discoveries of the late 1990s laid the gash supposition to rest, and turned the forensic researchers' interest toward the contribution to the sinking to the metallurgical qualities of the plate and rivets. Such matters are also discussed in Chapter 16.

Another important discovery was made in the forensic examination of the "Big Piece." As remarked earlier, the drain pipes from lavatories and toilets that discharged overboard above the margin line were not equipped with check valves. These drains, therefore, became routes for flooding when the level of water covered the exit pipe. Also, curious passengers or the crew had opened portholes above "E" Deck to catch a glimpse of the iceberg, but did not close them. In these ways, the area for flooding increased even before the bow of the ship was submerged.

Events in Boiler Rooms Nos. 5 and 6.[7] As the *Titanic* was speeding through the night, there were eight firemen and four coal trimmers hard at work in Boiler Room No. 6. The engineer in charge when the accident happened was Junior Second Assistant Engineer Jonathan Shepherd, who happened to be in Boiler Room No. 5 at the time. However, in the moments before the encounter, Leading Fireman Frederick Barrett was talking to Second Engineer James Hesketh on the starboard side of Stokehold No. 10 (i.e., the after-firing aisle in Boiler Room No. 6) near the station where an emergency was first signaled by the clock-like mechanism that indicated the ship's ordered speed range by colored lights: a white light indicated full speed, a blue light indicated low speed, and a red light indicated stop. The light had been white since the ship departed Queenstown. With no hint of what was to occur, the light flashed to red. In a scene that was reproduced in the 1997 movie, *Titanic*, Barrett and Hesketh called out in unison, "Shut all dampers!" This act choked the flow of air to the boiler furnaces.[8] Before the dampers could be shut, there was a crash or a roar like thunder. Suddenly water came pouring into Boiler Room No. 6 two feet above the stokehold floor plates, and the ship's side was opened intermittently from the third stokehold (Stokehold No. 9, the forward-most firing aisle of Boiler Room No. 5) to a point 20 feet from the stem. The water poured very quickly into the starboard side of Stokehold No. 10 about two feet from where Barrett and Hesketh were standing, and at a point about two feet above the floor plates, which were six feet above the inner bottom, or eleven feet above the keel.

Aware of the danger and hearing the bell that warned of the closing of the watertight doors, Hesketh and Barrett ran to a passage under the coal bunker where the watertight door was located. They jumped through the door opening into Boiler Room No. 5 as the massive watertight door was descending. Once into Boiler Room No. 5, Barrett looked into the open empty starboard coal bunker at the forward end of that space. Looking into the door opening, he noticed a tear in the side shell about two feet above the floor plate on which he was standing. The tear also extended longitudinally only two feet into the bunker along the side shell. He also noticed that there was no damage to the main transverse bulkhead. He estimated that the flow was at about the rate to be expected from an ordinary fire hose. However, it was not flowing out of the bunker into the boiler room nor was it flooding at the rate he had experienced forward. Barrett lowered the nonwatertight bunker door to its closed position.

Meanwhile, Fireman George Beauchamp was part of the crew working the furnaces in Stokehold No. 10 when the collision occurred. He noticed the red light that signaled STOP and heard Barrett's and Hesketh's orders to shut the dampers. Soon after, the watertight door behind him came down. Water was already flooding into the boiler room from the side of the ship and through the open coal bunker. Within seconds, an order was given to draw the fires from the furnaces. Beauchamp and others moved to comply, but when the water began to approach waist level, the incomplete drawing process had to be abandoned. Everyone in the room fled up the ladders.

Once into Boiler Room No. 5, Barrett consulted with Engineers Shepherd and Hesketh about the tear in the shell and the consequent flooding. Barrett and Hesketh were already quite aware of dangerous flooding in Boiler Room No. 6. In Boiler Room No. 5 the three men observed the lesser flooding—comparable to a fire hose, as Barrett later testified—entering through the room's forward coal bunker. Since Barrett's duty station was in the forward boiler room, he and Engineer Shepherd left to make an inspection of that room. Hesketh, convinced that no immediate danger threatened the crew or Boiler Room No. 5, then shouted out "all hands stand by your stations," an order for them to remain at their posts in the firing aisles. Meanwhile Shepherd and Barrett began their return to Boiler Room No. 6 via ladder from Boiler Room No.5 and the "E" Deck passage. However, they could not return to the working level in that space because there was 8 feet of water above the floor plates. This covered the alleyways that led to the watertight doors at either end of the boiler room. Fortunately, they thought, the watertight doors at both ends were closed. Shepherd and Barrett then returned to Boiler Room No. 5, retracing their steps on the access ladders. Junior Second Assistant Engineer Herbert Harvey and Senior Second Assistant Engineer Bertie Wilson met them at the foot of the ladder to ask what they had found in the room just forward. These two engineers had been tending the pumps in Boiler Room No. 5, but with the forward bunker doors shut no trace of water had yet appeared in the boiler room itself. The ship had stopped so that the water was not flooding into the coal bunker as rapidly as it had been when Barrett first entered this space.

The telephone rang; the caller in the Reciprocating Engine Room told Engineer Harvey to send all the Boiler Room No. 5 stokers above to the "E" Deck passage (Scotland Run). Barrett, however, was asked to remain. If Harvey needed anything, then Barrett could go for it. The only access was up the emergency ladders as both watertight doors were closed. At this time, only Barrett, Harvey, Wilson, and Shepherd were left in Boiler Room No. 5. The main transverse watertight bulkhead between Boiler Rooms Nos. 5 and 6 was holding as was the nontight coal bunker bulkhead at the forward end of Boiler Room No. 5. Just after the last stoker left, the lights in this room went out. Mr. Harvey sent Barrett for flashlights. Barrett went to the top of the escape ladder to "E" Deck, where the lights were still burning, and sent two firemen he found in Scotland Run to the Reciprocating Engine Room for the flashlights. One of these men was Leading Fireman Charles Hendrickson who returned with another man with an armful of flashlights. Hendrickson took one and went to check Boiler Room No. 6. He found the space dark and flooded. He then went below to Boiler Room No. 5, where Engineer Shepherd asked if he had the flashlights. Receiving confirmation, Shepherd instructed Hendrickson to place a lighted flashlight at each water gage of the boilers. After Hendrickson completed that task, he was asked to draw the fires.[9] With 30 furnaces in the room, this was going to be an overly long job for one man. Realizing this, Engineer Harvey told Hendrickson to find men to help. The latter then left for the Firemen's quarters on this mission. He would not return.

Meanwhile, Barrett, who had been on "E" Deck, went below to find that the lights in Boiler Room No. 5 were on, though they had been out for ten minutes. Barrett noted that the space was still clear of water. He went to the boilers to check the water gages and found no water showing in them. (This was the consequence of feed valves being shut and safety valves lifted to reduce pressure.) After Barrett had completed his inspection, and with men still unavailable to draw fires, he was asked to find men and put them to this work. He went back up to the "E" Deck passage, and there enlisted the help of 15 to 20 firemen standing around there. With the 30 furnaces requiring this process, the drawing required about 20 minutes. The trim of the ship became noticeable during this time, making the task difficult for those men taking the coals from the furnaces that faced aft.

When the drawing was completed, Barrett sent them back up to "E" Deck. The Boiler Room was now filled with steam from the quenching of the smoldering coal that had been raked onto the floor plates. The ventilation fans had been shut down and the draft through the furnaces blocked by closure of the dampers and furnace doors so that the steam did not dissipate. It hung there like a fog, making seeing and breathing increasingly difficult.

Around 0045, Engineer Harvey asked Barrett to lift a manhole cover from the floor plates so that he could have access to valves controlling the bilge pump located on the starboard side of Boiler Room No. 5. With steam clouding the room's atmosphere, Engineer Shepherd did not see the open hole as he hurried across the boiler room a short time later. He fell down into the hole, breaking his leg. Hearing his cry for help, Barrett and Harvey lifted him from the hole and carried him into the ash ejector room that was located on the port side in the alleyway to Boiler Room No. 4. Harvey and Barrett attended Shepherd there for about 15 minutes, during a time when there was still no sign of water in Boiler Room No. 5 above the floor plates. Nonetheless, flooding was obviously continuing in the compartments farther forward, since the ship's trim by the bow was becoming worse by the minute.

Despite problems with the trim, the engineers and Barrett were so busy that they had no time to consider the problems that the trim was exacerbating. But suddenly a green foaming rush of water came through the forward end of a pass, a space between the boilers giving access between firing aisles. The time was just before 0110. Engineer Harvey told Barrett to go up the ladder at once; he would go to get the injured Shepherd.

However, the flooding of the boiler room was so fast that Harvey, burdened with his crippled comrade, was unable to escape. Both men were lost.

Barrett believed that the flow of water came when the coal bunker doors failed, or when the bunker bulkhead plating itself failed under hydrostatic pressure. However, Barrett was so intent in seeking safety that he could not pause on the emergency ladder to confirm what was happening.

The only unknown aspect of this tragedy is just how Boiler Room No. 5 flooded. Barrett had lowered the starboard non-watertight coal bunker door when he entered the room from Boiler Room No. 6. The port bunker door must have also been shut, otherwise the boiler room would have had traces of water on the floor plates. In any case, Boiler Room No. 5 was free of water until 0110. By that time, the bow of the ship was deep in the water as evidenced by Barrett's description of the bow trim's effects in the boiler room. Meanwhile, water had accumulated in the forward coal bunker (empty of coal) until its hydrostatic force probably caused the structural failure in the bulkhead structure or the coal bunker door(s), allowing the flooding that Barrett described as a rush. Our surmise is that the failure was that of a bunker door rather that of the bulkhead itself.

Transverse Bulkhead between Boiler Rooms Nos. 5 and 6. There had been a coal fire in the aft starboard bunker of Boiler Room No. 6, beginning before the ship reached Southampton from Liverpool. For several days thereafter, continuing into the maiden voyage, coal was shoveled out of the bunker in the effort to extinguish the fire. As a precautionary measure, the coal in the adjoining bunker in Boiler Room No. 5 was also removed. Leading Firemen Frederick Barrett and Charles Hendrickson were the only surviving crew who took part in the effort to extinguish it. By Barrett's accounting, Chief Engineer Bell had given him the order for coal removal, accompanying it by the remark, "The Builder's men want to inspect that bulkhead."

Frederick Barrett saw the transverse watertight bulkhead from Boiler Room No. 6 on Saturday, 13 April. He thought that the bulkhead was damaged at the bottom. Some of the bulkhead plating at the bottom of the watertight compartment was bent aft while another part was bent forward. Hendrickson also had a similar observation. In fact, he noted that the paint had been consumed by the coal fire, exposing bare steel. He applied oil to protect the exposed steel.

Water may have seeped through the riveted joint connecting the inner bottom plating to the bulkhead between Boiler Rooms Nos. 5 and 6.

A distortion in the plating and loosening of rivets, caused by the bunker fire, may have allowed the seepage. Barrett's evidence rules out any major damage to this bulkhead at its intersection with the shell. The hull lies above the obscuring sediments in this vicinity, and could be given a very close examination. Such a survey would be of profit for far more than just this one spot, since its findings could be extrapolated to the other points of failure. However, available resources have not allowed this interesting step to be taken at the time of this writing. Absent this examination, there does not seem to be enough evidence that the coal fire that burned in the *Titanic* for several days was a factor in her loss.

The Mersey Report, Reprised. The Mersey Report of 1912 may well be called a document above criticism from its contemporaries, but the information gleaned from the marine forensic investigations of the *Titanic* wreck site in 1996 and 1998 would surely cause an updating to be somewhat different. We offer, following, the parts of the report covering flooding and damage as it might be rewritten, though with the language of the report preserved as far as possible. The particular sources for the revisions are noted.

Iceberg Damage. In assessing any aspect of the flooding in the *Titanic* or her sinking, one must define the damage from the collision that initiated the process. Table 15–1 identifies the major considerations that

TABLE 15–1 *Length of Damage, Frame Spacing, Type of Rivets, and Number of Rivets in Longitudinal Seams*

Space	Length of Space (Feet)	Frame Spacing (Inches)	Length of Frame Spacing	Length of Damage (Feet)	Rivet Size (Inches)	Rivets in Row
Forepeak	46	24	46	Trace	—	—
Cargo Hold No. 1	46.25	27	33.75	5	—	—
	46.25	30	12.50	6	—	—
Cargo Hold No. 2	50.50	30	17.5	15	—	—
	50.50	33	33.00	—	1⅛ Dia.	2
Cargo Hold No. 3 and Reserve Coal Bunker	51.00	36	51.00	33	1⅛ Dia.	2
Boiler Room No. 6	54.00	36	54.00	40	1⅛ Dia.	2
Boiler Room No. 5	57.00	36	57.00	2	1⅛ Dia.	3

the marine forensics investigation evaluated, such as the length of damage, the frame spacing, the type of rivets, and rows of rivets in longitudinal riveted joints. An unknown factor in the data presented in Table 15–1 is the amount of rust accumulation between the plates. Rust will act to increase the friction between plates.

The Friction between the Plates. An important consideration in the strength of riveted connections is the frictional resistance in the joint that depends upon the tightness of the parts that are being joined and the roughness of the faying surfaces. When this resistance is not overcome, the rivets are not overstrained and the joint behaves like a solid plate. The frictional resistance of the joint should be such that it will not be overcome by the working stresses in the vessel. The rivets will not be subjected to any undue stress. Surface scale and the gradual formation of rust between the faying surfaces of the riveted joint will tend to improve the frictional resistance as the ship ages.[10] Consequently, the rivets will not be subjected to any undue stress under normal working conditions. The exact mechanism by which a riveted joint transmits loads when it is stressed to the point of failure is less well known. Gayhart,[11] who instrumented a number of sample riveted joints under loads sufficient to cause failure, noted that frictional slip occurred early in each test, before any other kind of failure. He did not report any significant increase in the breaking strength of the joints due to friction. However, it must be kept in mind that his tests were conducted under laboratory conditions, with specimens made especially for his experiments.

In the case of the *Titanic*, she was new and the amount of rust was probably miniscule. After the small frictional resistance was overcome, the rivets tended to fail in shear and plates will break across a line of rivet holes. There was much evidence of the latter in the area of the hull fracture at the end of the bow portion of the wreck and in the "Big Piece" brought up in 1998.

All the shell plating from the keel plate to the margin plate (turn of the bilge) in way of the double bottom was hydraulically riveted. These rivets were mild steel. The rivet holes had the ordinary countersunk configuration in accordance with Lloyds Rules and had full points 0.25 inches above the plates.

The configuration of the holes above the margin plate is not as easy to determine. Photographs of the ship clearly show the smooth appearance characteristic of rivets in countersunk holes, up to roughly the level of "D" Deck. (Even in a countersunk hole, the point of a rivet would generally protrude about 1/16 inch beyond the surface of the plate. This may have been done for increased strength or to allow for some corrosion of the rivet material. As a result, the outlines of countersunk points are still visible in photographs. Over time, as successive layers of paint are applied, the rivets become less visible.) The holes above this point were not countersunk beyond what was caused by the punching of the plates, and these rivets had snap or "button" points. (The factory-formed end of the rivet is properly called the "head," while the end formed during construction is called the "point.") The contrast between the countersunk points and the more rounded shape of the snap points is clearly visible in any photograph that shows both.

The exact style of countersinking used is also difficult to determine. While Lloyds Register Rules called for countersinking in holes of this size to extend nine tenths of the way through the plate, the countersunk holes in the "Big Piece" show a shallower countersink, roughly consistent with the shape of a "Liverpool point." These points had a 45-degree countersink extending about halfway through the plate. We cannot be certain of the type of countersinking used in the rest of the side plating. This is probably because the *Olympic* class was classed with the Board of Trade whose rules were slightly different.

The holes above the margin plate were not countersunk beyond what was caused by the punching of the plates and these rivets had snap heads. Thus, the seam that came in contact with the underwater spur of the iceberg did so at the riveted joint, just above the margin plate, which is about 13 feet above the baseline in way of Boiler Rooms Nos. 5 and 6. The damage continued forward to Cargo Hold No.2. Forward of there, the damage was to the upper connection of that same strake above the margin plate. In way of the Forepeak Tank, this riveted connection is just below the Lower Orlop Deck.

The most important factor contributing to the iceberg damage was the ship's speed. Table 15-2 is data extracted from the Bedford and Hackett analysis of the sinking of the *Titanic*[12] and relates that area of opening in the riveted plates with the length of damage, measured by Paul Matthias during the 1996 *Titanic* Expedition. Note as you progress aft, the average width of opening increases, as does the frame spacing. The hidden elements not tabulated are the thickness of plates and the force of impact. There was more force of impact as the ship's beam increased. Thus, the greatest impacts of the collision between the ship and the iceberg took place in way of Cargo Hold No. 3 and Boiler Room No. 6. These two spaces filled more rapidly than the others and the data in Table 15-2 gives an indication of why. This was a high-energy sideswiping encounter in rather cold water (28-31° Fahrenheit). Rivets failed and the force of the impact was sufficient to push in a steel plate a fraction of an inch that allowed water to force its way in. The force of that impact in way of Boiler Room No. 6 was carried into Boiler Room No. 5 where the same seam opened two feet into that space. The area of flooding for the Bedford and Hackett data for Boiler Room No. 5 appears too large based on Leading Fireman Frederick Barrett's testimony before the British Inquiry. He maintained that water flowing into the forward coal bunker of Boiler Room 5 was like that from a fire hose. Considering fire hoses of that period, the flow would be that from a 3-inch deck pipe. The large frame spacing and the double row of wrought iron rivets with a joint efficiency of only 27% were contributing factors. One of the factors not considered in the computation of this joint efficiency was the slag content of the wrought iron rivets. Based on tests of 49 of these rivets taken from plates from various parts of the ship indicate that the slag could be as high as 8 to 10% in 18 of them. Such a high slag content weakens the wrought iron rivet and will also decrease the joint efficiency. Therefore, the damage to the *Titanic* was spread over approximately 110 feet of her forward hull length and at a sufficient depth to cause the rapid inflow of water. We, therefore, estimate the area of initial flooding to be 11.88 square feet, based on an adjustment in the area of inflow in Boiler Room No. 5. Most of the initial flooding area was concentrated in Boiler Room

TABLE 15–2 *Frame Spacing, Area of Opening, Length of Damage, and Average Width of Opening*

Space	Frame Spacing (Inches)	Area of Opening (Sq. Feet)	Length of Damage (Feet)	Average Width of Opening (Feet)
Forepeak	24	0.60	Trace	—
Cargo Hold No.1	27	1.43	5	0.135
	30	—	6	0.135
Cargo Hold No. 2	30	3.30	15	0.208
	33	—	—	0.208
Cargo Hold No. 3 and Reserve Coal Bunker	36	3.44	33	0.104
Boiler Room No. 6	36	3.01	40	0.073
Boiler Room No. 5	36	0.10	2	0.052

No. 6 and Cargo Hold No. 3. These two areas of the ship flooded to the level of "G" Deck in the first twenty minutes.

The intensity of flooding in two forward cargo holds can be measured by what some of the crew remembered during their escape from their quarters there. Seaman John Bulley and others noted that the tarpaulin over Hatch No. 1 on "C" Deck was elevated from the air escaping from the spaces below. The sound of flooding water could also be heard.

It is also believed that the ship pivoted around the iceberg for an instant in way of Boiler Room No. 4 causing the failure of rivets there. However, that damage is considered very slight since the pumps there were able to contain the flooding to a level below the floor plates.

It is important to recognize that the damage the *Titanic* sustained in this encounter was sufficient to cause her demise. It was only a matter of time. The ship was designed to withstand the flooding of any two major watertight compartments and the four forward compartments in the bow. The intermittent damage was spread over six compartments. The only hope for those aboard was to contain the flooding as long as possible to an area forward of Boiler Room No. 5. That was the critical compartment. When Chief Engineer Joseph Bell was informed that Boiler Room No. 5 flooded around 0115, he was heard to say, "We are lost!"

There have been those who have speculated that the ship could have been saved by plugging the leaking seams. This was not possible with the means at hand for the crew of the *Titanic*. They also were not trained in rudimentary damage control. Thus, stuffing mattresses, sheets, or blankets into or placing sailcloth over the leaking seams was not possible without diving equipment that was not part of the ship's outfit.

The Sinking Process. From the description of damage, all six compartments forward of Boiler Room No. 4 had seams open to the sea. The damage existed at about 13 feet above the keel in Cargo Holds Nos. 2 and 3 and Boiler Rooms Nos. 5 and 6, and 21 feet in Cargo Hold No. 1 and the Forepeak Tank.

During the first 10 minutes, the Forepeak Tank was still taking water. Cargo Hold No. 1 had seven feet of water. In Cargo Hold No.2, five minutes after the collision, water was seen flowing in at the bottom of the Firemen's Passage on the starboard side. This evidence provided by Leading Fireman Charles Hendrickson indicates that there was a failure in a seam above the floor of the Firemen's Passage. There is an indentation in the ship's side abaft of the forward transverse bulkhead in Hold No. 2 that caused the structure of the Firemen's Passage to be pushed inboard. The outer bulkhead was 3.5 feet from the outer skin of the ship. A riveted seam was compromised, thereby flooding both the hold and the passage.[13]

Ten minutes after the collision the water had risen to about 14 feet above the keel in all these compartments except Boiler Room No. 5 and Cargo Hold No. 3. With the Mail Room flooding, the depth of water here was 24 feet. After ten minutes, progressive flooding continued in all these six compartments. However, the storerooms above the Forepeak Tank remained dry and did not begin to flood until around 0120 when Boiler Room No. 5 flooded, causing the vessel's bow to be submerged to above "C" Deck. At that time, water spilled into the scuttle on "C" Deck, forward of the collision bulkhead. This was the main access to these storerooms and flooded all compartments from "C" Deck down to the top of the Forepeak Tank on the Orlop Deck. It is known that these compartments were dry until a few minutes after midnight based on the testimony of Hemming and Haines.

By midnight, water was coming up in Hatch No. 1. It was getting into the firemen's quarters on "F" and "G" Decks and drove the firemen out. Reginald Lee coming off his fateful watch from the Crow's Nest shortly after midnight heard the water flooding into Cargo Hold No. 1. Going to the hatch opening he could see water flooding into the cargo hold several decks down. He also saw a number of firemen scheduled for the 12–4 watch bring their belongings up to "C" Deck from below. Third Officer Herbert Pitman met these men exiting from the forecastle and went to No. 1 Hatch. He saw water overflowing the coaming of Hatch No. 1 on "G" Deck and coming mostly from the starboard side. Therefore, in 20 minutes after the collision, the water had risen above "G" Deck in Cargo Hold No. 1.[14] At this time, according to the calculations of Bedford and Hackett,[15] 3,975 long tons of water had entered the ship.

In Cargo Hold No. 2 about 40 minutes after the collision the water was coming in to the Seamen's Quarters on "E" Deck through the collapse of a centerline wooden joiner bulkhead separating the third class from those quarters, opposite the Seamen's Wash Place. Thus, the water had risen in Cargo Hold No. 2 to about 3 ft above "E" Deck by 0030.[16]

In Cargo Hold No. 3, the Mail Room was flooded within 20 minutes after the collision. During Third Officer Joseph Boxhall's second inspection, he saw mail bags floating on top of the water about 2 feet below "G" Deck. The surface of that water was rising rapidly. The Mail Room was located on the Orlop Deck and was 24 feet above the keel.[17]

The manually operated watertight doors on "F" Deck at the fore and after ends of Compartment No. 3 were not closed then.[18] There was no water on "F" Deck at 0005–0010.

There is a stairway on the port side on "C" Deck, which led down to the first-class Baggage Room on "G" Deck and a first- and second-class Baggage Room immediately below on the "H" or Orlop Deck. There was water in this lower baggage room 25 minutes after the collision. Within a half hour after the collision, water was overflowing onto "G" Deck into the Post Office.[19]

The crew of Boiler Room No. 6 abandoned that space within five to ten minutes after the collision. Ten minutes later the water had risen to 13 feet above the top of the double bottom. The water probably covered the passages to the watertight doors at the aft and forward ends of that compartment. By 0040 the water level reached "E" Deck, completely submerging this space. By this time, there was over 7,450 tons of water in the ship.[20]

In Boiler Room No. 5 there was no water above the stokehold plates, until a flood of water came through the pass between the boilers from the forward end, forcing Leading Fireman Frederick Barrett to leave the space by the emergency ladder.[21]

During the first ten minutes, water was coming into Boiler Room No. 5 in its empty forward starboard bunker at two feet above the floor plates in a stream about the size of a deck hose. Barrett lowered the starboard bunker door when he first entered this space. This action caused the water to be retained in the bunker until it rose high enough to burst the starboard bunker door, which was weaker than the bunker bulkhead. This happened about an hour after the collision. The water rose on the starboard side of the bunker which was empty of coal and then reaching the level of the Orlop Deck. It began to flood over the portion of the bunker that formed the alleyway to the watertight door in the main transverse watertight bulkhead. There was probably some coal in the port bunker, so it took awhile to fill the port side. Between 0100 and 0110, a sufficient hydrostatic head forced the starboard coal bunker door to fail. The mechanism of that failure cannot be established at this writing.

In Boiler Room No. 4 there was some flooding through leaking or failed rivets. By 0120, water was coming in its forward end from underneath the floor plates in the forward part, in small quantities. The men remained in that space until ordered to go above. The damage in way of Boiler Room No. 4 occurred before the ship finally cleared the iceberg. As the helm was put over, her shell plating at that boiler room made very light contact with the iceberg, but enough to caused failure in some rivets. Since there were few survivors from Boiler Room No. 4, it is not known precisely when this space flooded. However, shortly after Boiler

Room No. 5 flooded completely, downflooding into Boiler Room No. 4 could have occurred from nontight openings on "E" Deck.

In Boiler Rooms Nos. 1–3 there was no flooding until after Boiler Room No. 4 flooded by 0150. When the men left Boiler Room No. 4 at 0120, some of them went through the three aft boiler rooms to proceed to the Reciprocating Engine Room. From there they went up to "E" Deck. There was no water in either the Reciprocating or Turbine Engine Rooms.[22] It is also known that after 0130 the flooding rate began to increase rapidly from 50 tons per minute to over 700 tons per minute after 0200.

There was no damage to the shell plating in the electrical plant or the starboard shaft tunnel. Thus, there was no flooding in these compartments until after the ship broke apart and sank.

All the watertight doors aft of the Reciprocating Engine Room were opened manually after the collision. These doors were opened to allow four men to go to the starboard shaft tunnel and carry a 10-inch leather suction-pipe forward. This pipe was used for drawing the water up out of the bilges. By 0010 all the watertight doors were also open from the Reciprocating Engine Room to the after stokehold of Boiler Room No. 4.[23]

Final Effect of the Damage. The later stages of the sinking cannot be defined with any precision, because the survivors were confused about the exact time particular events occurred.

At 0045, Officer Boxhall and Quartermaster Rowe began to fire the distress signals. For the next 45 minutes, the rate of flooding slowed in the ship until Boiler Room No. 5 flooded. This caused the forward end of the Forecastle Deck to go under water shortly after 0135. By this time, over 23,000 tons of water had entered the ship, and the rate of flooding began to increase dramatically. The forecastle head was not submerged then, though it was getting close to the water around 0150.[24] By this time, Boiler Room No. 4 had flooded and 32,000 tons of water had entered the ship.

When Collapsible D was lowered from the No. 2 Lifeboat davits, the forward end of "A" Deck was underwater, and water began coming up the stairway under the Boat Deck almost immediately afterwards.[25]

After this, the other port Collapsible (B), which had been stowed on top the Boat Deckhouse, was uncovered, the lashings cut, and she was swung round over the edge of the coamings of the Deckhouse on to the Boat Deck. Very shortly afterwards, the vessel, according to Mr. Lightoller's account, seemed to plunge by the bow, and he just walked into the water. When he came to the surface all the funnels were above the water.[26] Collapsible B floated off.

Baker Charles Joughin. The most intriguing testimony came from Charles Joughin. He had been in his stateroom on the portside of "E" Deck drinking from a bottle of cooking sherry that he taken from the Galley. He had taken this action, perhaps at the suggestion of the Ship's Doctor, F. W. N. O'Loughlin, who he had a conversation with early in the evacuation of the passengers. When water began to seep in under his door (circa 0200), he calmly left and went up to "B" Deck where he threw a number of deck chairs into the water. He then went up to the starboard side of "A" Deck. His recollection of events on "A" Deck were as follows:

> I went to the Pantry on (starboard side of) "A" Deck, and while I was in there I thought I would take a drink of water. While in the Pantry, I was getting a drink of water and I heard a kind of a crash, as if something had buckled. It sounded as if part of the ship had buckled, and then I heard a rush of people overhead on the Boat Deck. People were running and, when I got up on top, I could then see them clambering down from that deck. Of course, I was in the tail end of the rush to the Poop Deck.

When asked at the British Inquiry about the sound of buckling or crackling and whether it was loud, Joughin recounted that, "You could have heard it, but you did not really know what it was. It was not an explosion or anything like that. It was like as if the iron was parting."

Joughin kept out of the crush of people as much as he possibly could, and he followed the crowd down toward the Aft Well Deck. Just as Joughin reached that deck, the ship took a large port list and threw hundreds of people into a heap with the exception of Joughin. He managed to get to the starboard railing of the Poop Deck and climbed over it along starboard side of that deck. He went aft as far as he could and stood on the outside of the railing near the ensign staff. While the stern hung there, Joughin was wondering what to do next. He tightened his belt and transferred some things out of his side pocket into his back pocket. He was wondering what next to do when the stern began its decent into the water. He recounted that it was like riding an elevator down to the surface of the water. When he reached the water, he simply swam away as quickly as he could to avoid the suction.

When the ship sank, Joughin calmly swam away without getting his hair wet and finally reached the upturned Collapsible B. He managed to stand on this craft in freezing water until rescued. However, that cooking sherry helped to insulate him from the cold.

Hull Failure. The stern of the *Titanic* kept rising gradually out of the water, and eventually the propellers and rudder became clear of the water. The ship reached an angle of 13–15 degrees and then broke in two. She did not attain the perpendicular. By the time the second funnel reached the water, there were no lights burning, though they were on until almost the last.[27]

When reaching an angle of 13–15 degrees, there was a rumbling sound that survivors mistakenly attributed to boilers exploding or crashing through bulkheads. These noises can be attributed to the failure of steel plate and rivets, items in the ship not secured falling to the decks or toward the forward bulkheads, and the implosion/explosion of bulkheads well beyond their hydrostatic design head. With about 39,000 tons of water in the ship, stresses had built up to a point where the hull failure occurred between the forward end of uptakes for the third funnel and a point midway between the aft expansion joint and the fourth funnel. This was a top-down failure in the structure. The stern then began to fall towards the water surface, but rose even more sharply than before, perhaps to 35 degrees. The stern slowly slid down below the surface of the water, disappearing around 0220.[28]

Observations. The *Titanic*, as constructed, could not have remained afloat long with the flooding of five forward compartments. Her bulkheads were spaced to enable her to remain afloat with any two compartments in communication with the sea. She had a sufficient margin of safety with any two of the compartments flooded, which were actually damaged.[29]

In fact, any three of the four forward compartments could have been flooded by the damage received without sinking the ship to the top of her bulkheads.[30]

Even if the four forward compartments had been flooded the water would not have gotten into any of the compartments abaft of them though it would have been above the top of some of the forward bulkheads. But the ship, even with these four compartments flooded would have remained afloat. But she could not remain afloat with those four forward compartments and the forward boiler room (No. 6) also flooded.

The flooding of these five compartments alone would have caused the ship to sink. The waterline would eventually rise above the bulkhead at the after end of Boiler Room No. 6. Water would then flow into nontight openings in "E" Deck over Boiler Room No. 5. That boiler room would fill until in turn Boiler Room No. 4 would commence flooding in the same manner. There would come a moment when the weight of flooding water forward caused bending stresses in the upper half of the ship to reach a failure point. It has been shown that water came into the five forward compartments to a height of about 14 feet above the keel in the first ten minutes. This was at a rate of inflow with which the ship's pumps could not possibly have coped, so that the damage done to these five compartments alone inevitably sealed the doom of the ship.

The damage done to the shell plating and rivets in Boiler Rooms Nos. 4 and 5 was too slight to have hastened appreciably the ship's sinking. As was shown by survivor evidence, no considerable amount of water was in either of these compartments for an hour after the collision. The rate at which water came into Boiler Room No. 6 makes it probable that this compartment was filled within 30 to 45 minutes as the riveted seam was opened over a distance of 43 feet in a space 54 feet long. It was also shown in the evidence of Barrett that the leak in Boiler Room No. 5, caused by a two-foot seam opening, was only about equal to the flow of a deck hose pipe about 3 inches in diameter.[31]

The slight leak in Boiler Room No. 4 was only enough to admit about 3 feet of water in that compartment in 1 hour and 40 minutes.[32] Pumping there kept the water to a level below the floor plates.

Therefore, the leaks in Boiler Rooms Nos. 4 and 5 did not appreciably hasten the sinking of the *Titanic*. The pumps were being worked in Boiler Room No. 5 soon after the collision. It is likely that the 10-inch leather special suction pipe, which was carried from aft, was more likely to have been carried for use in Boiler Room No. 4. An attempt was being made to pump out this water which the engineers found coming in. They sent for the special 10-inch leather suction pipe because they had only one pump in Boiler Room No. 4. The watertight doors were opened soon after the collision when water was known to be coming into Boiler Room No. 5. There is no evidence that the pumps were being worked in Boiler Room No. 4.[33]

The only evidence possibly favorable to the view that the suction pipe was required for Boiler Room No. 4 is the evidence of Frederick Scott, a greaser, who said that he saw four engineers dragging the suction pipe forward about one hour after the collision.[34]

The importance of the question of the damage to Boiler Room No. 5 is small because the ship as actually constructed was doomed as soon as the water in Boiler Room No. 6 and all compartments forward of it entered in the quantities it actually did.[35] It is only of importance in dealing with the question of what would have happened to the ship had she been more completely subdivided.[36]

It was stated in evidence that if Boiler Room No. 4 had not been damaged or had been damaged only to an extent within the powers of the pumps to keep under, then, if the bulkheads had been carried to "C" Deck, the ship might have been saved. Further methods of increased subdivision and their effect upon the fate of the ship are discussed later.

Evidence was given showing that after the watertight doors in the engine and boiler rooms all had been closed, except those forward of No. 4 group of boilers, they were opened again, and there is no evidence to show that they were closed again. It is probable that the engineers who remained below would have closed these doors as the water rose in the compartments aft of Boiler Room No. 4. However, it was not necessary to close them because each door had an automatic closing arrangement, which would have been activated as soon as a small amount of water came through the door.[37]

The ship would not have survived longer if all the watertight doors had been left open to allow the water to flow aft. The free surface of the flooding water would have significantly reduced the transverse stability of the ship, causing her to capsize around 0145. Such an action was also unnatural to Chief Engineer Joseph Bell and the engineers, entrusted with the safety and survival of their ship. It is important to point out the importance of preventing the flooding to the boiler rooms, particularly Nos. 2, 3, and 5, that supplied steam to the two emergency dynamos. Flooding of these spaces not only would have caused additional stress on a weakening hull, but would have been a loss of electrical power, creating a problem of evacuating a sinking ship. Electrical power was needed particularly in lowering the lifeboats. Therefore, such a proposition, even if advanced to Captain Smith or Thomas Andrews, was impractical.

It can be seen that the efficiency of the automatic arrangements for the closing of the watertight doors, which was questioned during the inquiry, had no important bearing on the question of hastening the sinking of the ship. The engineers would not have opened the doors after they were closed unless they had been convinced that the ship was doomed. There is no evidence that they did prevent the doors from closing.

The engineers had the pumps engaged when Barrett left Boiler Room No. 5, but even if they had succeeded in getting all the pumps in the ship to work they could not have saved the ship or prolonged her life to any appreciable extent.

Table 15–3 summarizes the flooding, drafts, and trims of the *Ti-*

tanic. Some of the data are based upon data from Bedford's and Hackett's paper.[38] The times and lists are approximate and based on survivor accounts. It should be noted that as the ship gained flooding water, the GM (metacentric height) increased. When Boiler Room No. 5 flooded, there was a pronounced change. The GM began to decrease and with that there was a dramatic decrease in the range of stability. As the flooding rate increased to 700 tons per minute, the range of stability also decreased, reflecting the fact that the Fore Deck was submerged and that the transverse and longitudinal metacenters were decreasing also. After 1:30 A.M. the officers and many of those passengers on the Boat and "A" Decks realized that the ship was going to sink. No passenger needed to be cajoled into boarding a lifeboat. It was at this time that some of the officers, like Lowe, had to brandish their pistols to keep men from boarding crowded lifeboats. The shift in list from starboard to port was gradual during the first hour of flooding. The port list represents the fact that flooding progressed down the port side more freely than the starboard side due to a watertight bulkhead and door fitted at the forward end of the first-class accommodations on "E" Deck that forced water along the long and wide passageway popularly known as "Scotland Run."

Effects of Suggested Additional Subdivision. A watertight "E" Deck may have prolonged the ship's stay on the surface. Such a watertight subdivision would have complicated passenger access, although openings for

TABLE 15-3 *Summary of Flooding, Drafts, Trim, GM, Range of Stability, and List as a Function of Time*

Time	Flooding Water (Tons)	Mean Draft (Feet)	Trim By Bow (Feet)	GM (Feet)	Range of Stability (Degrees)	List (Degrees)
11:50 P.M.	3,975	34.60	9.38	1.77	97.4	5 (Stbd)
12:00 A.M.	7,450	36.58	21.69	2.44	96.1	?
12:30 A.M.	14,131	40.29	41.90	3.17	85.0	?
1:10 A.M.	23,607	45.25	66.71	4.88	77.8	1.0 (Port)
1:20 A.M.	26,438	46.85	75.88	3.29	76.7	?
1:50 A.M.	32,741	58.46	141.08	1.60	63.1	2.1 (Port)
2:00 A.M.	35,312	78.88	247.77	0.12	16.3	?
2:20 A.M.	39,542	SINKS	—	—	—	—

pipes, cables, and ventilation trunks would have watertight sleeves fitted to the deck penetrations. Curious crew and passengers, eager to catch a glimpse of the iceberg may have compromised such a deck. They left portholes open and failed to secure them. As the ship sank deeper into the water, these provided additional access for flooding water. Also, the plumbing drains discharging above that watertight deck had no check valves fitted. This allowed water to back up into sinks and toilets.

Without discussing the advantage of a watertight deck, it is desirable to form an opinion in the case of the *Titanic* as to whether making the bulkhead deck watertight would have been an advantage in the circumstances of the accident, or in case of accident to ships of this class.

All the compartments certainly known to have been flooded were those forward of Boiler Room No. 4. The ship would have remained afloat longer if "E" Deck had been a watertight deck. If, however, Boiler Room No. 4 had also been flooded the ship would not have remained afloat unless, in addition to making the bulkhead deck watertight, the transverse bulkhead abaft of that boiler room had been carried up to "D" Deck. To make "E" Deck effectively watertight, it would have been necessary to carry watertight trunks around all the openings in the Bulkhead Deck up to "C" Deck.

If the bulkhead abaft Boiler Room No. 5 had been carried to "C" Deck, the ship would have remained afloat if the compartments certainly known to have been damaged had been flooded. Fitting a watertight deck below "E" Deck would have been impractical for the accommodations of the Steerage. Entrance to some of their quarters might have been necessary through watertight hatches. There is also the question of the vessel's transverse stability when those compartments below that deck flood on one side. This would lead to a list and possible capsizing.

Longitudinal Subdivision. The advantages and disadvantages of longitudinal subdivision by means of watertight bunker bulkheads were discussed before the British Inquiry. The flooding from the raking damage of the iceberg encounter could have been thwarted by the use of longitudinal watertight subdivision. The *Mauretania* and *Lusitania* were spoken of as examples of such an arrangement. Edward Wilding testified that one of the greatest of objections to longitudinal bulkheads in coal-burning ships was that each coal bunker was a watertight compartment. That meant that the coal had to be worked through watertight coal bunker doors. It was very difficult under such conditions to ensure that these watertight bunker doors would work when they had to. The other objection he voiced was that such an arrangement led to unsymmetrical flooding unless some means of counterflooding were provided in the opposite bunker.[39]

Extending Double Bottom Up the Side. It was also shown during the British Inquiry that there would be increased protection to the ship against raking damage if the double bottom were carried higher up the side than was done in the *Titanic*. Boiler Rooms Nos. 4 through 6 would not have flooded. The problem with this arrangement was that the ship took an initial list of 5 degrees because of unsymmetrical flooding on that side. Large wing compartments in those two boiler rooms would have produced a much larger list until intervention by the crew flooding aft port double bottom spaces. Although Edward Wilding was positive the ship would have survived with such an arrangement, flooding double bottom spaces would have increased the draft, making downflooding beyond Cargo Hold No. 3 a possibility.

Watertight Doors. As far as the watertight doors of the ship are concerned, there does not appear to have been any appreciable effect upon the sinking of the ship caused by either shutting or not shutting the doors. There were no difficulties in working the watertight doors. They were shut in good time after the collision. The ship had sufficient transverse stability to resist the massive amount of flooding and free surface that was generated during the last 90 minutes of this ship's existence above the surface. This would have been drastically reduced if the flooding had been allowed to continue aft. There would have been a massive loss of steam power that was keeping the lights lit and the lifeboats lowering. This maintained calm until those last few minutes when all the lifeboats were gone. However, the engineers would have had a better chance to escape with an earlier abandonment of the engine and boiler spaces instead of trying so hard to keep the ship afloat to the point of the hull failure. None of these men survived, yet their sacrifice allowed 706 persons to survive.

In other cases of damage in ships constructed like the *Titanic,* it is probable that the efficiency of the closing arrangement of the watertight doors may exert a vital influence on the safety of the ship. Consideration could be given to bulkheads with no openings below the margin line and the necessity of fitting watertight doors there. If there is a need for watertight doors, then the question of automatic operation needs to be considered.

Notes for Chapter 15

1. Chris Hackett and J. G. Bedford, "The Sinking of SS *Titanic*, Investigated by Modern Techniques." The paper was presented to The Northern Ireland Joint Branch, The Institute of Marine Engineers and The Royal Institution of Naval Architects, Belfast, March 26, 1996 and April 15, 1997. Bedford and Hackett used

the elementary orifice-flow formula that has flow rate proportional to the square root of the pressure difference across the orifice, multiplied by a flow coefficient of 0.485 that was stated in the main text. The coefficient value was an estimate based on a perceived configuration of the openings in the hull.

2. The description of the condition of Boiler Room No. 4 was extracted from the testimony of Trimmer Thomas Patrick Dillon, Questions Nos. 3816–3829.
3. This was extracted from the testimony of Fireman George Cavell before the British Inquiry, Questions Nos. 4214–4290.
4. This description of flooding was extracted from Able Seaman John Poindestre's testimony before the British Board of Inquiry, Questions Nos. 2843–2872.
5. The horizontal extent of the six openings was clearly shown by the sonar imaging, but the widths were below the resolving power of the sonar profiler.
6. Chris Hackett and J. G. Bedford, "The Sinking of SS *Titanic*, Investigated by Modern Techniques." See Note 1 for additional details.
7. This was extracted from the testimonies of Leading Fireman Frederick Barrett, Questions Nos. 1851–2361, Leading Fireman Charles Hendrickson, Questions Nos. 4899–4910, and Fireman George Beauchamp, Questions Nos. 661–680 at the British Board of Inquiry.
8. The dampers regulated the flow of combustion air to the boiler furnaces. Shutting them, therefore, effectively precluded further combustion. However, this desired effect could not occur instantaneously because the mass of coal would remain at near-combustion temperature even after the combustion reaction ceased. It is for this reason that "drawing fires" might be necessary in an emergency.
9. Drawing fires. The contents of the furnaces were raked out onto the floor plates of the firing aisle, where the still-burning coal was extinguished, preferably by being smothered by wet ashes. If no such resource was available, then resort was had to quenching by fire hose. In either event, a great deal of steam was produced. Drawing fires was usually an emergency procedure, as it surely was in this instance.
10. Bedford and Hackett. See Note 1.
11. Gayhart, CDR E.L., "An Investigation of the Behavior and of the Ultimate Strength of Riveted Joints under Load," *Transactions*, SNAME, 1926.
12. Chris Hackett and J. G. Bedford, "The Sinking of SS *Titanic*, Investigated by Modern Techniques." See Note 1 for additional details.
13. British Inquiry, Questions Nos. 2821–2, 4859, 4856–66.
14. British Inquiry, Questions Nos. 2455, 2479, 2485, Officer Pitman 14962, 14965.
15. Chris Hackett and J. G. Bedford, "The Sinking of SS *Titanic*, Investigated by Modern Techniques." See Note 1 for additional details.
16. Bedford and Hackett. See Note 1
17. British Inquiry, Questions Nos. Poigndestre 2844, 2845, 2861.
18. British Inquiry, Questions Nos. 14948, 14949, 15374, 15380.
19. British Inquiry, Question No. 15757.
20. Chris Hackett and J. G. Bedford, "The Sinking of SS *Titanic*, Investigated by Modern Techniques." See Note 1 for additional details.
21. British Inquiry, Question No. 3397.
22. Bedford and Hackett. See Note 1.
23. British Inquiry, Questions Nos. 1969, 2038, 2039, 2041.
24. British Inquiry, Thomas Dillon, Questions Nos. 3795, 3811.

25. British Inquiry, Question No. 5599.
26. British Inquiry, Questions Nos. 2541, 17685, 17689, 167.
27. British Inquiry, Questions Nos. 14081, 14023, 14035.
28. British Inquiry, Questions Nos. 14052, 14071, 14072.
29. British Inquiry, Questions Nos. 14078, 14075, 14084A, 14089, 2558, 10454.
30. British Inquiry, Question No. 14097.
31. British Inquiry, Questions Nos. 20320, 20286, 20293.
32. British Inquiry, Question No. 20364.
33. British Inquiry, Question No. 2255.
34. British Inquiry, Questions Nos. 4265, 3811.
35. British Inquiry, Question No. 1961.
36. British Inquiry, Question No. 5602.
37. British Inquiry, Questions Nos. 20311–4, 20320.
38. Chris Hackett and J. G. Bedford, "The Sinking of SS *Titanic*, Investigated by Modern Techniques." See Note 1 for additional details.
39. British Inquiry, Questions Nos. 20355, 20371–4.

In 1998 the RMS Titanic, Inc. and Discovery Channel sponsored an expedition to the *Titanic*'s wreck site. One of the major goals of this expedition was to photograph the location and analyze the damage caused by the ship-iceberg collision. This photograph shows the collision damage outside Boiler Room No. 5. Note the indentation of the one-inch plates, and the very small opening in the riveted seam—less than one half inch. (*Photo from W. H. Garzke collection*)

CHAPTER 16

Titanic *Hull Fracture*

If the *Titanic* could have been viewed from a distance at 10 minutes after the collision, the consequences of her fatal damage would scarcely be evident. A slight list and trim might catch the eye of an expert observer, but the lifeboats were snug under their davits, no unusual activity could be seen on the open decks, and, moreover, the ship was brilliantly lit. In view of such comforting evidence, a casual observer would probably be completely unaware of trouble, while the expert would attribute the trim and list to some mundane operational circumstance. Of course, the ship was stopped in mid-ocean, and great and noisy plumes of steam rose from the three forward stacks, but the presence of ice easily accounted for the absence of way, and the escape of steam was a natural and inevitable consequence of stopping suddenly. The *Titanic* must have received the ice warnings that had been flying from ship to ship all day, and the carefully alerted lookouts must have been alarmed by that haze ahead. Warned, alerted, and aware that the *Californian*, a little less than 20 miles to the north, had stopped to await daylight, a cautious Captain Smith would have elected to do the same.

The evidence of a ship in the process of sinking slowly increased as the inexorable flooding weighed down the bow, but even an hour after the collision the casual observer might be unaware that the *Titanic* was doomed. The bow would be depressed at an angle of three degrees, but the ship was still brilliantly lit, the sea was calm, no flame or smoke signaled danger, no frightening sounds of boiler explosion or the rending of steel signaled an accelerating progression toward foundering. The firing of eight distress rockets was surely a sign of something amiss, but they were *white*—a casual observer would expect *red* if there was life-threatening trouble aboard. But expert observers were at hand, for some of the crew of the *Californian* was watching, and they, though baffled, did not dismiss the rockets as a meaningless display. Unfortunately, they could not detect any confirming circumstances: their vantage was some 10 to 20 miles away, and at that distance only the rockets were clearly evident. Their wireless operator had gone to bed. The watch officer, Second Officer Herbert Stone, did react to the rockets by awakening Captain Lord, but upon being told that the rockets were white, the captain went back to sleep after muttering that they must be "company signals."[1]

Rebuffed, Stone returned to the bridge. He carried no order from his captain to respond to the eight rockets he had seen, but he did resume his interest in the ship that had fired them. Together with Second Mate James Gibson, he maintained his vigil from 0045 to 0200, frequently trying to make sense of the distant activity by using his binoculars. Around 0145 he remarked to Gibson, "Look at her now. She looks very queer out of the water; her lights look queer."

Aside from the unmistakable sight of bursting rockets, darkness and distance made it difficult to ascertain from the bridge of the *Californian* what was actually happening. Later, the two watchers did not agree completely on what they had seen, but neither saw the blinking Morse light, nor were they aware of the *Titanic*'s increasing trim. They were aware that the distant array of lights was diminishing in size, and that it finally disappeared, but they attributed this phenomenon to a turning and steaming away. No one on the *Californian* became aware of the terrible drama occurring so close at hand until their wireless operator came back on duty after daylight.

By the time the now-aware *Californian* got underway, the *Carpathia* had already arrived to pick up the 706 who had survived, and the calmly dying ship with the "queer lights" had sunk to an eternal rest some 12,500 feet below the surface of the Atlantic. Of course, its rest was not to be eternal, for all that finding and visiting the wreck was recognizably impossible in 1912. It would take 73 years to elapse before advancing technology and the questing spirit of its users brought discovery and the subsequent forensic analyses.[2] The most immediate and most dramatic discovery announced by Dr. Robert Ballard and French scientist Jean-Louis Michel in 1985 was that the wreck was lying in two well-separated pieces. *Titanic* canon stated that the ship had sunk intact, and here was contrary evidence. But did that rupture take place somewhere in that two-mile descent? Did it occur because of impact with the seabed or did it occur at the surface? When Dr. Ballard's second expedition in 1986 revealed that some 60 feet of the ship was missing, debate expanded: did the ship break into two pieces or three?

After the discoveries of the 1980s, fracture of the hull was accepted as incontrovertible fact, and the on-site evidence soon led to the conviction that the fracture had occurred at or near the surface. The 1912 conclusion of an intact sinking came only from a weighing of contradictory evidence in the form of eyewitness accounts. The American investigator, Senator William Smith, and his British counterpart, Lord Mersey, agreed on this point. Senator Smith wrote in the American report:

> There have been many conflicting statements as to whether the ship broke in two, but the preponderance of the evidence is to the effect that she assumed an almost end-on position and sank intact.

Lord Mersey wrote:

> Her stern was gradually rising out of the water, the propellers were clear of the water. The ship did not break in two, and she eventually attained the perpendicular, when the second funnel from aft reached the water. There were no lights burning then, though they kept alight practically until last.

Their conclusions prevailed until Dr. Ballard's images showed the parted hull seventy-three years later. Now, the testimony behind the conclusion of 1912 was not unanimous beyond agreement among survivors that there was an interval of loud noises from within the ship shortly before it disappeared. Several likely events, seemingly more plausible than hull fracture, could account for those noises. On the other hand, among the likely events only hull fracture would have been visible to survivors. Analysis of testimony that was discounted in 1912 shows that the breakup *was* actually seen by many of the passengers and crew. Apparently their statements were thought less reliable than the contradicting testimony by surviving officers, led by the senior survivor, Second Officer Charles Lightoller. This officer's testimony was obviously given more credence than it deserved.[3] Their view was that large objects, particularly engines and boilers, were breaking from their foundations and plunging forward through the bulkheads. That event had to happen if the ship actually reached the extreme trim angles that a few survivors reported, and was believed to account for the noise of the ship's demise.

Other survivors attributed the noises to boiler explosions, such as passenger Charles Stengel. This was based only on the general belief that boilers often exploded from the gross derangements of their operating conditions during sinkings.

Then there was the now-confirmed rupturing of the hull. The survivors who attested to this had seen the stern abruptly fall back from the hull's steep trim angle, and had assumed (correctly) that this seemingly anomalous event could occur only if an after section of the hull had separated from the forward.[4]

These three possibilities were not mutually exclusive; boilers could have exploded, engines could have crashed through bulkheads, and the hull could have failed, all around the same time. From post-discovery evidence we know that a top-down failure of the hull did occur, that the second seems unlikely, and that the first might have happened, but also seems unlikely.[5]

What the Survivors Said. The survivors' testimony is conflicting, as well we might expect it to be. Beyond the inevitable tendency of the most sincere witnesses to give inconsistent descriptions of events, there remains the circumstance that no one present had experience with the sounds of boilers exploding, of heavy objects crashing through bulkheads, or of ship hulls rending into two pieces. But let's see what those who *thought* they could distinguish among these possibilities had to say.

Some of their stories, appearing here, are from testimony at the American or British Inquiry, or are found in a collection of survivor stories by Jay Henry Mobray[6] and Walter Lord.[7] The first of these, from Seaman John Buley, seems the most credible because the witness was an experienced crewman, and recently discharged from the Royal Navy had become familiar with damage control procedures.

> The bow settled down and she went down as far as the after funnel. Then there was a little roar as though the engines had rushed forward and she snapped in two. The bow part went under and the after part came up and stayed afloat for about five minutes before it went down.
>
> The ship was horizontal at first and then went down. We could see the after part afloat and there was no forepart to it.
>
> We were close (200 yards) from her and the lights went out before she parted.

Seaman Buley's account is the best description of the breakup we have found from an experienced crew member, and it fits the failure of the hull as we believe it did happen. His testimony on the settling of the stern agrees with what young Eva Hart could vividly remember as late as October 1993 in an interview with the Discovery Channel.

Elmer Z. Taylor, a 48-year-old mechanical engineer in Lifeboat 5, was the most emphatic among survivors in his description of hull fracture. He later wrote:

> The cracking sound, quite audible a quarter of a mile away, was due, in my opinion, to tearing of the ship's plates apart, or that part of the hull below the expansion joints, thus breaking the back at a point almost midway the length of the ship.

Quartermaster Walter Perkis in Lifeboat 4 noted a cracking noise at about 0200 as he rowed by the area of the second expansion joint, which was located between the third and fourth funnel. He thought this to be the sound of dishes breaking in the galley as the stern began to up-end. However, both brittle and ductile fracture of steel is a noisy phenomenon. It was remarked by witnesses to the sudden failure of brittle steel in the hull of the tug-barge *Martha Graham* at Port Jefferson, New York, in February 1974.[8] Since this incident occurred in port, witnesses were numerous and scattered over a wide area. As with the *Titanic*, stories differed, but breaking of china plates or even a rifle shot is consistent with most of them. As shown by even a casual inspection of the wreck lying in full view at Port Jefferson, the hull had cracked apart; boiler explosions and crashings of boilers and engines through bulkheads had not occurred.

But can the noise of fracturing steel deceive a listener into believing that an explosion has occurred? Very possibly: fractures of steel can be observed in any laboratory where the fracture properties of that material is being tested. When a large structural element is being loaded to destruction, the noise can be quite loud and frightening.

George Weikman, one of the *Titanic*'s Barbers, managed to escape in a rather unusual manner that is worth recounting. As we see, he heard explosions, which he attributed to the boilers, but also reports seeing the hull break in two.

> While I was helping with the boats, there came explosions below deck and the ship took an awful lunge, throwing everyone in a heap. These explosions were caused by the rushing in of icy water on the boilers. A bundle of deck chairs, roped together, was blown off the deck with me, and struck my back, injuring my spine, but it served as a temporary raft. The time was 0150.
>
> The bow went clean down, and I caught the pile of chairs as I washed up against the rail. Then came another explosion, which blew me fifteen feet. While laying on the bundle of chairs, I could see the ship split in two. There was a rain of wreckage and a big piece of timber fell on me,

> striking my life vest. If it had not been for this, I probably would have been killed.
>
> After the water filled the forward compartments, the ones [compartments] at the stern could not save her. They did delay the ship's going down. If it wasn't for the [stern] compartments hardly anyone would have gotten away.

Mrs. J. Stuart White, a first-class passenger, was in Lifeboat 8 off the *Titanic*'s port side when the ship's death throes began. Even at that time, no one in the boat thought the ship was going to sink, but she saw the ship break apart.

> In my opinion the ship when it went down was broken in two. I think very probably it broke in two. I heard four distinct explosions, which we supposed were the boilers. Of course, we did not know anything about it.

Asked how loud the explosions were, she replied, "They were tremendous."

Olaus Abelseth, a steerage-class passenger, was standing on the aft end of the Boat Deck talking with his brother-in-law and cousin, and recalled that:

> We could see the water coming up, the bow of the ship going down, and there was a kind of explosion. We could hear popping and cracking, and the deck raised up and got so steep that people could not stand on their feet on the deck. So they fell down and slid on the deck into the water that was right on the ship.

Abelseth grabbed one of the dangling boat falls on Boat Deck until he could steady himself, then jumped overboard. His description of what he heard is very revealing from what we now know exists on the bow wreck. There are a number of missing rivets in the longitudinal seams and butt joints. The butt joints are the strongest riveted connection in the ship with as many as four and five rows of rivets. The stress levels had become so intense that plates also tore apart. What Abelseth heard were the rivets failing (popping noises) and cracking (plates failing). He was fortunate to be out of the zone of failure, otherwise he might not have survived as all the ship structure on the Boat Deck between the second and fourth funnels disintegrated.

Able Seaman Frank Oliver Evans was in Lifeboat No. 10 that had been tied with boats 12 and 14. Although busy at his oar with the task of station keeping, he managed to watch the final minutes of the *Titanic*. From

150 yards away, he saw the ship break in two between the third and fourth stacks. The stern fell back toward an even trim after the break, but then rose again and sank. This observation tracks with what we know at the time of this writing from a finite element modeling of the stress. It also agrees with Eva Hart's October 1993 recollection of the final moments of the *Titanic.*

Mrs. Arthur Ryerson, a first-class passenger, to port in Lifeboat No. 4, one of the last to leave the ship, watched it sink.

> The great ship took a plunge toward the bow, the two forward funnels seemed to lean and then she seemed to break in half as cut with a knife, and as the bow went under, the lights went out. The stern stood up for several minutes, black against the stars, and that too plunged down, and there was no sound for what seemed like hours. Then began the cries of people drowning, all around us, which seemed to go on forever.

Mrs. Ryerson's testimony agrees very well with what we now know happened. When the forward funnel came down after the forward expansion joint opened, the ship took a sudden plunge by the bow. This resulted in a wall of water surging along the Boat Deck, engulfing some passengers as it moved aft before dissipating. Colonel Archibald Gracie was swept overboard by this surge of water.

Greaser Thomas Ranger, who was in the same boat as Mrs. Ryerson, agreed with her affidavit with one exception: he saw the stern slowly settle and then plunge.

Lucille Carter was another first-class passenger in Lifeboat No. 4.

> It was cold and we had no time to clothe ourselves with warm overcoats. The rowing warmed me. We started to pull away from the ship. We could see the dim outlines of the decks above, but we could not recognize anybody.
>
> We had pulled away from the *Titanic* for a distance of a city block, that is about all, I should say, when the *Titanic* began to shake to pieces.
>
> She had exploded. There was a rumbling noise within her, she gave a lurch, and started to go down. We realized what it meant. That the sinking ship would suck us under with her. It was a moment later that the suction struck us. It was all we could do to keep from being caught, so strong was the drag that followed *Titanic.*

Other survivors remarked on the absence of suction as the stern went down. The Carter testimony apparently refers to suction from the bow section. Many of the spaces in that section must have been still filling with water as it went down; the inrush of this water created the "suction."

Steward George Crowe was in Lifeboat No. 14, drifting away from the port side of the *Titanic* as the ship went down. When his boat first drew away from the ship, the bow of the *Titanic* was already deep in the water. However, the lights were still burning brightly, though the bow was continuing to go lower and lower. Then the stern rose out of the water and the lights dimmed. Thereafter,

> She broke clean in two, with probably two thirds of the ship under water.

The aft third of the ship with the fourth funnel seemed to float back again with the forward portion disappearing. Crowe then heard an explosion and the aft portion began to sink. He also noted several muffled explosions before the ship broke apart.

Major Arthur Peuchen, first-class passenger in Lifeboat 6, recalled that while the T*itanic*'s lights still burned the bow was down and the stern up, but not in a perpendicular attitude. He also remembered hearing a rumbling sound, as of an explosion, followed by several more rumbling noises, with the lights being extinguished shortly after. Later, he watched the sea from the deck of the *Carpathia* as it passed through field of wreckage. He spotted a floating barber pole that he knew must have come from "E" Deck, low in the hull of the *Titanic*. He reasoned that only a calamitous rupture of the hull could have freed it to float away from the ship.

Appearance on the surface of wreckage such as Major Peuchen's barber pole or a piece of the carved oak paneling over the entrance to the first-class Lounge on "A" Deck that is now in the Maritime Museum in Halifax, Nova Scotia would be unlikely if the ship had sunk intact. Innumerable buoyant objects from the ship's *exterior*—deck chairs are an obvious example—would have littered the surface, but objects from the interior should have been rare. But like the barber pole, much of the debris was obviously from the interior. The area of the ocean in which the ship sank was covered with cork, an insulation material used in the ship's interior. Other debris, such as cabin fittings, wooden drawers, and cork insulation from ruptured bulkheads, were clear evidence of the violent nature of the hull breaking apart. On the other hand, it was not incontrovertible evidence that the breakup had occurred at the surface.

Quartermaster Arthur Bright saw the final minutes of the dying liner from the vantage of Collapsible D on the port side. From 100 to 150 yards away he saw the ship break in two, with the stern settling briefly toward the surface before it plunged.

Charles Emil Stengel, a first-class passenger in Lifeboat 1, reported the loud noises, but attributed them to boiler explosions.

> I saw her first row of port lights go under the water, and finally all of the bow was dark. She was going down by the bow most all of the time, and all of a sudden there were four sharp explosions, about that far apart, just like this [he indicated by snapping his fingers four times], and then she dipped and her stern stood up in the air.

When questioned about the character of the explosions, Stengel replied:

> I should judge it would be a battery of boilers going.

When questioned further as to the possibility of the explosions being the sounds of bulkheads giving way, he replied:

> I do not know. I have never been familiar with bulkheads giving way.

The 1998 RMS *Titanic*/IFREMER expedition obtained views of the after-most boiler room that was in use at the time of the sinking, Boiler Room No. 2. Being the after-most space in service (boilers in Boiler Room No. 1 were not fired), this was surely the last boiler room fired, and, hence, the one where explosions would have occurred when the hull was rupturing. Yet all boilers in this room are on their foundations, and show no evidence of explosion. Indeed, if any boiler had exploded in the manner of a bomb, the evidence of the happening would almost certainly be apparent on the wreck. Stengel's four sharp explosions, therefore, must have been the sounds of a fracturing steel structure.

Seventeen-year-old Jack Thayer and another first-class passenger, Richard Williams, saved themselves at the last possible moment by diving overboard as a plunge of the bow sent a wave rolling toward them along the Boat Deck. Before hypothermia could claim them, they managed to reach and climb aboard the overturned Collapsible A, washed overboard by that same wave. From that uneasy vantage they watched the ship buckle and break apart before it sank. Thayer even believed that he had seen the second stack lifted off.

His judgment is probably correct, since the stays supporting the stacks would surely have failed as the upper structure of the ship disintegrated. Aboard the *Carpathia*, his description was sketched by L. D. Skidmore, and later published in the Philadelphia *Bulletin*. The sketch shows the stern parted from the forward hull.

John Thayer was haunted by the events of the *Titanic* and in 1941 he committed suicide at the age of 46. He gave presentations and wrote articles on his experiences and what he had seen. Each time his recollection changed. The latter is an illustration of why a marine forensic investigator must get the testimony of the survivor of a ship casualty as close to the event as possible. The other tragedy of John Thayer is that the experience of surviving a ship sinking is an event that can cause terrible nightmares and severe depression.

Taken without rebuttal, the evidence cited previously seems to be conclusive support for the contemporary assertions that the ship broke at the surface before sinking. But there was strong support for the opposing viewpoint. Colonel Archibald Gracie, in numerous speaking engagements before his death in December 1912, insisted that the *Titanic* had sunk intact. His explanation for the contrary testimonies was that the falling of the forward stack, accompanied by a shower of sparks, had given witnesses the illusion of the ship breaking apart.

Colonel Gracie's view was shared by Second Officer Charles Lightoller and Third Officer Herbert Pitman. At the American hearings, Pitman testified to hearing explosions, sounds like those of a big gun firing in the distance, but he believed the source to be bulkhead failures after the ship was below the surface. He firmly rejected the notion put to him that the sounds indicated boiler explosions, asserting that the boilers had been fed steam for almost 2½ hours, and that their water levels had been reduced. Authorities that agreed with him that boilers did not explode accepted his testimony to support the notion of internal rendings and crashings rather than a breaking of the hull. It should also be noted that Gracie, Lightoller, and Pitman had a number of conversations and interviewed survivors aboard the *Carpathia* before this ship arrived in New York on 18 April 1912.

Some survivors, to be quoted in Chapter 26, judged the noises to be evidence of boiler explosions. These anti-breakup views prevailed before both Senator Smith and Lord Mersey. The popular belief for the next seven decades was that the ship lay intact on the ocean floor except for that "gash" made by the iceberg (well, maybe those boilers that crashed through bulkheads were piled in a heap somewhere forward). Some believers in a scarcely-damaged ship down there put forward schemes to salvage the *Titanic.* A few among these believed a salvaged *Titanic* might be cleaned up and put back in service.

All opinions, all judgments by the official inquiries, were superceded when the wreck was found to be in two distinct pieces. Later during the 1998 RMS *Titanic* /IFREMER Expedition it was found that the ship broke into four pieces. A section of the double bottom was found upside down, but that measured only 26 feet in length. Another 34 feet of the missing hull had not been found at the time of this writing. A die-hard advocate of an intact sinking might say that the fracture occurred as the hull struck the bottom, but the three identified parts are too far apart to support that scenario.

In a later chapter we shall explain what evidence from the wreck shows about the events on the surface in the last few minutes of the sinking.

Notes for Chapter 16

1. Conflicting stories were told by the officers and men of the *Californian.* Our summary is a reconstruction from several testimonies of likely events.
2. Dr. Robert Ballard, 1990, *The Discovery of The Bismarck,* Warner Books, New York.
3. Weight may have been given to the officers' testimony because they were seemingly better qualified than any other witnesses. A century later, one might agree that they were probably better qualified to analyze what they saw, but disagree that they were better qualified to know what they were seeing. The sinking of the *Titanic* was an unprecedented event—neither its Captain nor the least of its steerage passengers had seen such a thing before.
4. Analysis of the wreck suggests that the broken halves of the hull were still connected by a section of bottom/inner bottom structure in the last moments, so that a quibbler might maintain that the ship didn't break *completely* in two at the surface.
5. Boiler explosions are discussed in Chapter 25. We note there that a boiler might explode in a manner quite analogous to the ignition of dynamite, but that in 1912 much less devastating rupturings of boiler shells, or even of pipe joints, were classified as explosions. Explosions of the most extreme kinds did not occur during the sinking, but lesser explosions cannot be ruled out.
6. Jay Henry Mobray, 1912, *Sinking of The Titanic: Most Appalling Ocean Horror with Graphic Descriptions of Hundreds Swept to Eternity Beneath The Waves*, memorial edition, National Publishing Company, Philadelphia. The author names his survivors, but says nothing about how he collected their stories.
7. Walter Lord, 1955, *A Night to Remember*, Henry Holt, New York.
8. The *Martha Graham* failure was widely reported in the marine trade press of the time.

The *Lusitania* is shown arriving in New York Harbor during her maiden voyage in 1907. The size of her four funnels can be seen in this photo. (*Courtesy of the Mariners Museum*)

CHAPTER 17

Lusitania: *The Event*

World War I began on 4 August 1914, just as the *Lusitania* was about to depart from New York. The decade before had seen a frantic buildup in sea strength by the two forces that now faced one another: the Home Fleet of Great Britain and the High Seas Fleet of the German Empire. The Germans' battleline strength was still inferior, for they had not been planning to reach parity with the Royal Navy before 1920. However, the ships of the High Seas Fleet had an inherent advantage over their rivals. In spite of the grandiose "High Seas" name, the German heavy units were designed for a confrontation in the North Sea. On the other hand, their expected enemies were perforce designed for service world-wide. An obvious consequence: without the need for long-voyage fuel capacity, the German designers could devote more of their ships' displacement to armor or armament. If the Home Fleet were to make a descent upon the German coast, then the concentrated High Seas Fleet might well be victorious. The War to End All Wars would then have been mercifully short.[1]

The Home Fleet did not weigh its anchors from the mud of Scapa Flow, however. Instead, the Royal Navy promptly moved to impose a distant blockade on German maritime trade. This measure was far less spectacular than would have been a Nelsonian thrust through the minefields of the Jade to bombard Wilhemshaven. The blockade was, in the long run, far more successful as a measure of war. The Royal Navy also imposed a distant blockade on its enemy's ships in certain neutral ports, such as New York, where three cruisers stood patrol and a number of German liners were stranded.

There was the submarine, however. Its powers were not appreciated on the day the War broke out. They soon were, however. During September 1914, as the Germans continued their advanced through Belgium toward France and Paris, three British cruisers were sunk in the North Sea by a single *U*-boat. A new and frightening dimension had suddenly been added to naval warfare.

The involvement of merchant shipping was seemingly a different matter. International law in 1914 was quite explicit: no merchant ship could be sunk without first being warned, inspected, and given time to evacuate its people into boats. Further, the safety of the evacuees had to be guaranteed. The requirement for warning and inspection was a severe handicap for a submarine essaying an attack on a merchant ship. It required surfacing, and thereby abandonment of the submarine's invulnerability to counterattack; even if not armed, a merchant ship could deal a fatal blow to a surfaced submarine by ramming. Even more unreasonable from the submariner's viewpoint was ensuring the safety of the merchant ship's people when the encounter took place far offshore. Such assurance was essentially impossible; there being scant room for guests aboard the *U*-boat.

International Law had been made without thought for submarine warfare. There was no way it could be amended to accommodate an effective use of submarines against non-combatants. But the unrestricted use of the submarine was too potent a weapon to be foregone. Although controversy shadowed its use throughout World War I, acceptance eventually spread, even among naval powers whose principal strengths lay on and above the surface—witness the order issued to the U.S. Navy on the afternoon of December 7, 1941 to commence unrestricted naval (and air) warfare against Japanese commercial shipping.

In the beginning, acceptance did not come easily or quickly.[2,3] The British were shocked, and vociferous in their condemnations, when the Germans sortied their *U*-boats to begin a counter-blockade in the waters around the United Kingdom. Initially, their successes were few and submarine losses actually exceeded enemy surface ships sunk by the *U*-boats. Despite the "customary usage of war" that was nominally still obeyed at the end of 1914, the masters of British steamers had been given blanket secret orders to try to escape immediately on sighting a hostile submarine. The same secret orders instructed the British ship to ram or attempt to ram the German *U*-boat, if such tactics would best contribute to the safety of the British ship.[4]

Where was the *Lusitania*? After war was declared in August 1914, the *Lusitania* returned to Liverpool and entered the Canada Dry Dock there because the Royal Navy at first planned to use her as an auxiliary cruiser. Although there was some concern with the amount of coal she would burn, the Director of Naval Construction, Sir Eustace Tennyson-d'Enycourt was more concerned about unarmored longitudinal bulkheads in battleships and cruisers exposed to submarine attack without suitable escorts. Although the *Lusitania*'s conversion had begun, it was decided to place her back into service as a passenger ship on 24 September 1914. Therefore, she made two transatlantic round trips in October 1914, and in November resumed a monthly regular scheduled passenger service. To conserve coal and reduce required manning in her machinery spaces, one boiler room (Room No. 4, the aftermost) was not used. This measure reduced her maximum speed to about 21 knots.

There were some problems with this situation among the engine room crews who felt it was unsafe with a war at sea for the *Lusitania* to have a speed restriction. Finally Captain David Dow decided to take action on the matter and told Albert Booth, Chairman of the Cunard Board of Directors that he would not make an eastbound voyage without full bunkers of coal. Dow pointed out that coal was cheaper in New York than in Liverpool and a study by Cunard also determined that Dow was correct. He also pointed out that full bunkers of coal out of New York would aid in the trimming of coal since some parts of that system were defective. Furthermore, naval architects would also have told him that coaling in New York would improve the stability of *Lusitania* in her passages through the Irish Sea where the *U*-boat threat was greatest. During the January 1915 voyage, coaling the *Lusitania* began in New York for the round trip of the vessel. On 16 January 1915, during that westbound passage from Liverpool, the *Lusitania* created a widely publicized international incident. En route to Queenstown, the liner encountered heavy seas that delayed her transit. Commodore Dow, fearing the delay would cause his arrival at the Irish port during a time favorable to *U*-boat attack, hoisted the American flag. With the United States still neutral, he believed the Germans would be reluctant to attack a ship under American colors. The*Lusitania* was also fortunate during that voyage because the German raider *Karlsruhe* was off the Grand Banks of Newfoundland. A fog prevented an encounter.

On February 20, 1915 the *Lusitania* began another westbound voyage. The danger from submarines began to increase as during her passage through the Irish Sea two merchant ships were sunk about ten miles from her track. During the eastbound voyage another cargo ship was torpedoed by *U-20* just as the *Lusitania* entered St. George's Channel. This incident was too much for Captain Dow who relinquished command on 8 March 1915 saying that he did not mind being a master of merchant in these waters, but he did not want the responsibility of carrying passengers with contraband or munitions aboard. For her westbound voyage that left from Liverpool on 17 April, Captain William Turner relieved Captain Dow as master of the *Lusitania*.

By May of 1915 the submarine controversy was far from settled. After all, it was to be a major causus belli for the United States in 1917. But the reality of submarine stealth in attacks on merchant shipping had reached fact-of-life status, though the Germans were still mindful of American attitudes. They also had found out that since the liner had been sailing since October 1914, she had been carrying contraband on her eastbound voyages. They apparently had a sincere desire to avoid causing American casualties, since on 23 April 1915, the German embassy sent warnings about sailing on ships belonging to Great Britain and her allies. They would be entering a war zone around the British Isles and would be legitimate targets for *U*-boats. Americans were informed that travelers who would ignore this warning risked their lives. The warning was apparently composed with the *Lusitania* especially in mind. Notices were sent to 50 newspapers around the United States but only the *Des Moines Register* carried the warning. The U.S. State Department had contacted the *New York Sun* and told them that no advertisement should be carried by any publication for any belligerent embassy without approval of its legal department. On 1 May a similar warning was carried by the *New York Herald Tribune*.

When the *Lusitania* left New York Harbor on 1 May 1915, the event was cloaked in a controversy largely stirred by these warnings to the American public. Secretary of State William Jennings Bryant had conferred with President Woodrow Wilson several days before the liner's departure, and recommended that no American passengers be allowed to sail on her. The President knew of the consignment of ammunition being shipped aboard her, and spent several days agonizing whether he should pursue Bryant's recommendation. Although the *Lusitania* was known to have a consignment of small arms ammunition and empty shell cases aboard, this material was not likely to endanger passengers by exploding, nor was it cargo that was in violation of American laws on carrying ammunition aboard passenger ships. The 3-inch shrapnel shells were stowed in the special magazine for 6-inch shells in Cargo Hold No. 2.

On the other hand, there is still a persistent unresolved controversy whether the liner carried bulk guncotton, clearly dangerous contraband prohibited by American law. Suspicion was stirred by a shipment from Du Pont, the well-known American chemical company that included explosives in its product line and by suspicions over the validity of the ship's cargo manifest. For example, that manifest listed a large shipment of "furs." The suspicious items may indeed have been furs, but they seemed to be an incongruous item for a wartime cargo.

She made her final sailing from Pier 54 in New York on 1 May 1915, with some 1,959 passengers on board. The cargo was entered on the manifest as foodstuffs, fur, cheese, metal rods, ingots, and boxes of cartridges. Controversy about the true nature of the cargo would persist for many years. The cheese was consigned to the Admiralty Research Establishment at Shoeburyness, a place where ordnance was designed and tested.

After the ship sailed, a *Lusitania* officer encountered three young men lurking in a storeroom. They were not passengers and they were not members of the crew. Were they stowaways, saboteurs, or what? From items confiscated from them it is believed that they intended to take pictures of the *Lusitania*'s secret gun emplacements or its cargo. No one knew, and the men themselves offered no answer, for they had no documents and they refused to speak, or couldn't speak. Captain Turner ordered the men confined to a cell (dooming them to death by drowning, as matters worked out) and ordered a search of the ship for explosives. No explosives were found, and the voyage continued.

The *Lusitania*'s 202nd crossing was uneventful, with unremarkable weather broken only by to-be-expected patches of fog once the ship had reached the southern Irish coast on the morning of 7 May 1915. As with the *Titanic*, there had been no lifeboat drill during the eastbound crossing. Thus, much of the crew and most of the passengers were unaware of their lifeboat locations.

At 1408 on 7 May, while passing the Old Head of Kinsale, Ireland, 12 to 15 miles offshore, the *Lusitania* sank within a span of 18 minutes after being hit by a single torpedo fired by the German submarine *U-20*. Of the 1,959 persons aboard, 1,198 perished, including 124 of the 159 American passengers. The time of the torpedoing was significant because with the beautiful weather and lunch concluding, the passengers and some of the crew had opened portholes. This action should not have been allowed in a war zone.

A torpedo . . . an explosion . . . that much is incontrovertible. But a second explosion followed seconds after the first and Kapitanleutnant Walther Schweiger of the submarine *U-20*, denied vehemently that he had fired a second. The testimony of survivors is so heavily weighted toward a second explosion its occurrence from a torpedo or some other cause is regarded as fact in some quarters. Beyond that, controversy dominates the history of the unfortunate ship. The second explosion *might* have been that of a second torpedo. Lord Mersey,[5] in his report on the sinking, boldly said that it was, and implied that it had been fired because the Germans acted "not merely with the intention of sinking the ship, but also with the intention of destroying the lives of the people on board." Mersey's judgment was almost certainly distorted by wartime animosity toward the Germans. When that is discounted, the source of the second explosion could have been coal dust in the bunker that was hit by the torpedo. The explosion also could have been that of boilers shocked by the torpedo explosion, or by the sudden impingement of cold seawater on the steam piping or boilers. It could have been an explosion of small-arms ammunition that was admittedly part of the cargo. It could have been "munitions," i.e., high explosives in bulk, which might have secretly been part of the cargo.

Further, the whole affair may have been the climax of a plot, masterminded by Winston Churchill, to gain political advantage by having some Americans killed as a result of a torpedo attack by a German submarine. In fact, Churchill was away in Paris the entire time the *Lusitania* approached the danger zone. Plot or not, it may have been the consequence of inadequate precautionary instructions sent by the Admiralty to the *Lusitania*'s Captain William Turner. Turner, on the other

This is a view of the *Lusitania* at speed passing the Old Head of Kinsale on one of her peacetime westbound Atlantic voyages. She would be torpedoed and sunk in this area on 7 May 1915. Note the large stacks and funnel smoke. These were distinctive features that led *U-20* to her. (*Courtesy of the Mariners Museum*)

hand, told the passengers that he expected a naval escort as he began to make his final approach toward Liverpool. In the meantime, the *U-20* had been active the previous day sinking several ships in the Irish Sea, but Captain Turner claimed until his death that he was not warned of these events. There were instructions sent to Masters approaching or in the Irish Sea regarding *U*-boat activity, but special ones were sent to Turner. These have not been divulged. Also, the instructions may have been quite suitable, but in either arrogance or ignorance, the captain (who survived) did not adopt courses that would have avoided the submarine, and speeds that would have made Schweiger's targeting much more difficult.

Here is a rich source for the conspiracy mavens, but even if conspiracy is ruled out a priori, there remains a wealth of viewpoints to be espoused by authors who would add to the literature of the *Lusitania* event. Two that expound contradictory viewpoints while telling the story are cited here.[6,7] Our recounting, following, borrows from sources such as these, but is composed only of elements that are not in dispute.

During World War II, the fast passenger liners of the North Atlantic crossed the great open expanses of that ocean unescorted, successfully depending on high speed for their protection from the *U*-boat. In 1915 the primordial *U*-boat was slower, of shorter range, and car-

The *Lusitania*, being prepared for one of her eastbound voyages to Liverpool. Note the activity at the pier, indicating departure day.

ried fewer torpedoes, but the speeds of its liner victims were almost as high as they were to be a quarter century later. The *Lusitania* had been built as an auxiliary cruiser for the Royal Navy, but her speed had been reduced by Cunard as an economy measure deemed necessary by the sharp decline in passenger revenue since August 1914. Therefore, Boiler Room No. 4 had been shut down to conserve coal and stoker wages, and also because of the wartime shortage of stokers. In fact, Captain Turner had a difficult job in finding recruits to man his ship on return voyages to the United Kingdom since many men would jump ship in New York. As a consequence of these events, maximum speed was limited to 21 knots. Despite this casting-away of the protection inherent in the ship's record-breaking speed, the 21-knot maximum should have been sufficient protection against the *U*-boat—if it were maintained through dangerous waters.

The fateful ocean crossing was uneventful, but as the ship approached the Irish coast, Captain Turner felt the need to make a landfall in order to confirm his position. He slowed his ship, for a while moving at only 15 knots in a dense fog. He also testified that he had slowed so that the *Lusitania* would arrive at the mouth of the Mersey at a moment when he would not have to stop—and perhaps really be exposed to torpedoes—to wait for the proper tide. He also asked for an escort that morning and that escort started out of Crosshaven, but was recalled. Perhaps he felt such a delaying action would be covered by the fog off the Irish coast. Once out of the fog, he later increased speed, but to only 18 knots, which was not enough to disconcert his adversary, the skipper of the *U-20*. However, Captain Turner cannot be accused of ignoring the *U*-boat threat, since he posted extra lookouts, ordered portholes and skylights covered, and had the lifeboats uncovered and swung outboard. Guidance from the Admiralty to merchant ships was clear:

- Stay away from headlands (where *U*-boats tended to lurk)
- Stay near the middle of sea lanes (more than 60 miles offshore, in this location)
- Steer a zigzag course
- Operate at maximum speed near harbors

For reasons never satisfactorily explained, Captain William T. Turner, the experienced Master of the *Lusitania,* ignored all this guidance. He plotted his course to pass Ireland about a dozen miles offshore, near the Old Head of Kinsale. At 1200 orders were sent to the engine room to maintain the steam pressure at maximum (196 psi) in case it was necessary to increase speed to the maximum (21 knots). Around this time, the fog that had shrouded the *Lusitania* during all the morning lifted and the ship's speed was increased from 15 to 18 knots. Land now came into sight on a heading of 087. At 1220, Turner ordered a course change to 067° after receiving a message from the Admiralty. The port turn was done rather quickly and it caught some passengers off balance. When the Old Head of Kinsale came into view at 1320, Turner changed course back to 087°. The liner continued straight on the latter heading, to make a straight run for a period of 40 minutes starting at 1350. Captain Turner testified at the Mersey Inquiry that his intentions with all these maneuvers were to bring his ship into Liverpool at high tide at 0400 on 8 May 1915.

The *U-20* had reached the Irish Sea via the North Sea and a passage southward around the west coast of Ireland. German intelligence had received information of the departure of a number of transports for the eastern Mediterranean, carrying troops for the Gallipoli Campaign. In

The *Lusitania* begins her last eastbound voyage on 2 May 1915. Her funnels are painted black as she moves down the Hudson River and a number of passengers are lined at the rails. A total of 1,198 passengers and crew would lose their lives in this tragedy of the sea and 128 were Americans who were warned by the German Embassy in Washington not to sail on the ship. (*Courtesy of the Mariners Museum*)

turn, British naval intelligence was aware of the German intentions, having broken the German naval code. In the Irish Sea between England and Ireland, Schweiger had sunk several merchant steamers, prompting the warnings that were sent by the Admiralty to the *Lusitania*. Suddenly, heavy black smoke appeared on the horizon. At the moment of a coincidental encounter with the *Lusitania,* the *U*-boat captain was beginning the return passages by retracing his outbound route around Ireland. The German submarine submerged, increased speed, and headed northeast for an interception that would provide him with a good torpedo firing position.

The encounter with the *Lusitania* was fortuitous. Schweiger's thoughts were on the long passage home, but while he was still in sight of the Old Head of Kinsale, there was his famous victim-to-be with speed and course that would give him no trouble in reaching a good firing position. If the target continued on the present course or altered it slightly to port, the *U-20* would have had trouble overtaking it. Recognizing that the ship was either the *Mauretania* or *Lusitania*, he believed he was about to sink a troop ship that he had been instructed to attack. Actually the *Mauretania* was at Avonmouth loading troops for the Dardanelles Campaign. However, Captain Turner made a course alteration to starboard that brought the *Lusitania* on a heading of 087° and directly across the path of the *U*-boat.

This early in World War I German torpedo attacks were failures more than half of the time, often for reasons unknown. The torpedo warhead might be a dud, the ejection equipment was unreliable because of faulty triggers or other problems in the mechanism, torpedo speeds were modest, the torpedo's steering mechanism could take it on erratic courses, and torpedo fire control techniques were decidedly primitive. From February to September 1915, only 40% of torpedoes fired found their targets, a situation that would be repeated in German and American submarines in World War II. However, there was to be no failure this time. The torpedo warhead made direct contact with the hull of *Lusitania*, insuring that the trigger would perform in igniting the tytrol. Discounting the human tragedy that was involved, then the *U*-boat's attack would be called a resounding success.[8]

Shortly after sighting the *Lusitania*, Schwieger fired a single torpedo at a range of 500 meters, with a depth setting of three meters. Only two torpedoes remained on board for possible use on the long passage home.

Aboard *Lusitania,* Turner had ordered double lookouts to watch for hostile activity. One of the bow lookouts, Able Seaman Leslie Morton, spotted the torpedo wake and shouted to the Bridge, "Torpedoes coming on the starboard side!" Able Seaman Thomas Quinn, aloft in the Crow's Nest, also spotted the approaching torpedo and reached for the telephone to alert the Bridge. However, their sightings were too late for any evasive maneuver to be effective. Second Officer P. Hefford heard Morton's warning and repeated it to the Bridge personnel. Captain Turner was on the lower bridge taking a four-point bearing on the Old Head of Kinsale. When he heard Hefford's warning, he turned and took one step toward center of the lower bridge. He had a quick glimpse of the white wake left by the torpedo. The torpedo struck the starboard side at a point that survivors' consensus placed somewhere near the mid-length of the hull. Captain Turner ordered a port turn that would take the stricken liner towards shore where he could possibly beach her.

There were loud hissing noises from below—the sound of broken steam pipes. As they burst they sounded to some of the survivors like the "rattling of a Maxim gun."[9] The engine telegraphs no longer worked and the electric lights began to flicker as the steam pressure began to fall in the steam line to the turbo-generators.

Marconi Operator Robert Leith was eating in the second-class Dining Room when the torpedo struck the ship. He instantly left his place and ran up the stairs and through the passages to his Radio Room. He relieved his junior assistant and sent out a message:

"Come at once, big list, 10 miles south of Old Head Kinsale MSU." As he typed out the message, he noticed that the electric power was beginning to fail. Only several more messages were sent.

First-class Passenger Charles Lauriat was on the Promenade Deck and described the torpedo impact as a "heavy, rather muffled sound." He also witnessed the shower of coal, steam, soot and coal that James Brooks had seen. He also saw gratings and other debris fall onto the superstructure. Soot was sprinkled on his white shirt. He immediately saw the ship list to starboard and going down by the bow. His first thought was that the ship would sink, but she steadied herself, lulling him and others that somehow the ship might still survive.

The torpedo explosion was followed very quickly by a second violent explosion. This explosion apparently doomed the ship. Some survivors testified that it followed about 30 seconds after the first. To Charles Lauriat it sounded like a boiler explosion. Lord Mersey, apparently wishing to impute as much evil to the Germans as possible, simply stated in his findings that the second explosion came from a second torpedo hit. Those wishing to impute as much evil to the British as possible declare that it was the secret cargo of military explosives, touched off by the first and only torpedo. Since the torpedo hit in way of a coal bunker, the explosion of coal dust has been suggested. Since this was also the locale of boilers, the explosion of one or more boilers has been suggested. At first, a steam explosion seems to be the mostly likely phenomenon. Reasons are to be outlined in Chapter 25, devoted to the question of boiler explosions.

The torpedo struck at 1408 GMT on 7 May, announced by a heavy detonation and a large plume of water up the starboard side. The ship itself vibrated in a severe whipping response to the explosion. Witnesses report that a lifeboat, deck planking, coal, and coal dust were flung skyward. The lifeboat might seem an improbable part of the geyser of seawater, but William Holton, a bellboy whose lifeboat station was Number 5, reported seeing it immediately after, drifting away astern of the ship. Some other survivors testified that this lifeboat was damaged when ripped away from its davits by the plume of the torpedo explosion.

The tearing-away of a lifeboat without severe damage within the ship in the immediate vicinity of the boat is strong evidence that the cause was the water plume from the torpedo explosion, i.e., not the second explosion (unless it too was the discredited second *torpedo* explosion). The identification of the boat as Number 5 is a convincing indicator of where the torpedo struck, i.e., at a place forward of that lifeboat's location. This is particularly true because all the lifeboats were in lowering position because of a submarine threat, making them vulnerable targets to a rising or descending plume from a torpedo detonation. It would not be a descending plume because these are not free-floating masses that go up and then down. Plumes from a torpedo explosion are continual streams of water that go up and then fall down. How could a plume go up, skip Lifeboat Nos. 1 and 3 and then knock No. 5 off the ship?

Eyewitness accounts of sudden, life-threatening events are typically muddled and contradictory, but an analysis of Holton's testimony concerning Lifeboat No. 5 helps to pinpoint the location of the torpedo explosion and supports the theory that the torpedo hit just after the Bridge. Subsequent chaos led to many contradictory accountings of the consequences of that explosion, but the outline of events is beyond contradiction: the ship quickly took a heavy list to starboard, flooded rapidly, listed to a point that allowed seawater to pour down the stacks, and was gone within 20 minutes (18 minutes is the consensus figure).

Some witnesses spoke of the torpedo hitting between the third and fourth stacks, while Schwieger, peering through his periscope in the *U-20*, observed the torpedo to have hit near the Bridge. There is enough evidence to validate those claims, including Captain Turner, that the torpedo hit between the third and fourth funnel. The most immediate and severe damage below decks was in Boiler Rooms Nos. 1 and 2 which agrees with what Schwieger saw. There was a total of four surviving crew members from the two forward boiler rooms. Both testified before Lord Mersey, describing narrow escapes that involved scrambling up through ventilators. Water poured into the two forward boiler rooms through any opened coal bunker doors, and in Boiler Room No. 1, through tears or failed riveted seams in the longitudinal coal bunker bulkhead. According to the operating instructions given to the Mersey *Titanic* Inquiry, Naval Architect Leonard Peskett testified that these coal bunker doors were closed each day at 12 noon as a precautionary measure. He also testified that there were no problems in opening and closing these doors with coal clinkers or dust in the sills. The longitudinal bulkheads, while regarded by some authorities as contributors to watertight integrity, served in the diametrically opposite sense—they limited the cross-flow of water so that the ship quickly listed to starboard. On the decks above, particularly the Lower Deck, a few feet above the waterline, many portholes were open, either opened by passengers or the torpedo explosion fractured their glass. As the starboard side dipped low, many of those on "E" Deck were immersed, exacerbating the peril by allowing rapid flooding much higher in the hull than the boiler level. Soon those on "D" Deck also were immersed. The flooding of these spaces outboard on the starboard side added to the unsymmetrical flooding that occurred in the coal bunkers below. The nontight subdivision on "D" and "E" Decks acted as a temporary-flooding boundary until a sufficient hydrostatic head caused them to rupture.

Torpedo Hit and Damage.[10] The torpedo struck just forward of the midpoint between two main watertight transverse bulkheads, but closer to the one that divided Boiler Room No. 1 and the Coal Bunker.[10] The point of impact was also just forward of a discontinuity in the longitudinal coal bunker bulkhead. Due to the fine lines of the hull needed to make speed the designers had found it necessary to step this bulkhead inboard to provide sufficient bunkering capacity in the forward end of Boiler Room No. 1. Evidence of this location as the epicenter of the torpedo explosion is the ripping away of Lifeboat No. 5 from its davits. Accounting for the ship's forward speed, this would place the torpedo impact point about 90–120 feet (see Figure 17-1) forward of the center of Lifeboat No. 5. James Brooks, a passenger who was near the point of the torpedo impact, was sent sprawling to the deck by the effects of the detonation. Mr. Brooks recalled:

> There was a dull explosion and a quantity of debris and water was flung into the air beside the Bridge. The waterspout knocked me down by the Marconi office. This explosion seemed to lift the ship hard over to port and was followed soon after by a second rumbling explosion entirely different from the first.

The debris that Brooks referred to came from a large ventilator on the starboard side of the forward funnel. Steam, water, smoke, coal and other debris spewed out of this ventilator towards an advancing Mr. Brooks. The water from the torpedo's plume, knocked him down on the deck. Suddenly he was enveloped with a cloud of steam and its concentration was almost suffocating. However, the ship's forward speed of 18 knots was unabated at that instant and the air quickly cleared. However, he was covered with soot from head to shoes. Brooks quickly got up and ran across the Marconi Deck down to the portside of the Boat Deck below.

The explosive charge of the torpedo was about 300 pounds of trotyl (a variant of TNT). This amount of explosive was much less than was used in World War II torpedoes. However, observations[11] of torpedo and mine damage to merchant ships in World War I established 40 feet as the typical fore-and-aft length of hull damage from a single torpedo hit. The hole created by the torpedo explosion was very near the waterline.

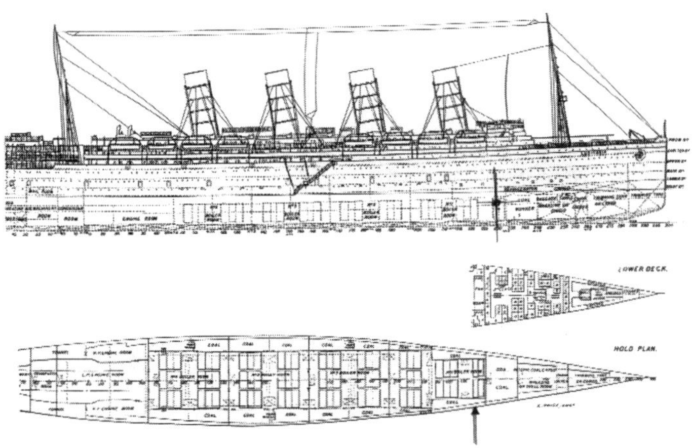

FIGURE 17-1 These plans show where we believe the torpedo may have hit. The Baggage Room sustained heavy damage and a number of able-bodied seamen were killed there. The actual location of the torpedo hit is difficult to establish because the ship is resting on her starboard side in sediments.

Further, rivets were probably sprung along longitudinal seams for as much as 20 feet on either side of the 40-foot penetration. Experience with the *Titanic*, plus experience gained from the forensic examination of the carrier USS *Yorktown* (CV 5) in 1998 suggests that these opened seams could also be sources of flooding that contributed to her sinking. The hole created by the torpedo's explosion was equivalent to that of a peacetime collision, and the *Lusitania*, like other British liners built in that period, were two compartment ships; that is, they could sustain flooding of two compartments and still survive.

Further trouble had its source in the location of the torpedo hit near a discontinuity in the longitudinal bulkhead. At this location, this bulkhead stepped inboard to allow additional space for coal in the forward boiler room. An explosion at or near that point could cause enormous structural damage. The shallow setting of the torpedo placed its detonation close enough to the Lower Deck to cause the rupture of that deck's plating in way of what had been third-class passenger quarters as the explosion vented itself inward and upward. Overall, the damage produced asymmetrical flooding relatively high in the ship, seriously compromising the ship's transverse stability. This was also a case of high flooding and low buoyancy intact that is very detrimental to a ship with a low metacentric height (GM).

The explosive penetration into the ship might have been approximately 10 feet, a distance that over most of the boiler-room length would not have ruptured the inboard bunker bulkheads, generally 16 feet inboard from the shell. However, in the area of the explosion, these bulkheads were much closer to the shell. Further, the inboard step of the bunker bulkhead in the vicinity of the explosion may have experienced flooding through fractured plates and failed riveted joints or opened or not-tightly-closed coal bunker doors. It was fortunate that the torpedo hit high in the coal bunker because the presence of coal and air space allowed some dissipation of the energy of the explosion.

It appears from crew survivor testimony that the gas jet from the torpedo's explosion did not penetrate into Boiler Room No. 1 or into the stokehold that was immediately forward of the boilers. However, the testimony of Leading Fireman Albert Martin, a survivor from Boiler Room No. 1 and one of the few surviving the sinking, indicates that there was some failure in the longitudinal bunker bulkhead. He maintained that water burst through this bulkhead in the manner of water bursting a paper bag. Water also poured in through the access doors through the bunker bulkhead, and the tear in the plates or separation of riveted seams in that bulkhead. The water was approximately 18 inches deep on the floor plates when Martin began his escape up ladders and through ventilation ducts. The list and the current of water pouring in made him lose his footing, but eventually he found his way to an escape ladder and climbed up to the Boat Deck through a ventilating shaft.

Fireman Thomas Madden and Trimmer Frederick Davis were working between the two forward boilers in the forward stokehold at the forward end of Boiler Room No. 1 when the torpedo hit. The concussion of the explosion threw him to the grating and the noise temporarily made him and some of his mates deafened. Madden maintained to a Board of Trade inspector at Liverpool on 28 and 19 May 1915 that the torpedo explosion seemed to be centered forward of his location. According to his testimony no water had come into the boiler room for 2–3 minutes and he had enough time to go aft to the watertight bulkhead but found it shut. Steam began to collect in the space and Madden was afraid of a boiler explosion. The lights in the boiler room then went out. Going forward in the boiler room, he noticed that water was pouring into the space through open coal bunker doors. He and Davis made their way to a ventilating shaft to make their escape. On the way, there was about 18 inches of water on the grating and with the list to starboard, keeping their footing was very difficult. Madden maintained that the boilers in Boiler Room No. 1 did not explode while he was in the boiler room or the ventilating shaft.

As noted earlier, studies of riveted ships subjected to explosive forces indicate that leakage through riveted seams can take place as far away as 40 or more feet from the point of detonation. Therefore, the ingress of water would not be confined to the hole ripped into the shell plating by the torpedo's explosion; sheared rivets and parted seams would have allowed water to pour in over a wider area. In Boiler Room No. 1, at least, the flooding would have been at a rate that precluded any attempt to draw fires, or even to lift the boiler safety valves to reduce steam pressure. However, since the wreck lies on its starboard side, the damage to the hull has never been confirmed by examination. To the date of this writing, the difficulty of forensic examination makes it impossible to know exactly what transpired on the starboard side in way of the torpedo hit on the *Lusitania*.

The severe whipping response of the hull to the explosion may also have been a contributing factor in the flooding, causing failures of rivets and parting of riveted joints. Also, portlights may have also shattered from the shock response. Such failures only increased the area of flooding.

Since the torpedo was set for a shallow 3-meter depth, much of its energy was vented into the atmosphere, as evidenced by the large water plume.

"Good commercial design practice" from a later era of ship technology advises that for best potential damage resistance, wing compartments on merchant ships should encompass about 20% of the maximum beam on each side. At best, if this doctrine had been followed in design of the *Lusitania,* the effects of the torpedo detonation would have been contained within the wing bunker system. The 16-foot width of the bunkers near amidships was close to that standard (20% of beam was 17.6 feet). Unfortunately, the longitudinal bulkheads forward, especially in way of Boiler Room No. 1, were much closer to the side shell than that. In addition, the longitudinal bulkhead was stepped inboard about midway of that space. The damage to the hull led to the flooding of the forward boiler room and outboard passenger quarters on the Lower Deck over this space, as well as slow flooding in Boiler Room No. 2.

The explosion took place ten feet below the Lower Deck where there had been a number of Lower Deck third-class passenger staterooms over the coal bunker and forward boiler rooms. When that deck dipped below the water line, the joiner bulkheads that bounded the passenger staterooms just inboard from the shell acted as a flooding boundary for a short period of time until the hydrostatic forces became large enough to bring about their failure. However, by this time there was enough water trapped between them and the side shell of the ship to make a significant contribution to the list caused by the flooded bunkers.

Unable to flow to the port side, the water confined to the starboard side and led to a buildup of water that caused a list of 15 degrees that would soon increase rapidly—and fatally—to the point of the ship's capsize. Such a list made footing difficult and compromised the safe launching of lifeboats.

In addition to causing the flooding of Boiler Room No. 2, the parted seams and loosened rivets caused by the shock effect of the torpedo explosion may have allowed some slower flooding into the cargo hold that contained some of the contraband munitions.

Some water came in from the starboard bunker of Boiler Room No. 2 through the open bunker doors. This flooding probably originated from the failed rivets and parted seams in the shell or in the main transverse bulkhead structure between Boiler Rooms Nos. 1 and 2.

Trimmer Ian McDermott was thrown to the grating in his workplace when the torpedo exploded, but he claimed that he did hear Chief Engineer Archibald Bryce order that everyone should go inside at once. He was the only one from Boiler Room No. 2 to survive the sinking. Like Albert Martin, Thomas Madden, and Frederick Davis in Boiler Room No. 1, three of the survivors from Boiler Room No. 1, he was forced to escape through a ventilation shaft because the list and the pressure of the water prevented the overriding mechanism for the watertight doors from operating. As he made his way to the escape ladder in darkness, he was waist deep in water.

The watertight doors that gave access to the coal bunkers were almost certainly open when the torpedo hit. These doors, perhaps, were difficult to close because the accumulation of coal dust and particles prevented an efficient closure, even though Leonard Peskett testified otherwise in the 1912 Mersey Inquiry into the *Titanic* sinking. This situation was a prime factor in the loss of the Cunard liner *Oregon* in 1886 off Fire Island, New York. Another problem with stowing the coal in outboard bunkers was its permeability when the bunkers were flooded. Not only did water fill the interstices among the coal, but the coal itself also absorbed water. This outboard arrangement of coal was very detrimental to transverse stability from raking, mine, or torpedo damage in way of the propulsion plant without some cross connections between bunkers in the same subdivision. The *Lusitania* had none of these features. The loss of transverse stability was particularly detrimental in passenger ships, whose operating GMs were generally much lower than warships.

Second Explosion. All survivors commented on a second, duller-sounding explosion that occurred almost immediately following the torpedo detonation. It is difficult to conclusively establish the time interval between the two explosions, although one survivor believed it to be about 30 seconds. The sound of the secondary explosion was markedly different from the initial torpedo detonation. One survivor likened it to "a million-ton hammer hitting a steel boiler," while another likened it to "a ship grounding on a rock." In less than 10 seconds, the *Lusitania* listed 15 degrees to starboard and much of the ship was plunged into darkness as the electrical power failed. This power failure was probably the consequence of severing the steam lines to the turbo-generator engines. The main steam line to the turbo-generator room may have originated in Boiler Room No. 1. This line was probably ruptured, if not by the torpedo hit, then by events that took place in the engine room.

Around the time the second explosion occurred, a mixture of steam, soot, coal dust, and smoke burst from a starboard ventilator at the base of the first funnel. James Brooks, a surviving passenger testified that this explosion seemed entirely different in character from the first, and filled the air with a dense steam cloud that made breathing difficult. He was sure that the second explosion was not from a torpedo, but was steam induced.[12] What James Brooks probably saw was steam, soot, and other debris that came up from below because of the venting of some of the torpedo explosion up the starboard ventilator that was very close to the area where the torpedo impacted.

Senior Second Engineer Andrew Cockburn was on duty in the forward engine room on the upper level of grating. There was a deafening concussion and a blinding flash of light. Within seconds the lights in the engine room went out, followed by the sound of escaping steam. The air became difficult to breathe as it was filled with coal smoke.[13] This combination indicates steam pipes rupturing. As Cockburn held onto the railing, he then heard water flowing into his space. He guessed that the damage had occurred in the forward two boiler rooms and those were quickly flooded. Cockburn also had seen Chief Engineer Archibald Bryce and much of the engineering staff in the forward engine room. None of these men would survive.

The second explosion also hastened the end that was likely to have been inevitable even without this added blow. Prior to that Captain Turner noticed that the indicator lights for the Pearson fire and flooding alarm board for the forward compartments went berserk. He made a vain attempt to head the ship toward shore, as the Old Head of Kinsale was only twelve miles away. He then asked Second Officer Heppert to have the yeoman carpenter to check the forward compartments for flooding. He then shouted to Heppert to give him the angle of list from the list indicator. Staff Captain John Anderson sent a message to the Bridge suggesting the flooding of the port trim tank.[14] Captain Turner also saw the bow trim also increasing, so he ordered the port-side trim tanks ballasted to stabilize the trim and list so that the lifeboats could be safely lowered. He knew that this would decrease his buoyancy, but the list and trim were alarming indicators that his ship was probably doomed. However, the engineers who were to open the valves had left the space because of the large list. This is not surprising since any list over 15 degrees makes walking difficult and strongly signals the need to abandon ship. Unable to have those spaces ballasted, Captain Turner raced up to the navigating bridge where he could view the status of the lifeboats. The list to starboard caused the port boats to swing in, while the starboard ones were swung out, making it difficult to launch these boats successfully and safely. Also, he noticed smoke and coal dust poring from the ventilators and blowing toward him. Recognizing a need to slow the ship's forward motion, he ordered a FULL SPEED ASTERN and the engine rooms dutifully complied.

Down in the Engine Room, Third Engineer George Little received the engine order telegraph of setting of FULL SPEED ASTERN. When Little applied pressure to the astern turbine something disastrous happened to the safety valves. Before the *Lusitania* sailed Captain Turner requested that the Board of Trade inspect the ship; Inspector Laslett told Captain Turner and Chief Engineer Bryce that FULL SPEED ASTERN should not be used in an emergency because this would place too much strain on the safety valves. Many of the Royal Navy's battleships and battle cruisers of that time had this problem and still put to sea in wartime. One of the two main steam pipes was claimed to have fractured and it was reported to have blown off the top of the condenser on the Boat Deck. (We discount this report for reasons stated in Note 7.) It does appear that in engaging the smaller astern turbine resulted in an overabundance of steam into the astern turbine before valves regulating that steam could be properly set. The propellers and shafting had not yet stopped their forward motion and the smaller astern turbine could not overcome that motion. In his haste to take the steam pressure off the astern turbine, Little pushed the engine room telegraphs to FULL AHEAD to reduce the steam pressure surge caused by the astern maneuver. However, the steam safety valves could not take the accumulation of steam in the lines and these valves burst open, sending a cloud of steam to the Boat Deck. The steam pressure then dropped from 196 to 50 psi. In any case, continuing the ship's forward speed with the severe list already developing would have increased the risk in launching the lifeboats. The loss of the steam lines also cut the source of power for the turbo-generators and the steering gear. As a result, the lights failed and there was no steering function.

Meanwhile, when Captain Turner saw the steam bursting from the safety valves, he ordered a turn to port that would head to shore. The helm would not respond. There was nothing left to do but evacuate the passengers and crew.

The ship was already plunging by the bow and taking a list greater than twenty degrees. In four minutes after the torpedo exploded, the foredeck was under water. Chief Officer Piper told Captain Turner that he was going to the Fore Deck and secure the hatches, saying, "I am going down to the forecastle to help Hefford with the hatches. She is sinking fast by the bow, perhaps we can slow her down a bit." Turner nodded agreement. However, water began to pour into the hatches. Piper and Hefford were not seen again. Fifteen minutes after the torpedo struck, the list was 25 degrees and the water was coming over the starboard side of the Bridge Deck. The ship kept plunging by the bow until her rudder and propellers were clear of the water. Just before the ship finally sank, the bow struck the granite seabed, and the *Lusitania* began to pivot around to starboard (apparently there was still some forward motion at this time). The combination of list and trim allowed seawater to pour into the funnels. Two surviving passengers, Margaret Gwyer and William Pierpont, were struggling in the water nearby, and reported being swept into the third funnel by the inpouring water. Both were saved after being suddenly expelled by abrupt outrushes of warm water, apparently steam mixed with air and soot. (The likely cause of their miraculous escapes is discussed in Chapter 27.) The force of this explosion detached funnel number 3 from the ship. Without question, given the time at which this event must have occurred, this could not have been the disputed second explosion.

Shortly after the torpedo exploded, Captain Turner asked Third Engineer Albert Bestic to go down to the forward Baggage Room where many of the able-bodied seamen were working to bring the passenger baggage topside. Turner wanted these men to go topside and handle the boats. Bestic decided to go to his quarters first and put on his work clothes before going to the Baggage Room. When he got there, he found out that a number of these able-bodied seamen had been killed and access to the Baggage Room was blocked. He then went to the portside of the Boat Deck to begin the evacuation of the passengers.

The time span between the torpedo hit and the foundering of the *Lusitania* is usually cited as eighteen minutes. There were lifeboat accommodations for 2,605 persons, nominally quite enough for the 1,959 persons that might have been saved (less those trapped or killed by the explosions, of course). Lord Mersey states:

> They [passengers who testified] speak in terms of the highest praise of the exertions made by the crew. No doubt there were mishaps in handling the ropes of the boats and in other such matters, but there was, in my opinion, no incompetence or neglect, and I am satisfied that the crew behaved well throughout, and worked with skill and judgment.

One might question whether 18 minutes would have been sufficient to place almost 2,000 persons safely afloat in lifeboats, even in the absence of list and forward way on the ship. An important factor of doubt is the fitting of only 22 pairs of davits, with the boats they primarily served having a total capacity of 1,323. After these boats were in the water and released from their falls, those falls would have to be recovered before the second tier of boats was launched. As the event unfolded, the ship had a severe list to starboard long before any boat was even loaded with its 60-person capacity. Launching of portside boats was extremely difficult, and soon became impossible when the list went past 20 degrees. The difficulty was further compounded by the lack of enough able-bodied seamen, many of whom were lost while retrieving passenger baggage just before the torpedo hit. Many of the ordinary seamen were not familiar with the lifeboat gear. Boarding of the starboard-side boats was also difficult because when lowered to deck-edge for access, they tended to hang too far away from the deck for the safe entry of the passengers. Furthermore, the *Lusitania* and *Mauretania* had a history of rolling in a seaway so that their lifeboats would rock back and forth. The inward motion brought these lifeboats into contact with the collapsibles, damaging the latters' canvas gunwales. To stop this motion, a six-foot chain was attached to the deck and secured to the boat with a wooden peg. A wooden mallet was provided to remove the peg. However, some of the seamen attending the boats were unaware of the procedure to unfasten the chain. In the late stages of the sinking, the list had become so severe that work on deck of any kind was impossible. Very few of the survivors escaped without spending time afloat in their lifejackets.

The water temperature was about 45° Fahrenheit, not as chilling as it had been during the *Titanic* disaster, nor was rescue very far away. However, the water was cold enough to cause hypothermia after a period of time. This must have accounted for the drowning deaths of many who would otherwise have survived. There was no *Carpathia* to risk its own existence in a heroic attempt to reach the sinking ship before it was too late. No need, it would seem, since the Irish shore was in sight, and vessels of many sizes and descriptions were available in nearby Queenstown, about 25 miles away. Wireless quickly informed them of the event, but getting underway, and traversing the seemingly short distance was a matter of several hours. However, enough vessels arrived before dark to rescue 761 persons.

Cause of the Second Explosion. As one author, an advocate[15] of guilt lurking in the Admiralty, described the disaster as thus:

> The result of Schweiger's torpedo was the flooding of the starboard coal bunkers and a 15-degree list. The result of the second explosion was the sinking of the *Lusitania*.

The question that had yet to gain an incontrovertible answer some 90 years later is "what caused that second explosion—coal dust . . . munitions . . . gun cotton . . . boilers?" At first, we believed the source to have been a steam explosion. Given that a fairly lengthy technical background is needed to support that conclusion, the discussion and advocacy is reserved for a separate chapter (Chapter 25).

Another prime candidate is guncotton. It is known that the British government purchased quantities of guncotton from Du Pont for the explosive filler for their mines. If this material was carried in the converted spaces over the boiler rooms or in the forward magazine, the devastation wrought by the explosion of guncotton could have destroyed the vessel's ability to survive. Guncotton was made in an infinite number of grades. However, the most important grades had a specific ratio of nitro groups to carbon groups. This provided optimum power as well as stability.

Dry guncotton is quite unstable but was used as a primer. If dry guncotton were heavily wetted, chemistry and explosion testing background would possibly indicate that the heat generated would be small relative to the suppression effect of the water. If some of the dry guncotton became damp and bacteria started its decomposition, then over a period of days or weeks an explosion could happen much like spontaneous combustion of overly damp hay or sawdust. Dry guncotton can also be readily ignited by projectile ignition or from hot fragments as well as fire or a nearby explosion.

Wet guncotton would need properly sealed containers to be shipped because it must retain its moisture. Otherwise, it would be very dangerous to let wet guncotton become dry. In its wet form, guncotton is highly resistant to projectile ignition. However, wet guncotton can be ignited by fire or a nearby explosion.

Based upon the chemistry and type of guncotton, the following scenarios are offered as possibilities if some form of guncotton were present on the *Lusitania* near the area where the torpedo hit.

- If dry guncotton were stored in the forward magazine (the converted forward coal bunker), it could have been set off from a flying piece of metal from the torpedo explosion or from the ammunition, which might have also fired off in a nonspecific manner in reaction to the torpedo. However, the dry guncotton simply becoming wet would have had no explosive effect.
- If wet guncotton were stored in the forward magazine, it could be set off by the torpedo explosion. Even if the Germans had attempted to corner the market on the normal shipping containers for wet guncotton, it is assumed that Du Pont would have the ability to find adequate replacements for them.
- The influx of water to the guncotton would have had no bearing on the explosion.
- Since the forward magazine was near where the torpedo exploded against the hull, anything in the magazine would have been very close to the torpedo explosion.

Until the wreck of the *Lusitania* is thoroughly investigated to determine the plausibility of the involvement of guncotton, it can be considered as another possibility for the second explosion. However, we believe that a combination of exploding guncotton and hot boilers and piping coming into contact with cold seawater certainly would be enough to cause multi-compartment (more than two major watertight subdivisions) flooding for which the ship was not designed to survive.

The munitions answer is favored—indeed, insisted upon—by the those who believe that the *Lusitania* was secretly carrying an explosive cargo of war material, and that the fact was to be concealed by the British at all costs to avoid the inevitable rage in America, homeland of 124 victims. The contention is that the torpedo hit in way of the cargo hold that extended between 100 and 140 feet (approximately) of the bow, and that its blast front touched off the munitions stowed there. Support is claimed from the state of the wreck over that length of the hull. The port side—opposite the stricken side—appears to show plates blown outward by an internal explosion.

Critics of the munitions advocacy believe that the port-side hull damage is an outward rupturing of this narrow forward part of the hull that occurred as the bow of the stricken ship pivoted on the granite seabed. They also question why the ship heeled so severely and so suddenly to starboard if the two explosions were both far forward, and blasted completely through the hull.

Coal dust, when suitably mixed with air, is explosive. It is a likely event in coal mines, but has rarely, if ever, been reported on a ship at sea. Near the end of an Atlantic eastbound crossing, coal might well have been low in the side bunkers, leaving an air space filled with a potent mix of fine coat dust. The reserve coal bunker would still be filled during an eastbound voyage. The fiery torpedo blast through the hull plating would presumably ignite the dust. Kapitanleutnant Schweiger, observing events through his submarine's periscope, commented on the unexpected magnitude of the torpedo explosion. Perhaps he was seeing a torpedo blast magnified by a near-simultaneous explosion of coal dust. The "second" explosion (i.e., the one reported to have followed about 30 seconds later) then would have actually been a third explosion. If so, it is unlikely to have been coal dust because the torpedo explosion would have brought a quenching surge of seawater into the bunker.

It is very unlikely that most of the sound from the second explosion came from the detonation of some or all of the shrapnel shells in the forward magazine that was located below the baggage room that Third Officer Bestic found to be inaccessible. The casualties in that space point to some event happening below. In addition, this magazine was located within 50 feet of the torpedo's explosion raising the distinct likelihood of a sympathetic detonation. It was claimed by British authorities that these were empty shell casings, but that could be false.

Before World War I, coal bunkers were regarded as part of a warship's protection against explosive attack. Tests were performed, particularly by the Royal Navy, by firing shells into both full and partially full bunkers. No coal dust explosions were recorded.

Fires in coal bunkers, ignited by spontaneous combustion, were an occasional occurrence in the era of coal as the dominant sea-going fuel. Indeed, the *Titanic* was subjected to a stubborn bunker fire over several days after leaving Belfast. These fires (which also can occur in shore-side bunkers) received some attention from investigators, so that a modest literature on the phenomenon exists. But coal dust *explosions*—there's no possibility of stating unequivocally that they did not occur at sea, but the absence of references in the marine literature indicates that they were quite rare. Further, the belief that a dust explosion ignited by a torpedo would occur almost simultaneously with the explosion of the torpedo itself leads us to declare against coal dust as the source of the widely recognized second explosion. A strong argument against a coal dust explosion is that the coal bunkers were located abreast of the boiler rooms, and hence the heat and humidity that characterized those spaces would also be present in those bunkers. With 45-degree seawater cooling the outside of the shell plating and the warm air coming into contact with this moist air, puddles would form in the floor of the coal bunkers. Such conditions were far from ideal for the dry air needed for a coal dust explosion.

Many ships sank without an accompanying explosion of their cylindrical boilers, but such events *did* occur, and with sufficient frequency that the possibility was recognized whenever a ship sank. For an instance quite pertinent to our discussions, boiler explosions during the *Titanic* sinking were reported by some survivors. We now believe that those reports were false, that the explosive sounds the survivors heard came from the rending of steel structure as it failed during the late stages of that ship's sinking. The point of argument at the moment is that the phenomenon of boiler explosions during sinkings was well enough known that some survivors thought "boiler explosion" when they heard loud noises during the late stages of the *Titanic* disaster.

As recounted earlier, two *Lusitania* passengers reported being ejected from the ship by sudden eruptions of hot water that could have originated in boiler explosions. However, these lucky events happened late in the sinking, and hence are unlikely to be linked to the legendary second explosion.

The boiler spaces of the *Lusitania* have never been examined, so that no certainty can be assigned to explosions from that source. Even if the boiler rooms could be accessed, the hull has begun to collapse and the present beam of the ship on the seabed is but a fraction of what it was on the day the ship sank. The forces involved in the collapse could have masked the damage done by the aftermath of a steam explosion. We nonetheless believe that the origin of the second explosion was probably caused by a combination of steam and munitions aboard the *Lusitania*, happening soon after the torpedo hit. Our justification is that such events are known to have occurred in other cases, and that such an event aboard the *Lusitania* is less unlikely than the alternatives of coal dust or munitions explosions. The only other plausible alternative is the detonation of some of the 3-inch shrapnel shells in the forward magazine. We also believe that there was some collateral damage from steam-related explosions at the same time. A final assessment of what actually caused the second explosion has to wait until a full marine forensics analysis of the wreck can be accomplished.

Further discussion of boiler explosions is found in Chapters 25 and 26.

Captain William T. Turner.[16] After the passing of eight decades punctuated by world wars, cold war, and other conflicts of varying intensity, the role of Captain Turner in the *Lusitania* tragedy remains a puzzle and an enigma. There are some who believe that he did not reveal all that he knew at the Mersey Inquiry.

In May of 1915, Will Turner was 50 years old, a veteran of more than 40 years of life as a seaman. A Cunard employee since 1878, Turner had achieved his first Cunard command in 1903. Famed for his proficiency as a Master and uncomfortable with the socializing with the elite expected by the Master of any great ocean liner of the day, William T. Turner was in his glory on the Bridge of his ship. At the end of 1909, Captain John T. Pritchard, who had commanded the *Mauretania* since her maiden voyage, retired and was succeeded by Turner.

When the *Lusitania* departed New York City on 1 May 1915, Turner was in command for the second such voyage since replacing Captain David Dow, who had complained of ill health and the strains of command in time of war. This was also Turner's second assignment by Cunard as Master of the *Lusitania*. He had previously been Captain of the *Mauretania*, but she had been laid up since the start of World War I.

Tension or friction between officers of the merchant marine and their naval counterparts has proven to be the norm, especially relatively early in wartime when the exigencies of naval requirements conflict with the peacetime practices and experience of the civilian mariners. With the passage of time, the harsh lessons of war gradually ameliorate this friction, leading to improved understanding of the reasons for naval requirements.

By May of 1915, Royal Navy wartime experience had been codified by a series of secret advisories given to the masters of merchant ships. Experience had already shown that compliance with these procedures could enhance the likelihood of safety of these ships. Such procedures, of course, could not *guarantee* safety. Ignoring the hard-learned practices, conversely, enormously increased the likelihood of disaster. Captain Turner, in part because of habit and concern for the comfort of his passengers and in part because of policy direction from Cunard management, ignored all too many of the Admiralty advisories, setting the stage for an epic tragedy. However, Captain Turner was not well served by the Admiralty either. Several ships had been torpedoed within 24 hours of the *Lusitania*'s arrival off the Old Head of Kinsale. Perhaps an escort should have been provided to escort the ship into Queenstown. However, such an escort would have made the ship more at hazard to a legitimate German submarine attack.

Notes for Chapter 17

1. The sources for a recounting of the naval history of World War I and the prior building race between Germany and Great Britain are overwhelming in number. For example, the catalog of the University of Michigan libraries has 628 entries under the subject heading "Great Britain—Naval History." German sources (particularly in English) are much fewer in number. One of them that serves quite well is Admiral Alfred von Tirpitz, 1919, *My Memoirs*, Dodd, Mead & Co., New York.
2. A large part of the Tirpitz memoirs (see Note 1) is devoted to justification for the unrestricted warfare against enemy shipping. His thesis held that the German measures were a response to the behavior of their enemies: they armed their merchant ships, and (in particular) those ships, even if unarmed, sent alarms by wireless and attempted to ram surfaced submarines that were obeying the "stop and search" provisions of International Law.
3. The issue of submarine versus merchant ship appeared to have been settled by the conduct of the combatants in World War I, but witness this excerpt from the British *Defence of Merchant Shipping Handbook 1938*: "On the outbreak of war it should be assumed that the enemy will act in accordance with International Law, and fire should not be opened until he has made it plain that he intends to attempt capture." However, in a later paragraph of the same document this declaration appears: "No British merchant vessel should ever tamely surrender to a subma-

rine, but should do her utmost to escape." In his *Memoirs* (English edition 1990, U.S. Naval Institute Press, Annapolis) Grand Admiral Karl Doenitz pointed out that International Law did not forbid "capture," but required the merchant ship to submit if it wished to escape harsh consequences.

4. Hoehling, A.A., 1992, *Ships that Changed History*, Madison Books, Lanham, New York, and London.
5. Shipping Casualties—Loss of the Steamship *Lusitania* ("Report of a Formal Investigation into the circumstances attending the foundering on 7 May, 1915, of the British Steamship *Lusitania*, of Liverpool, after being torpedoed off the Old Head of Kinsale, Ireland," Lord Mersey, Wreck Commissioner). Referred to here as "the Mersey Report."
6. Bailey, Thomas and Ryan, Paul, 1975, *The* Lusitania *Disaster*, The Free Press, New York.
7. Colin Simpson, 1972, *Lusitania*, Longman Group, London.
8. From the Mersey Report. See Note 5. Unless noted otherwise, all subsequent descriptions come from this report or from Garzke et al., Note 9.
9. Hoehling, A.A. and Hoehling, Mary, *The Last Voyage of the* Lusitania. Henry Holt and Company, New York, 1956.
10. William H. Garzke, Jr., Robert O. Dulin, Jr., Peter K. Hsu, Blake Powell, and F. Gregg Bemis, *The Saga of the RMS* Lusitania, *A Marine Forensic Analysis.* The New York Metropolitan Sections of the Society of Naval Architects and Marine Engineers and the American Society of Naval Engineers, January 1998.
11. Professor Edward Welch, 1921, *Transactions* of the Royal Institution of Naval Architects.
12. Archibald Hurd, "A Merchant Fleet at War," Cassell and Company, Ltd., London, New York, Toronto, and Melbourne, 1920.
13. Hoehling, A.A. and Hoehling, Mary, *The Last Voyage of the* Lusitania. Henry Holt and Company, New York, 1956.
14. Port trim tank? This term seems to be an oxymoron, since "trim" designates only fore-and-aft inclination. However, the ship did have a large trim tank, far aft, divided by a centerline bulkhead.
15. The "condenser explosion" is difficult to explain, since condensers typically operate at pressures below atmospheric. Further, they are unlikely to be found high in a ship. We conclude that this must be a case of mistaken identification.
16. Hoehling. See Note 4.

CHAPTER 18

Britannic

The third ship of the *Olympic/Titanic/Britannic* trio was laid down at Harland & Wolff on 30 November 1911 as the yard's hull 433. Erection on the slipway had been underway for only 4½ months when loss of the *Titanic* provoked immediate demands for improved safety at sea. The upgrading invoked for the *Olympic*—to be discussed in Chapter 24—obviously could be applied easily to a ship in such a nascent stage of construction, and it was. In the case of the lifeboat outfit, the changes were more radical than those possible with the older ship.

Meanwhile, a change of considerable significance already had been instituted for the new ship: an increase in the beam of the hull. *Titanic*, since it never developed a significant list as it sank, could not be faulted for lack of transverse stability during that 2½-hour ordeal. However, it was known by its designers to be less stable than the *Olympic:* in formal terms her metacentric height[1] was six inches less than that of her sister. This deficiency was the consequence of the changes made in the *Titanic*'s passenger accommodations, those previously mentioned changes that increased her gross tonnage by about 600 tons over that of the *Olympic*. This volume increment came at the cost of a weight increase, and being high in the ship, the increase did adversely affect stability: metacentric height was that estimated six inches less.

The *Titanic* changes had, of course, been instituted at the Harland & Wolff Shipyard while the ship was under construction, but a meaningful inclining experiment to get an accurate determination of the vertical center of gravity could not be performed until the ship was completed (or very nearly so). However, a weight estimate showed the 6-inch de-

The *Britannic* was unceremoniously launched on 26 February 1914 as were the other two sisters. She would not be completed until December 1915 as Harland & Wolff had a number of orders to fill for the Royal Navy.

crease in metacentric height from the *Olympic* value. This reduction was acceptable in the *Titanic*, but had it not been acceptable, the remedy surely would have been addition of fixed ballast low in the ship.

If, in April 1912, the *Titanic* inclining experiment demanded a corrective measure in the form of added ballast, the same remedy could be applied to the new ship. But for a ship still on the ways, other options were possible. One of them was an increase in beam—but the *Britannic* design had to be frozen in October 1911. The beam increase would be preferable to an addition of ballast. Rather than gamble on the chance that neither increase in beam nor ballast increase would be needed, White Star opted for the beam increase. The change was the small, but not insignificant, increase in beam by 18 inches.

In mid-April 1912 another imperative suddenly arose. White Star found itself with two ships on hand whose supposedly unsinkable sister had just sunk from an accident that opened six forward compartments, including its two forward-most boiler rooms. To regain the trust of passengers, the surviving ships *had* to be refitted so that no similar accident could sink them. Lifeboats or life rafts would be added. Watertight bulkheads would be carried watertight to higher levels. An inner skin would be fitted to contain the kind of flooding that had sunk *Titanic*. The last of these, in particular, would be difficult and expensive for the existing ship (*Olympic*), but was nonetheless accomplished (Chapter 24). However, the *Britannic* was being built with the added beam that would give the serendipitous benefit of space for a watertight inner skin that would thwart any raking-type damage.

Given the early stage of its construction when the decision was made, the difficulty of implementing these changes to the *Britannic* was minimal. The 18-inch increase in beam made the inner skin—much like that added to *Olympic*—less intrusive. In the event, the inner skin was added from the forward bulkhead of Boiler Room No. 6 to the after bulkhead of the Generator Room, extending upward from the inner bottom to a point 6.5 feet above the load waterline. Five of the bulkheads were extended from Upper ("E") Deck (8 feet above the load waterline) to "B" Deck (36.5 feet above the load waterline) (see Figure 18-1). A watertight bulkhead was added between the turbine engine room and the generator room, giving the *Britannic* 17 watertight compartments rather than the 16 in the earlier ships. For all that these additions were effected more readily than those of the *Olympic*, there was a cost beyond the obvious monetary one: displacement was increased by about 1,000 tons.

After the Mersey Inquiry in the summer of 1912, a bulkhead committee and a load line committee (sub-units of the Board of Trade) considered matters under their jurisdictions in light of the *Titanic* experience. They found that the lost ship was in compliance (as had the Mersey Inquiry) with the existing regulations, but the consequences of raking-type collisions had not been considered in those regulations. Even with the new experience in hand, the committee members were unable to offer suitable amendments to their previous work. Regardless of regulations, the inner skin was a seemingly effective measure to a company that needed so badly to counter the public's new fear of ship collisions with icebergs. The White Star Line decided to use the inner skin concept without waiting for any regulatory compulsion.

Other features of the *Britannic* that were different from her sisters were the arrangements for closing the watertight doors and pumping out flooded compartments. The latter was a direct result of the *Titanic* experience. The longitudinal riveted seams amidships had an additional rivet row added to strengthen the hull. Like her sister ships, the flat keel was 1½ inches thick and the bar keel below was 3 inches thick and 19½ inches wide. The plate for the center vertical keel was 1 inch thick and was connected to the flat keel and to the horizontal plates, forming the inner bottom or tank top by heavy angle bars. In the *Britannic*, however, there were six corresponding longitudinal girders between the center vertical keel and the bilge, four of these were intercostals, the fifth one was the margin plate, and a sixth intercostal plate outside the latter in the wing tank. The double bottom height was increased by one foot.

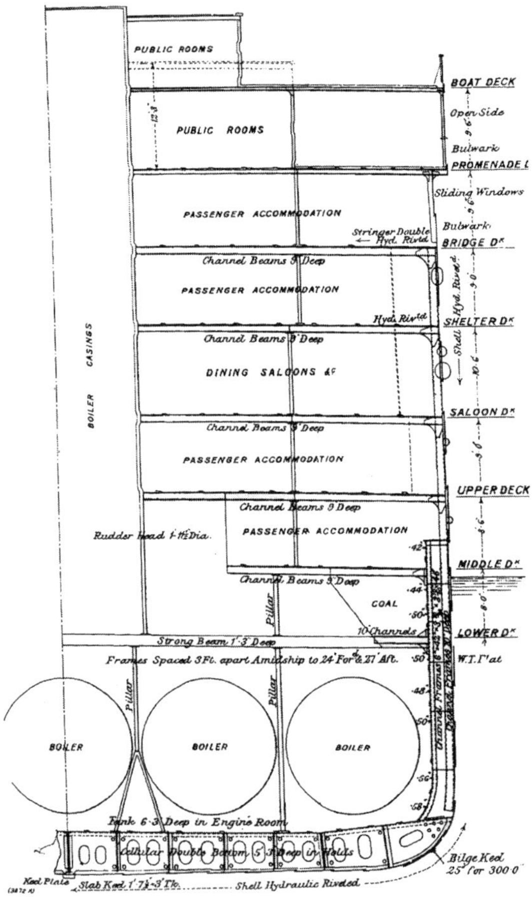

Fig. 1. Half Mid-Ship Section.

FIGURE 18-1 Half-section of the *Britannic* hull. A particular item of interest is the inner skin, extending from the inner bottom to a point above the waterline. This was intended to limit flooding from raking types of damage. The existing double bottom was also increased in depth from five to six feet. (*From* Engineering, *vol 94, 1913*)

Immediately after the *Titanic* disaster, Bruce Ismay decreed that all ships operated by International Mercantile Marine should be equipped with lifeboat capacity sufficient for everyone on board. That decision was not implemented until after the crew of the *Olympic* went on strike because they felt the lifeboat situation for that ship was inadequate for safety considerations. To satisfy the crew, White Star adopted the immediate expedient of adding additional wooden lifeboats to the *Olympic* before its first post-*Titanic* voyage began. During the addition of the double hull, the multiple-boat-per-davit plan of Axel Welin and Alexander Carlisle (Chapter 27) was implemented.

The *Britannic* was unquestionably to be provided with the boats, but the launching apparatus also was a radical improvement over that of the earlier ships. The main boat-launching outfit of the *Britannic* was intended to consist of eight sets of gantry davits (see illustration) that could pick up loaded boats from the deck, and could launch them in the face of lists and trims of up to 15 degrees. Those fitted between the funnels could reach across the deck to pick up boats from the opposite side of the ship if the list were too severe to allow launching from both sides. In all, each set of cranes was expected to handle as many as six boats. Each gantry was electrically powered and equipped with a lighting system for launching lifeboats at night. The ease of boarding the boats, and the safety of the always precarious lowering of them, was surely enhanced by this system. However, the time to launch six boats in succession might be

This is a starboard view of a model of the *Britannic* as she would have appeared when completed as a transatlantic liner. Note the position of the gantry-crane davits. Only one other passenger ship was so equipped, the *Pittsburgh*, which was completed in the 1920s by Harland & Wolff. (*Courtesy of Harland & Wolff*)

questioned. That time was never stated publicly. The aesthetic qualities of the *Britannic* were not improved by these rather large and industrial appearing cranes. One is reminded of the landing craft carried on amphibious assault ships that appeared during World War II.

The outfit actually provided included only five of the eight planned sets of the gantry davits because of the urgent need to get the *Britannic* into service as a hospital ship. The shortfall was made up by six sets of conventional Welin radial davits amidships on either side of Boat Deck. Two additional sets of these davits were installed on the Shade Deck, one on either side. In all, the outfit of boats consisted of 45 34-foot boats, one 26-foot cutter, and two 26-foot boats.

Another important safety feature of the new ship was a pneumatic tube that could whisk navigational messages from the wireless room to the bridge.

Some differences in passenger accommodations distinguished the *Britannic*. In first class, the spaces aft of the second funnel were similar to those of the *Titanic*; forward, they were similar to those of the *Olympic* in that the staterooms did not extend to the side of the ship. In detail, a Hairdressing Salon (a first among transatlantic liners) was added next to

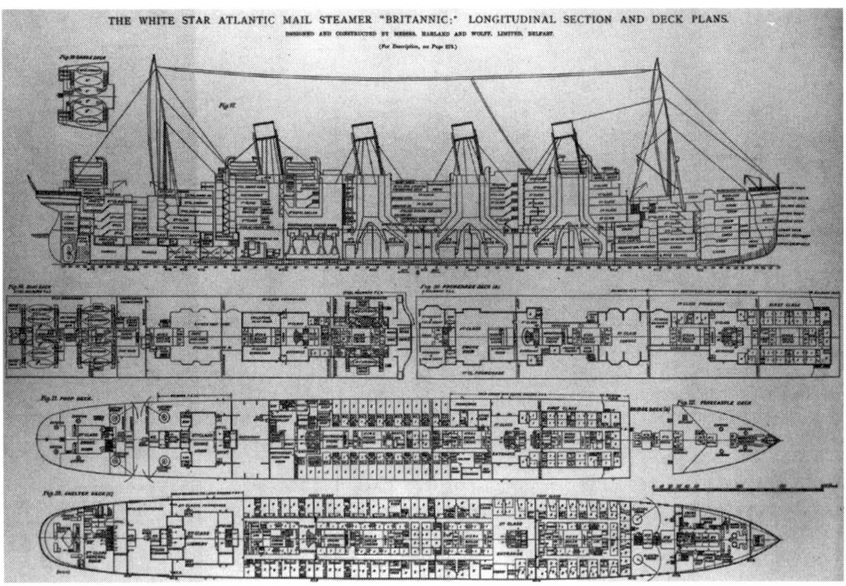

This is the inboard profile and the superstructure decks of the *Britannic* as she was to be completed as a passenger ship. (*Courtesy of the Mariners Museum*)

the Barber Shop, the a-la-carte Restaurant was enlarged at the expense of the Café Parisien, the number of staterooms with private baths was increased, a children's playroom was added on the Boat Deck opposite the first-class Gymnasium, and a fourth elevator was added amidships to give access to "E" Deck.

The differences for the second- and third-class passengers were somewhat greater. The after Well Deck was enclosed to give third-class passengers a covered promenade, and to give space for a second-class Gymnasium. The covering deck was an extension of "B" Deck, and was assigned to second class as a promenade. A third-class Smoking Room was added also at the after end of this deck.

There were other minor changes. The hospital was relocated, dog kennels were added on Boat Deck at the port side of the fourth stack, and the fresh water pumping system was improved so that hot water would reach the taps more quickly.

The changes within the superstructure, and the addition of gantry-crane davits, required a third expansion joint, just forward of the fourth stack. The forward expansion joint ran down the side of the deckhouses on "A" and "B" Decks near the officers' quarters, at the same location it did on the *Olympic* and *Titanic*. The second expansion joint was moved forward between the second and third stacks. A fourth expansion joint was added to the extended "B" Deck in an area that had been an open Well Deck in the *Britannic*'s sisters.

The respacing of expansion joints within the deckhouse is believed to have been in response to Second Officer Lightoller's testimony that the *Titanic*'s first expansion joint opened enough to cause failure of stays supporting the forward stack.[2] However, the *Britannic*'s stack stays for the forward funnel also were anchored in positions that required them to pass over the forward expansion joint.

Some changes were made to the propulsion plant. The 24 double-ended boilers were each approximately one foot longer than in the first two ships. The reciprocating engines were given minor modifications that allowed higher power. The turbine engine was built by Harland & Wolff rather than by John Brown, Clydebank, and was 70 tons heavier than the John Brown product. However, this approximately three percent increase in weight of machinery allowed a 2,000 horsepower increase in output, or approximately 13%.

The Britannic *as a Hospital Ship.* The *Britannic* was launched on 26 February 1914 with as little ceremony as her sisters. Her maiden voyage was scheduled for the spring of 1915, but the advent of World War

I in August of 1914 caused suspension of construction for the duration of hostilities even though it was believed that the war would be of short duration. The hull languished at its outfitting dock as the war dragged on until the ill-fated Gallipoli campaign (the event that had caused the *Olympic* to be pressed into service as a troopship) made the *Britannic* a resource that the War Office could not afford to leave idle. As winter approached in late 1915, the mounting casualties of the distant campaign demanded increased hospital ship capacity. On 13 November the ship was requisitioned for conversion to that service.

Very little of what could be called design work was required: furnishings not required for hospital ship service were simply stripped from the ship and replaced by the necessary medical equipment, stowages, and beds for patients. Elegant public rooms such as the first-class Dining Room was equipped as the main hospital with the adjoining first-class Grand Reception Room becoming an emergency operating room. Other dining areas and main public spaces were converted to wards, operating rooms, and treatment areas. The first-class entrance area on "D" Deck became a triage area because the nearby elevators made access to the other decks easier for quick movement of the critically wounded to sites appropriate for their treatment.

Modifications to the living quarters featured bunks for the wounded and staterooms for the hospital staff. The essential medical personnel were assigned to the first-class accommodations on "B" Deck, while the cabins on the lower decks were allocated to medical orderlies and any wounded who could act as stretcher bearers in emergency. In all, the ship was outfitted to carry and care for 3,109 wounded personnel instead of the 2,579 passengers which she was designed to carry. Her medical staff consisted of 52 officers and doctors, 101 nurses, and 336 orderlies. A complement of 675 officers and crew provided the engineering, stewarding, and navigating functions.

The *Britannic* was delivered to the Department of Transport on 8 December 1915, although brief trials were held on 11 December. She arrived in Liverpool on 12 December and left Liverpool for the Mediterranean as a hospital ship eleven days later. She was fitted with the standard hospital ship markings: a white hull with a green stripe, running from stem to stern, and interrupted by a large red cross at the bow, at the stern, and at amidships. The buff funnel color of White Star was replaced by yellow. For identification at night, a line of green lights ran beneath the Promenade ("B") Deck, and illuminated red crosses were positioned on the outer bulkheads of Boat Deck, just aft of the forward stack. (The lights of the forward starboard red cross were to be turned off during conditions of poor visibility so that its glow would not cause misidentification of the green starboard running light.)

During her first voyage back to the United Kingdom in January 1916, she carried approximately 3,300 wounded and sick personnel. By April HMHS *Britannic* had made three voyages from Southampton to the Mediterranean, touching at Mudros (on the island of Lemnos), Naples, and Augusta (Sicily). The island of Mudros was a central collection point of wounded from campaigns in the Dardanelles, Gallipoli, Macedonia, Mesopotamia, and Palestine. Then, because the withdrawal from Gallipoli in the spring of 1916 reduced the need for hospital ships, HMHS *Britannic* returned home to the Isle of Wight where she remained offshore as a floating hospital for a period of four weeks. She was released from hospital ship duty in June and returned to Belfast to be converted back to a mail and passenger ship. In July 1916, however, new Allied offenses in Mesopotamia, Palestine, and the southern Balkans was producing a new tide of overseas casualties that called the ship back into service on 28 August 1916.

The Fateful Voyage. HMHS *Britannic* made two voyages to the eastern Mediterranean during the early autumn of 1916. Then, on 12 November 1916 she departed Southampton for Mudros, where she was to pick up a full complement of casualties. Crew and Royal Army Medical Corps staff totaled 1,125 men and women. After taking on coal and water at Naples, and being delayed briefly by a Mediterranean storm, she sailed for Mudros on 19 November.

Meanwhile, the forces that would imperil the second *Olympic* sister had already been at work. There was no drifting iceberg this time, but instead a German mine-laying submarine in the Aegean, bent on the destruction of shipping that supported the Allied encampment at Salonika. The submarine was the *U-73*, commanded by Kapitanleutnant Gustav Siess. His boat was armed with an 88-mm deck gun and two 50-cm torpedo tubes, and was capable of carrying 34 mines that could be deployed through a 100-cm mine tube in the stern. The mines each contained an explosive charge of 150 kg (330 lb) of TNT or equivalent, and were moored to anchors on the seabed that allowed the mines to float about 10 feet below the surface. (Actual depth at any time would depend on the state of tide and current.)

Commencing in April 1916, *U-73* laid mines from Lisbon to Malta. On the morning of 28 October the *U-73* laid two mine barriers of six mines each off the island of Kea, just southeast of the Gulf of Athens in the Aegean Sea. These mines were laid across the main shipping channel

This is a port view of the HMHS *Britannic* at Mudros. Note the large gantry-crane davits abreast of the forward and after stacks. She was painted white with a green stripe that extended from bow to stern, broken by three red crosses, painted on the port and starboard sides. At night the ship was illuminated by a string of green lights running beneath the Promenade Deck and two crosses on the Boat Deck. (*Courtesy of the Imperial War Museum*)

This is a starboard quarter view of the HMHS *Britannic* at Mudros on 3 October 1916, seven weeks before the ship hit a mine and was lost. The illuminated red crosses can be seen just aft of the forward funnel and between the aft sets of gantry-crane davits. (*Courtesy of the Imperial War Museum*)

after Siess had observed shipping movements. British patrols never detected the mines' presence.

Early in the morning of 21 November, the *Britannic* was in the vicinity of St. Nicolo, the principal port of Kea. Many of the hospital staff were at breakfast. At 0812, while a change in boiler and engine room watches was being completed, the ship was shaken by a tremendous explosion. While the change in watch was underway, many of the watertight doors were opened. The official inquiry into the incident described a single explosion taking place low on the starboard side of the ship, centered between Cargo Holds Nos. 2 and 3. It should be noted that few survivors actually saw anything of particular relevance, as the vast majority of the crew and medical personnel were at breakfast at the time of the explosion. Nurse Violet Jessup, who was eating her breakfast, described the explosion as "a dull, deafening roar," and added that "the ship responded with a shiver that was a long drawn-out shudder from stem to stern, shaking chinaware on tables, breaking things until it subsided." Nurse Sheila Macbeth Mitchell also recounted that the ship shook for a long time.

Two of the three survivors on deck claimed to have seen torpedo tracks and a periscope, leading, of course, to stories that the Germans had deliberately torpedoed a clearly marked hospital ship.[3] These sightings, however, were at different ends and on opposite sides of the ship. There was a vague testimony by lookout Seaman J. Conelly that, shortly before leaving his watch at 0800 he saw "two suspicious objects." It is not surprising that once the final report was completed some three days after the sinking, it remained inconclusive. It concluded by saying:

> The effects of the explosion might have been due to either a mine or a torpedo. The probability seems to be a mine.[4]

Evidence is now conclusive that the ship was not torpedoed. German records show that no submarine was present at the time, and that the *U-73* had returned to base at Cattaro (later Kotor in Montenegro) in the Adriatic. Dispassionate analysts are convinced that the fatal instrument was one of the mines laid on 28 October. Commander Gustav Seiss admitted in 1928 that the ship probably struck a mine that his submarine had laid.[5]

The explosion ruptured the *Britannic*'s hull over an area estimated to be about 600 square feet.[6] The extent of damage into the hull is difficult to evaluate. With the ship at a 33-foot draft at the time, and with the mine exploding 10 feet below the surface, immediate damage to the centerline Firemen's Passage is unlikely, but that passageway, leading into the boiler rooms, may have been a significant factor, as we shall see. The immediate consequence was massive flooding of the two cargo holds, followed by a small list to starboard.

Two compartments were flooded in an area where the *Britannic* should have survived flooding of her five forward compartments. With the watertightness of the *Britannic* bulkheads extending higher than those of the iceberg victim, flooding of the entire forward end of the hull, including Boiler Room No. 6, should have left the ship safely afloat. The resolution of the mystery—why the rapid demise of the *Britannic* in the face of established naval architectural principles—apparently lies in three factors: the near immediate occurrence of an internal explosion triggered by the impact of the mine, open portholes, and shock damage that prevented the closing of watertight doors.

Although evidence for a secondary explosion is strong, inspections of the wreck by several diving teams has shown no indications of one. The mine explosion occurred on the starboard side, but a major rupture was evident on the port side of the ship, with hull plating and framing severed and bent outward. Given that the German mine could not have blown a hole all the way through the hull, an internal explosion may have been the only reasonable explanation for such damage. Whatever the cause (which we shall examine later), the bulkhead separating Cargo Hold No. 2 from Cargo Hold No. 1 was ruptured, as were the bulkheads bounding the Firemen's Passage. After this secondary damage, water poured into the Firemen's Passage, and then into the boiler rooms.

At the time, the watertight doors between boiler rooms, and between the passage and Boiler Room No. 6 were open to facilitate the changing of watch in progress at the moment of explosion. In 1998 American diver John Chatterton attempted to penetrate the broken fireman's tunnel, in the hope that he would be able to observe some of the forward watertight doors which failed to function. Unfortunately, the area was considered too hazardous to proceed too far inside and, although Mr. Chatterton observed what he thought was a partially open door, unfortunately, it was not possible to obtain any images.[7] If this observation is accurate, then it appears that the automatic closing feature of the door to Boiler Room No. 6 did not function as designed. It is very likely that the shock response to the explosion damaged the device and perhaps the closing mechanism. Thus, the door could not close and be sealed. These doors were fitted with means for quick closure by a control signal from the bridge. Failing that, a float-activated signal at each door would close them automatically in response to flooding. Local manual operation was also possible. When Murdoch on the Bridge of the *Titanic* threw the lever on the Bridge, the doors closed, but the gesture was wasted because the iceberg damage was too extensive. Now, when the damage, even including that of the supposed secondary explosion, was more limited, the doors did not close, water poured from the cargo holds (via the ruptured tunnel), and the ship was lost in consequence.

Why did they not close? A likely answer lies in the hull's whipping response—the "shuddering" noted by Nurses Jessup and Macbeth-Mitchell. Cast iron fittings were part of the door construction, and tests done on cast iron specimens recovered from the *Titanic* show a susceptibility to shock damage. In a more general sense, the very suddenness and violence of the event precluded damage-control action by either man or machine. The *Olympic*-class ships were just not designed for war service.

A highly significant contributing factor in the loss of the *Britannic* may have been open portholes on "E" Deck. Amidships these ports were about 20 feet above the waterline. Ignoring an order that they remain closed, the hospital staff had opened them to air out the wards before the boarding of wounded scheduled for that afternoon. The list and trim that had developed about 15 minutes into the event was sufficient to immerse the ports, allowing water to cascade through the openings. At this point, the non-tight wooden joiner work added to the problem of unsymmetrical flooding. Although unable to resist hydrostatic head, this structure impeded flow inboard, with the consequence being the development of a large listing moment.

The massive flooding along the starboard side penetrated the third-class Galley and Boiler Room No. 1. Until then, water had not penetrated aft of Boiler Room No. 5, and the ship might have survived had not this unfortunate bypassing of watertight boundaries ruined the best-made decisions of the Harland & Wolff naval architects.

Captain Charles Bartlett had recognized instantly the peril to his ship. He activated the switch that would close all the watertight doors. He ordered a distress call to be sent and the lifeboats to be made ready for launching. Then, when he realized that his ship would sink, he began a valiant effort to beach it in the shallow water offered by the nearby Island of Kea, about 8–9 miles distant. Heavy trim by the bow, list, and the failure of the rudder to respond to the helm (a possible consequence of shock damage) made maneuvering impossible. He attempted to maneuver by using the propellers, but that expedient was also futile because of diminishing steam pressure and the rising of the propellers to the surface as the trim by the bow and list to starboard increased.

During the course of the *Britannic*'s sinking process, her heading would have been affected by two factors, the currents in the Kea Channel and an exposed turning port propeller. One sentence from the report of the *Britannic*'s senior medical officer, Lieutenant Colonel Henry Stewart Anderson, is illuminating:

> The wake showed that the ship was moving in a wide circle to the right.[8]

This circle to the right (or starboard) may have been caused by the starboard list. That list was further compounded by the bow trim. As flooding continued unabated in the bow, the stern rose out of the water, exposing the port wing propeller. That propeller kept turning as evidenced by several lifeboats that drifted into it. With one propeller out of the water and one submerged, there was unequal thrust that caused the ship to keep turning to starboard. This combination of events was sufficient to turn the ship away from shore. Another interesting factor is that *Britannic*'s rudder has been found to be turned considerably to port (by at least 10°), almost as if the captain was trying to compensate for this drift to starboard.[9]

Another factor in the sinking process probably was the prevailing current in the Kea Channel. The closest estimation of the tidal conditions experienced in this area would probably come from the 1997 IANTD expedition organized by Kevin Gurr, as this expedition was carried out during the month of November, the same time of the year that *Britannic* was sunk. Mr. Gurr reported that some of the currents experienced were so strong that during one in-water decompression period (which usually lasted approximately three hours) the decompression platform drifted some seven miles from its exclusion zone—a dangerous situation in what is now a particularly busy shipping lane. This suggests that the prevailing northeasterly currents in the Kea Channel at this time of the year can be particularly strong and this would almost certainly have been enough to assist in turning the ship. However, once the *Britannic*'s bow had touched the seabed it would have been impossible for it to pivot around with the whole weight of hull and entrained water concentrated on the seabed. The bow structure also had been weakened by the effects of both the explosion and the high-speed dash for the shoreline. The stresses exerted in this weakened area were massive. A review of the video images tend to confirm that there was a twisting motion, causing the bow structure forward of Cargo Hold No. 3 to become almost severed from the main body of the hull. The hull, however, is still in one piece, but only because the deck structure at the level of the Well Deck appears to be intact. Interestingly, as on the *Titanic*, the two cargo cranes remain comfortably in place, which might suggest that the overall structural integrity of the surviving deck in this area is stronger than might be supposed. However, in that *Britannic* actually lies on her starboard side at an approximate angle of 85°, while *Titanic*'s bow is upright, the fact that the *Britannic*'s cranes are still in position is due to a slow rate of corrosion. The advance decay seen in the *Titanic* is brought on by the action of bacteria and fungi leaching out the steel in the hull and superstructure. This has caused these cranes to pitch inboard at an angle of 15 degrees by 1998. Eventually as the deck structure continues to deteriorate, these cranes will fall into the hull.[10]

During the final moments of the *Britannic* water rushed into the boiler rooms causing smoke to bellow forth from the ventilators. From within the hull there were the sounds of explosions. It is not clear whether these were from collapsing bulkheads or exploding boilers. However, considering the short time in which the ship sank, it is possible that some boilers did explode. As she continued to capsize to starboard, all four funnels came loose and fell into the sea. Finally with only her stern exposed, the great liner slowly slid below the surface.

No shortage of lifeboats sullied the reputation of White Star this time: with only the crew and medical staff aboard, boat capacity was about three times the number of lives to be saved. In spite of some difficulties in launching caused by list and trim, everyone aboard reached the placid surface of the Aegean safely. Unfortunately, however, two of the portside boats drifted into the still-turning propeller, exposed by the ship's trim and list. Thirty people were killed, and another 41 were injured enough to require hospitalization. A third boat avoided this fate when Captain Bartlett, having been told what was happening, ordered the engines stopped.

Just as with the *Titanic,* the late stages of sinking were marked by falling of the forward stack, because the opening of the forward expansion joint between the stack and the anchor points of their supporting stays. (A diving team in 1999 reported that the forward expansion joint had opened beyond the constructed distance of two inches.) Finally, the ship capsized to starboard as the forecastle submerged. The other three stacks would also fall off when the vessel capsized to starboard. At 0907 the stern disappeared below the surface. Survivors were picked up by the steamer *Heroic* and the destroyers *Foxhound* and *Scourge.*

Analysis of the Sinking. Both the *Titanic* and the *Britannic* sustained grievous damage forward and sank because the circumstances were different from those envisioned by their naval architects. The damage to the *Titanic* allowed initial flooding of six watertight compartments; the damage to the *Britannic* allowed very rapid flooding, possibly exacerbated by a secondary explosion, and was accompanied by a shock response that may have had devastating consequences to structure and appliances.

The *Titanic* sank slowly over 2½ hours because the inflow of water was slow, initially, at least, coming through an area of about 11.88 square feet. She sank with a slight port list because there were no major barriers to cross-flooding, and the slow rate of flooding allowed the water to be distributed across the hull before a list developed that might allow flooding high in the hull on one side (e.g., significant flooding through open portholes or non-tight deck openings). The *Britannic* did not sink with the disastrous speed of the *Lusitania*, but its time was rapid compared to that of its sister, only 55 minutes. Although the rate of initial flooding was far greater than that of the *Titanic,* the time difference may lie in the effects of shock that deranged the watertight doors and damaged structure that was part of the intended watertight integrity. It lies more obviously in the downflooding from the opened portholes on "E" Deck that not only bypassed surviving watertight bulkheads, but that exacerbated the list, which in turn increased the downflooding until the ship capsized, an event quite absent from the *Titanic* sinking.

Had the *Britannic* been completely free of list at the time the open ports were first immersed, the consequent flooding would have presumably been at an equal rate on both sides. Although the ship would still have sunk, its failure to capsize would have prolonged the event, perhaps even to the point where the attempt to beach would have been successful. As it went, the flooding of "E" Deck created a free surface high in the ship, and with the initial list, that free surface did its deadly work in increasing the list to starboard.

But why the initial list, perhaps as much as 10 degrees, significantly greater than that of the *Titanic*, and happening as a consequence of flooding from both sides? The answer may be the double skin, built into this ship as a protection against raking-types of damage. The explosions or the whipping response that followed them may have fractured wrought iron rivets and allowed seams to open through which the inner skin plating and stiffeners joined the forward bulkhead of Boiler Room No. 6. The protective space between inner and outer shells—on one side only—could then have flooded to produce a heeling moment. This protective feature, this inner skin that seemed so essential to protect the *Olympic* and *Britannic* from becoming a second *Titanic*, thus may have been key to the demise of the ship in the case at hand.

Marine Forensic investigations of the carrier USS *Yorktown* (CV 5), the *Titanic*, and the *Lusitania* have shown the propensity for rivet failure when ships of riveted construction are subjected to sudden impact loads. The USS *Hornet* (CV 8) and *Enterprise* (CV 6), sister ships of the *Yorktown*, also showed the same type of failure when these two carriers were damaged by bomb or torpedo attack. The British battle cruiser HMS *Hood* also was narrowly missed by German bombs in September 1939, resulting in a number of sprung rivets in her torpedo bulge. Also, HMS *Prince of Wales* sustained damage to her shell plating from a near miss bomb while completing at the Cammell Laird Shipyard in Birkenhead in January 1941. The plating was distorted, rivets failed and riveted seams sprung.

The diary of Kapitän zur See von Egidy, who commanded the German battle cruiser SMS *Seydlitz* was very emphatic about rivet failure when he analyzed the damage sustained by his ship during the Battle of Jutland:

> The torpedo bulkhead held, but it was seriously strained, as were parts of the armored deck. Where the rivets had gone completely, the holes could be stopped with wooden pegs. Where they only leaked, which they did in great numbers—more than enough for our needs—they became a distinct menace because there was no way to plug them effectively.[11]

The wrought iron rivets may have contributed to the extensive bow failure as most of these riveted seams in the area of the mine explosion were double riveted lap seams with wrought iron rivets. As was pointed out in a previous chapter, 19 of 48 rivets from the *Titanic* have an abnormally high slag content. The University of Sheffield did an optical microscopy on a wrought iron rivet retrieved from the wreck of the SS *Arabic*,[12] another White Star liner built by Harland & Wolff in 1902. The examination revealed that the volume fraction of slag was particularly high, ranging from 16% at the top of the rivet head, whereas the bottom of the rivet contained approximately 25% slag. Admittedly, the results of an examination on a single *Arabic* rivet cannot be regarded as commonplace, though further samples from the *Arabic* were tested at Sheffield that hopefully will widen the database. However, the percentage of slag is similar to the results found in 19 of the *Titanic* rivets.[13]

Even if a ship has no longitudinal watertight barrier, the numerous nontight bulkheads on the higher decks can delay cross-flooding above the bulkhead deck, with possibly fatal consequences. Ships afire have often listed dangerously, and even capsized, because water from fire hoses was added high on one side at a rate greater than it was able to flow away through nontight bulkheads and decks. This was the reason for the capsizing of the *Normandie* in February 1942 at a New York pier and the near capsize of the *Mauretania* in Southampton in June 1921. In the case of the *Britannic*, an array of open portholes on "E" Deck was the principal source of water that entered at a high level, and was trapped on one side in this way.

A Possible Second Explosion? The outward rupturing of the port side of the hull, opposite to the mine damage, could be explained by an internal explosion within the hull. The contrary view (i.e., that the mine explosion blasted a path completely through the ship) must be rejected on the basis of evidence from mine damage sustained and survived by many ships throughout two world wars. In the case at hand, a missing 2,000 to 3,000 square feet of hull structure can be attributed to the secondary event, leading to estimates that the equivalent weight of TNT would be from 500 to 2,500 pounds. Even the smaller estimate suggests a major explosion.

The *Lusitania* also suffered a secondary explosion within a few seconds following the torpedo strike, and the accusation was immediately leveled that the explosive agent was a secret munitions cargo. As might be expected, the Germans trumpeted this view, while the British vigorously denied it. By the end of the century, no clear case had been made either way, but the denials for both ships are accepted by most authorities.[14]

The *Britannic* case has one strikingly different factor: she was a hospital ship, painted with universally accepted and recognized symbols of that service, and was known to be operating as a hospital ship. That she also could be carrying munitions in her cargo holds is certainly physically possible, but if so, some unexploded items should appear in the debris field. None have been found. Having no positive evidence of such a violation of convention, we will be content with discussing the other possibilities.

Coal Dust or Methane.[15] Coal dust and methane, with the proper mixes of air can explode, and have been agents of death for many coal miners. If such an explosion did occur to add to the agony of the *Britannic,* the most likely source would be the 600-ton reserve coal bunker in Cargo Hold No. 3. A scattering of coal on the seabed near the wreck is the main evidence for a coal dust explosion, but that is weak evidence indeed. Several factors counter the coal dust hypothesis:

- The top of the bunker was below a deck that was in turn well below the likely point of detonation of the mine.
- Since the bunker in question had not been disturbed by recent use, the presence of dust suspended in air is unlikely.

- A bunker explosion of the severity indicated by the rupturing of the port-side shell also should have ruptured the bulkhead between the bunker and Boiler Room No. 6. However, boiler room personnel denied the presence of such damage, maintaining that flooding of the boiler room came only via the firemen's tunnel.
- Very little air was available in the bunker, being only that found in the interstices among the coal, so that the necessary air/fuel ratio for an explosion could have existed only in small pockets. The damage done therefore could not have been produced by the amount of oxygen available to participate.

In 1999 Jarrod Jablonski of Global Underwater Explorers explored the area around coal bunker hatch to the reserve coal bunker in Cargo Hold No. 3. That hatch was in position and there was no damage to the structure for the hatch opening in the Forward Well Deck. A coal dust explosion would have followed the least line of resistance and vented upwards. This would have destroyed the hatch cover that is not as strong as the structure around and over the bunker. Furthermore, none of the officers on watch at the time of the sinking reported any such occurrence. The latter combined with the fact that that hatch cover still seems to be intact, is a strong argument against the exploding coal dust theory.[16]

Methane might have been present from slow oxidation of coal, but the points raised against a coal dust explosion are also valid against this alternative.

Ether and Compressed Oxygen.[17] These fluids would be legitimate supplies for a hospital ship or cargo for field hospitals at Salonika. Though no cargo manifest is available to prove or disprove the presence of these substances, they may have been stowed as cargo in Holds Nos. 2 and 3, and thereby be at hazard from the mine explosion. Ether, an extremely volatile and flammable substance, forms an explosive mixture when in concentrations between 2.34 and 6.15% by volume in air. If carried, it would have been in carboys packed in straw-filled wooden crates, wicker baskets, or cardboard boxes. The mine explosion or the hull whipping response that followed easily could have fractured the carboys.

The level of damage found on the *Britannic*'s port side should require the energy released by 200 to 1,000 pounds of ether mixed with the necessary amount of air, which is on the order of 200 cubic feet per pound of ether. If oxygen cylinders also were stowed in the holds, and were ruptured by the mine or the whipping response, a much lesser volume of the suddenly enriched air would be required. However, no metal cylinders are to be found in the debris field around the wreck, making the presence of bottled oxygen at the site of the explosion unlikely. Without the enhancing effect of oxygen enrichment, an ether explosion of the expected intensity appears to be unlikely. One may still postulate an ether explosion, but should look to an additional cause for the hull damage evident.

Gasoline.[18] This seems to be an unlikely cargo, but the ship might have legitimately been carrying a supply for use in ambulances and in hospital stoves and lanterns at Mudros and Salonika. If so, the likely container would be the standard British four-imperial-gallon petrol tin, a container with a bad record for loss of gasoline through spillage, leakage, seam splitting, and evaporation through the cap. According to an estimate from World War II, they often lost 30% of the gasoline they transported. Also, ambulances might have been carried in the two cargo holds. Given that the danger of carrying vehicles below decks with gasoline in their tanks had not been recognized by 1916, such a cargo also could be a source of explosive vapors. However, if substantial leakages from any source were occurring during the passage from England or Naples, the odor is unlikely to have gone unnoticed. Further, without the unproven oxygen bottles in the holds, the amount of air confined there seems to limit the consequences of a gasoline explosion to a level far below that demanded by the hull damage. The conclusion must be that a gasoline explosion could have occurred, but not as the sole contributor to the secondary damage.

Nitrous Oxide.[19] Nitrous oxide is an anesthetic that would be needed in the normal functioning of a hospital ship, or that might be carried as cargo to field hospitals. The likely containers would be metal cylinders. As with the purported oxygen cylinders, the absence of any metal cylinders in the debris field at the wreck site appears to rule out the presence of this fluid as well.

In summarizing the evidence in hand at the time this is written, the evidence for the occurrence of the secondary explosion appears compelling, but its cause remains a subject of speculation. Part of the problem is the shallowness of the water in the area in which the *Britannic* sank. With a water depth of 400 feet, the damaged bow of the ship probably made contact with the seabed. As the ship continued its list to starboard that bow was probably anchored in the mud and a torsional-type failure of the hull could have resulted.

Divers on the 1999 expedition managed to penetrate the forward Cargo Hold and, as far as their video reveals, it is completely empty. In view of the generally excellent condition of the hold itself (even what looks like some reasonably well preserved decking), this might possibly raise questions as to the nature of the so-called "secondary explosion" which may have occurred. Even in the extensively damaged area of what was once Cargo Holds Nos. 2 and 3, parts of the hull look to be in extremely good condition. This evidence might help to strengthen the argument that most of the visual damage was caused during the sinking process, rather than because of the mine explosion.[20]

Condition of the Wreck. HMHS *Britannic* lies in water approximately 400 feet deep off the Aegean Island of Kea, in Greek territorial waters. Unlike the wreck of the *Titanic*, lying some 2½ miles down in the North Atlantic, this one can be visited by divers, albeit for times severely limited by the depth. Visits to the wreck are restricted and regulated by the Greek government, and by the owner of salvage rights, Simon Mills. Since the ship was a naval auxiliary at the time of its loss, her wreck has been declared a military grave site, severely restricting the removal of artifacts. Nonetheless, allowed visits to the wreck have determined its condition and garnered some forensic evidence.

When the French undersea explorer Jacques Cousteau discovered the wreck on 4 December 1975, he had difficulty in locating it from Admiralty charts for the area. By widening the area of exploration and using new side-scan sonar technology developed by Dr. Harold Edgerton of the Massachusetts Institute of Technology, he found the wreck lying on its starboard side some 6.75 nautical miles northeast from the recorded location of the sinking. Simon Mills found that the incorrect position was probably done during a survey of the wreck by the Royal Navy in May 1947.[21] Cousteau returned to visit the wreck in 1976. Some of the reports from his visit were of an expected nature: two or three inches of marine fouling on the steel surfaces, rotted wood decking and doors, broken portlight glass, some of the boat davits turned out with falls dangling.[22]

The most significant observation, in addition to the error in Admiralty charts, was damage to the hull that raised the suspicion of a secondary explosion. The ship is lying on its starboard side, exposing the severe port-side damage in way of Cargo Holds Nos. 2 and 3. In that area, the hull had the aspect of something ripped by a can opener. All of the shell, its frames, and some sections of keel were blown away. The remains of frames were bent and blown outward. The only structure surviving to hold the bow to the ship aft of Cargo Hold No. 3 was an intact Shelter ("C") Deck. Overall, the surviving bow structure was bent upward, an obvious consequence of its striking the seabed as the ship sank.

The wreck also was visited in 1995 by an expedition headed by Dr. Robert Ballard, using remote operated vehicles and the U.S. Navy research submarine *NR-1*. This expedition located all of the stacks, and found them to be in a condition that was surprisingly good, for all that they were non-structural items. An inspection of watertight doors to determine their open-closed status was on Ballard's agenda, but jagged debris, torn plating around the cargo hold rupture, and a swirl of snagged fishing nets appeared too threatening to the umbilical cords of the remotely operated exploration vehicles.

Dr. Ballard also scanned the seabed for the anchors and cables of mines, and for possible munitions scattered from the sinking ship. Results were completely negative.

The last expedition to visit the *Britannic* before press time of this book was that of Kevin Gurr in 1997, using human divers.[23] As in 1995, the hazardous condition of the damaged portions of the wreck thwarted plans to examine watertight doors and to look for evidence that might establish the source of the secondary explosion.

An examination of the wreck is expected to take place perhaps in 2003 by a diving team that will be working with the Marine Forensics Panel of the Society of Naval Architects and Marine Engineers. The wreck will be thoroughly surveyed and coal, steel plate, and rivets will be recovered for analysis.

Notes for Chapter 18

1. Metacentric height and its significance are discussed in Chapter 28.
2. Recall that the opening of the *Titanic*'s forward expansion joint as the hull was bent downward by the weight of flooding water caused stays to fail. The joint was forced open about 18 inches in some areas. In consequence, the forward stack fell, killing some of the till-then survivors.
3. One might speculate that those who reported evidence of a submarine attack had never seen a torpedo track or a periscope before, just as *Titanic* survivors who reported boiler explosions had never experienced a boiler explosion.
4. Official Inquiry into the Loss of HMHS *Britannic*, by Capt. H. H. Heard and Cdr. G. H. Staer, aboard HMS *Duncan*, 24 November 1916. Public Record Office, Kew, England. (File No: ADM 137/2171).
5. Commander Gustav Seiss stated in 1928 that his submarine *U-73* was probably responsible for the loss of the *Britannic*. He mentioned in his statement that mines do not pick their targets, but submariners that lay them bear responsibility for any ship striking them.
6. Professor Edward Welch, writing in the 1921 *Transactions* of the Royal Institution of Naval Architects, notes that the damage to British ships mined in World War I varied from 13 ft by 10 ft to 43 ft by 21 ft.
7. Report on the Preliminary Findings of Explorations to the Wreck of HMHS

Britannic, Kea Channel, Northern Aegean by Simon Mills. Report given to the Marine Forensics Panel on 1 March 2000.

8. The reasons for the error in the *Britannic*'s position are unclear, but Simon Mills' examination of the records at the British Hydrographic Office in Taunton, Somerset, indicates that the incorrect position was probably first recorded during a survey of the area carried out in May 1947.
9. *Calypso's Search for the Britannic.* Warner Home Video, The Cousteau Society, 1977.
10. Report on the Preliminary Findings of Explorations to the Wreck of HMHS *Britannic*, Kea Channel, Northern Aegean by Simon Mills. Report given to the Marine Forensics Panel on 1 March 2000.
11. *Ibid.*
12. Cheetham, Michael. *The Sinking of the RMS Titanic—Puncture or Brittle Fracture? Mechanical Properties of Early 20th Century Ship Steels and Ship Safety.* University of Sheffield, Department of Engineering Materials, England, 1999.
13. Report on the Preliminary Findings of Explorations to the Wreck of HMHS *Britannic*, Kea Channel, Northern Aegean by Simon Mills. Report given to the Marine Forensics Panel on 1 March 2000.
14. William H. Garzke, Jr., Simon Mills, Robert O. Dulin, Jr., F. Gregg Bemis, Dale Ridder, Dr. Timothy Foecke & David K. Brown. *The Saga of HMHS* Britannic: *A Preliminary Marine Forensic Analysis*. Proceedings from Research to Reality in Ship Systems Engineering Symposium, American Society of Naval Engineers & Society of Naval Architects and Marine Engineers, September 1998.
15. *Ibid.*
16. Report on the Preliminary Findings of Explorations to the Wreck of HMHS *Britannic*, Kea Channel, Northern Aegean by Simon Mills. Report given to the Marine Forensics Panel on 1 March 2000.
17. William H. Garzke, Jr., Simon Mills, Robert O. Dulin, Jr., F. Gregg Bemis, Dale Ridder, Dr. Timothy Foecke & David K. Brown. *The Saga of HMHS* Britannic: *A Preliminary Marine Forensic Analysis*. Proceedings from Research to Reality in Ship Systems Engineering Symposium, American Society of Naval Engineers & Society of Naval Architects and Marine Engineers, September 1998.
18. *Ibid.*
19. *Ibid.*
20. Report on the Preliminary Findings of Explorations to the Wreck of HMHS *Britannic*, Kea Channel, Northern Aegean by Simon Mills. Report given to the Marine Forensics Panel on 1 March 2000.
21. The reasons for the error in the *Britannic*'s position are unclear, but Simon Mills' examination of the records at the British Hydrographic Office in Taunton, Somerset, indicates that the incorrect position was probably first recorded during a survey of the area carried out in May 1947.
22. William H. Garzke, Jr., Simon Mills, Robert O. Dulin, Jr., F. Gregg Bemis, Dale Ridder, Dr. Timothy Foecke & David K. Brown. *The Saga of HMHS* Britannic: *A Preliminary Marine Forensic Analysis*. Proceedings from Research to Reality in Ship Systems Engineering Symposium, American Society of Naval Engineers & Society of Naval Architects and Marine Engineers, September 1998.
23. Observations of the wreck were reported by William H. Tatum, IV of the American Titanic Historical Society.

PART IV

Contemporary Analysis

After almost a century, there are important comments, especially those generated by forensic analyses, about the ships that sank. There were also contemporary critics, both expert and inexpert, that commented on the losses of these ships. The loss that elicited by far the most comment with respect to their design and construction was the *Titanic* tragedy. The *Lusitania* and *Britannic* events were clearly the consequences of wartime ordnance. Those sinkings may have caused shocks to public perceptions of what was allowable in a war at sea, but not to any lingering ideas of passenger ship invulnerability. On the other hand, the *Titanic*, with its aura of unsinkability, was destroyed by a thing so mundane as a block of ice. This event stirred those expert and inexpert critics to quick-draw condemnation and advice about ship design and construction. The expert critics went on to distill their criticisms into proposed regulations for the safety of passenger ships, and that, of course, had implications for more than just one lost ship.

The work of the expert critics of the *Titanic* was soon overshadowed by the outbreak of World War I, but their findings eventually found their way into international safety regulations. The work of the inexpert critics could be judged as of no consequence, since it did not influence ship design. From the professionals' viewpoint, it deserves oblivion. However, it casts a long shadow—the unwarranted criticisms that have been repeated in many of the innumerable books on the *Titanic* that have appeared since 1912. Authors from the professional class, writing in a later century, recognize this material as a cloud that taints the memory of a great ship. For that reason, Part IV begins with three chapters on that unattainable perfection, "unsinkability," that repeat—and condemn—the offerings of three critics whose analyses were published in the 1912–1914 period.

CHAPTER 19

Unsinkability—An Introduction

Three of our five ships sank, leaving their wrecks on the seabed for a later generation to study. The first ship, *Titanic*, was cursed with the oxymoronic label "unsinkable," bestowed inadvertently, it appears, by its principal designer, Thomas Andrews, in talking to the press. Andrews, wise in the ways of ships, took pride in the new White Star vessel as being well protected against the hazards that usually led to sinking, but he knew that no artifact floating on the water's surface can be incontestably immune to sinking. The enthralled reporters heard him say "practically unsinkable" but the attendant adverb was soon forgotten. The wreck had scarcely settled into the mud 12,400 feet down when its popular name became *The Unsinkable Titanic*.

The essence of the *Titanic*'s briefly held reputation of being unsinkable arose from its watertight subdivision. Transverse watertight bulkheads would limit any flooding to such a small volume that the consequent weight of water would not sink the ship. This feature was by no means a novelty at the time, having long since become common practice in passenger ship design. Because of the likelihood that a collision might rupture one of the transverse bulkheads, the practice was extended to make ships capable of surviving the flooding of two adjacent compartments. Also, an inner bottom would prevent flooding from any expected damage to the bottom of the ship was also part of the protection. However, at least one of the reporters who heard Thomas Andrews speak of these things in touting the new ship must have thought he was hearing of something clever invented for the *Titanic*. The *Olympic*, although almost identical to her sister, escaped that unfortunate label.

If a perceptive reporter had asked Thomas Andrews, or any other naval architect, "Can you conceive of *any* way that the *Titanic* can be sunk?" the reply might well have been, "Well, I suppose there could be some sort of glancing blow—two ships trying to turn away from each other, and not quite making it—that would open up a gash in the hull plating that extends over more than two watertight compartments. But that's so very unlikely, and . . . ." Unlikely, but make one of those ships a passive iceberg, and the unlikely event will happen.

Two sisters of three, one far from the tragedy, the other still building, survived that night to remember. They were just as unsinkable as the *Titanic*, but were suddenly illuminated as being unworthy of such a claim. Obviously they must be fitted with enough lifeboats or life rafts for their total complements, and the existing sister, *Olympic*, did not begin another voyage until that matter was resolved. The other sister still on its building berth, the *Britannic*, was not only given sufficient boats, but was fitted with boat handling cranes that appeared to be greatly superior to the boat davits of the two older ships. Beyond that, a sincere reiteration by White Star of its operating rule (paraphrased here) "Caution in navigation has a higher priority than arrival on schedule," should have been sufficient. But for public image, if nothing else, both ships were given an upgrading that would protect them from raking-type collision that sank their sister. They were provided with inner skins in way of boiler and engine rooms, and the height of their watertight bulkheads was extended one deck higher.

In spite of being made iceberg-proof, the two White Star survivors were only marginally more resistant to sinking than their lost sister. The inner skins, being only 30 inches inboard from the outer, added no protection against a ship-to-ship collision, an event far more likely than an iceberg encounter. The more unsinkable *Britannic* succumbed early in its life to the hazards of war, an underwater explosion of a German mine.

Lusitania and *Mauretania* entered service four years before the *Olympic*, first of the White Star ships. Unsinkability was just as much an objective at the time of their advent, and they were fitted with the same watertight bulkheads to limit the extent of flooding, leading the New York Times to conclude after her maiden voyage that the *Lusitania* was unsinkable. When the *Lusitania* was sunk in 1915, its sinking after having been hit by a single torpedo elicited little surprise. On the other hand, for a time after the *Titanic* went down, the two Cunard ships were pointed to as being closer to the ideal of unsinkability than *Titanic*. Their longitudinal bulkheads separating coal bunker from boiler room were regarded as providing a degree of subdivision greater than that of the White Star ships.

Leonard Peskett, designer of the Cunard ships, discussed the virtues of longitudinal bulkheads during the Mersey hearings on the *Titanic*. Edward Wilding disputed his advocacy there, and again in New York at the liability hearings in 1915. All those advocating the superiority of longitudinal bulkheads were to be discredited by events, prominently that of the *Lusitania* sinking.

Unsinkability is a worthy objective, surely, and is one that ship people of the early 20th century seemed to view as attainable. It seemed to be close to attainment in the *Lusitania/Mauretania* and *Olympic/Titanic/Britannic*, but the sinking of three of these ships seems to have destroyed that notion. Nonetheless, maximization of passenger ship (in particular) safety still remains an objective a century later. In the formal sense, the pursuit began as a response to the *Titanic* sinking, since the first attempt at international standards was a direct result of that shattering disabusement of the unsinkability concept.

We shall pursue the topic, especially as it flowered after *Titanic*, through several chapters. In the remainder of this chapter we shall look at a few instances of the responses, immediately following the sinking, by those who had the answer to the problem of unsinkability—if only the world would listen. There was a chorus of suggestions. Most of them can be dismissed by a statement from E. T. Chamberlain:[1]

> The files of the British Board of Trade following the loss of the SS *Titanic* contain quite as many and quite as foolish suggestions as those to be found in the department [Department of Commerce, Bureau of Navigation] records in Washington.

Beyond that, we shall discuss the comments of four critics, two who were amateurs, one who was a semi-professional, and one who was a practicing designer of passenger ships. The professional findings, those of Lord Mersey, and those of SOLAS 1914, appear in a following chapter. This chapter will concern critics who have not been trained in the ship design profession. The first of these will be Joseph Conrad, who was given a well-publicized opportunity to speak in 1912 by his stature as novelist and experienced man of the sea. That stature might be cited as making him something more than an amateur, and indeed, his public recognition gave an air of authority to his words. On the other hand, his words, and the temper with which he expressed them, count strongly against any professional competency.

The *Titanic* was carrying £40 belonging to Conrad, money that was being sent to America to fund a modest investment in art. He was so confident that such a ship would carry the money safely that he did not protect it with insurance. He tells of this in a letter,[2] claiming that given his circumstances at the time, the financial loss was a severe one. His distress signifies nothing in the arena of ship criticism, but hints at a possible source of the choler that surfaced later when his criticisms were published.

His two pieces devoted exclusively to the *Titanic* are essays appearing in the *English Review*, the first in the May 12, 1912 issue, the second in the following issue.[3] In both he unloads on every target in sight, including the ship, and in tones intemperate and satirical.

In the first article he targeted the right of the U.S. Senate to investigate the *Titanic* tragedy and termed the investigation a comic opera. He also stated that the British Board of Trade was asleep while ships grew in size past 10,000 register tons, the upper bound of its lifeboat regulations. He criticized the English press for accusing Captain Smith of desertion through suicide. He said that the Marconi Company was guilty for "a welter of lies that has not been sent vibrating without some reason, for which it would be nauseous to inquire too closely." Finally, he criticized the White Star Line for building a "45,000-ton hotel of thin steel plates to secure the patronage of, say, a couple of thousand rich people . . . a fatuous handful of individuals who have more money than they know what to do with" and "a perfect exhibition of the modern blind trust in material and appliances." Was all of that mere spleen over loss of £40? Perhaps not, but that more-money-than barb suggests that it might be.

On 7 November 1879 during a dark and foggy night, a transatlantic liner, the *Arizona*, much smaller and slower than the *Titanic*, hit an iceberg head-on at a speed of 15.5 knots, damaging 20 feet of her bow. On that occasion, however, the collision bulkhead held, allowing no flooding beyond the first compartment. The ship was saved and slowly steamed backward to St. Johns, Newfoundland. During repairs, 200 tons of wedged ice were removed from the area of damage. This incident was spoken of in 1912 as a way of hitting icebergs that might have worked equally well for the *Titanic*. Not for Joseph Conrad! His scornful satire imagines a "modern seamanship" in which young officers must demonstrate their skill in steering their ships head-on into any obstacle that threatens collision. More generally, he makes clear his belief that the *Titanic* was too big to be handled safely, either at sea or in port.

Conrad's second *English Review* essay was written during the Mersey[4] hearings in London, opening in May of 1912 and extending into late July. There he was as scathing, satirical, and contemptuous as he had been earlier, but with fresh targets from his apparent dismay at Mersey's witnesses:

> It is amusing, if anything connected with this stupid catastrophe can be amusing, to see the secretly crestfallen attitude of technicians. They are the high priests of the modern cult of perfected material and of mechanical appliances, and would fain forbid the profane from inquiring into its mysteries.
>
> Lately there came before it [the Mersey Inquiry] witnesses responsible for the designing of the ship. One of them was asked whether it would not be advisable to make each coal bunker of the ship a water-tight compartment by means of a suitable door. The answer to such an obvious question should have been "Certainly," for it is obvious to the simplest intelligence that the more water-tight spaces you provide in a ship the nearer you approach safety. But instead of admitting the expediency of the suggestion, this witness at once raised an objection as to the possibility of closing tightly the door of a bunker on account of the slope of coal [coal heaped in the doorway]. This with the true expert's attitude of "My dear man, you don't know what you are talking about."

Conrad, with his own air of Olympian wisdom, dismisses the expert's objection as "inane," and immediately proceeds to describe an infallible design for coal bunker doors, and the operational procedures for closing them without trapping the coal passers who were to be found within an in-use coal bunker. These bunkers were nearly always, as he reminds his reader, against the outer shell of the ship (the *Titanic* being an exception), and hence with his doors would constitute additional subdivision.

Longitudinal subdivison distinguished the watertight integrity of the *Lusitania* and *Mauretania*. Lord Mersey recommended study of its use, and other critics urged its post-*Titanic* adoption. As we shall see in later chapters, it was a naïve concept that never gained lasting professional favor, for all that some longitudinal subdivision might have saved the *Titanic*. With respect to this putative improvement in ship safety, Joseph Conrad was in good company in May of 1912, but his notion that watertight closure of coal bunker doors would save a ship was not supported by engineers.

Professor Fred Purvis[5] visited a number of ships to interview their chief engineers on this bunker door problem cited by Conrad. He asked (paraphrasing here) "how long would it take you to clear away the coal and get that bunker door shut?" Purvis distilled their varied answers into his conclusion—ten minutes. If an emergency like that befell the *Lusita-*

nia a few years later, that time would be tantamount to "forever." And suppose the *Titanic* had been given the same bunker arrangement as the *Lusitania*. There, the professor must have been thinking of Boiler Room No. 5, where water is said to have entered initially at a rate comparable to the flow from a fire hose of a 1912 vintage. That measure of flooding was not great enough to drive the boiler room crew into immediate flight. Suppose, however, that water had been entering the room via a coal bunker door. Shovel desperately to clear away the loose coal so that the door could be closed while the water continuously washes more coal into the doorway? Maybe the flow of water wasn't that strong, but in *Titanic*'s Boiler Room No. 6 the flooding was severe enough to send the crew into almost immediate flight.

Conrad's quoted slur at the "high priests" was surely aimed at naval architects and marine engineers, men facile at handling slide rule and drafting pen, but ignorant of the sea. A later generation relying on electronic calculators and computer-aided drafting, and whose navigational skills are limited to reading position from GPS or Loran, might still wither under the scorn of this veteran 19th-century seadog. Nonetheless, we would like to defend our profession.

We first concede that Joseph Conrad was unquestionably a person of high intelligence and extensive experience. Much of this experience must have come in moments when decisive action was required—when his instincts and training called for a particular decisive move. Naturally, and quite typically as a veteran seaman, he was impatient with those thoughtful and timid men who were targets of his scorn. In contrast, persons of the high priest ilk have and had, even in 1912, an access to written literature and to oral communication through technical conferences, all inundating them with ideas and analyses that they must take into account. They work under the duress of specifications and legal strictures. They must compromise to suit a shipowner's wishes, and maneuver from day to day as those wishes change capriciously. All of that requires them to work cautiously as they accommodate many inputs. And they know that a design choice that is so obviously suitable in *one* situation must also be suitable in many others. Their experience has shown them that many a good idea has failed egregiously in the hostile real world; as "true experts" they must evaluate all foreseeable objections before bestowing approval.

The same Conrad article contains the comparison that appears to have been given wide currency, namely the implied statement that a Huntley and Palmer biscuit tin[6] was much stronger for its size than the *Titanic*. Conrad admits that the hull plating of a big ship cannot be proportionally as thick as that of the biscuit tin. Other critics nonetheless took Conrad's words to mean that ships of the *Olympic* class are too big *because* they can't be as strong for their size as a high-quality English biscuit tin. This criticism also can be classed as naïve, but may be justified when viewed in the proper light; see our discussion of strength of these ships (Chapter 29).

After settling the coal-door problem, Conrad proceeds to a self-assured lecture on how ships must be outfitted in order to launch their lifeboats successfully in all circumstances, and outlines a method of doing so. He adds his scorn for those who cannot see the obvious merit in his idea. Hindsight tells that his idea was but one among many for reliable lifeboat launching. However, it also tells us that his idea was among the *unreliable* ones, since the problem of launching boats in all circumstances is one that has never been solved for passenger ships. Ships can sink in just too many ways, while passengers of wide ranges of age and physical ability exacerbate any problem of boarding and launching. The fleet of cruise ships that came to maturity in the late 20th century—recognized as successors to the transatlantic liners—are fitted with lifeboats of sufficient capacity for all aboard. Launching equipment has greater capability than that of 1912. The staff faithfully carry out lifeboat drills with every new batch of passengers. Little more can be provided.

About lifeboat drills for cruise ship passengers: with knowledge of a century's woeful tales of passenger ships sinking in war and peace, a critic might think of the drills in the scornful term "window dressing." Of course, these later-day ships also are outfitted with inflatable life rafts. In some circumstances they might be much more valuable than lifeboats, especially since they are designed to inflate and break free automatically if the ship sinks under them. Many hundreds of *Lusitania* victims, in particular, might have been survivors if that ship had carried such devices. But when you give your grandparents a cruise as a 50th anniversary present, can you picture them leaping from the sea from Promenade Deck and swimming to a life raft?

In early 1914, before World War I began, the Canadian Pacific liner *Empress of Ireland* was rammed by the Norwegian collier *Storstad*, and sank in the St. Lawrence River.[7] Like the *Lusitania,* to be torpedoed in 1915, the flooding was initially confined to one side, causing a severe and immediate list, which prevented the launching of half her lifeboats, and led to sinking within about 15 minutes. The loss of life was impressively large. Conrad also commented publicly on this tragedy in the *Illustrated London News.*[8] Unlike his *Titanic* articles, this one had nothing to say about the ship or its designers. (Perhaps he was chastened by the realization that the longitudinal subdivision he had advocated in 1912 was a major contributor to the rapid sinking of this ship.) Instead, his principal target was the seamanship of the respective crews: no one on either ship thought to rush a fender into place between the ships when the collision became inevitable. A cork fender dropped before the bows of the *Storstad*, he says, could have reduced the accident to one "with no more damage than bulged and started plates." It is apparently true that the *Storstad* was moving very slowly at the moment of impact, but the notion that a mere fender would have made the difference between a catastrophic rupturing of the hull and a harmless denting seems preposterous now as it did to many writers-to-the-editor who soon tasked the renowned author/authority severely. The tragedy of the *Empress of Ireland* is not discussed here, so no further comments will be made except to say that the fender notion does cast a shadow on the credibility of Joseph Conrad as a commentator on ship accidents.

On the whole, Conrad's criticisms of these unfortunate ships and the pertinent features of their design hold up poorly when illuminated with even small doses of hindsight, reinforced with a modest knowledge of naval architecture. He might well have left the whole topic alone, since his real target seems to be the pompous landsmen who so often infuriate the shoot-from-the-hip seaman. Indeed, he comes off rather poorly overall. Because he did wear the façade of authority, his criticisms gained currency, and hence they are proper targets for counter-criticisms here.

There were other literary figures who made public comment on the *Titanic* disaster. George Bernard Shaw and Arthur Conan Doyle are prominent examples.[9] However, they wisely stopped short of criticizing the design of the ship.

Another amateur naval architect is Walter Lord. He did write a splendid book, "*A Night to Remember,*"[10] that may be the best account of the *Titanic* event, and one that deserves credit for reviving a declining interest in the Great Ship long before Bob Ballard first pulled on his seaboots. For all its merit, the book does contain this passage:

> In technology especially, the disaster was a terrible blow. Here was the "unsinkable ship"—perhaps man's greatest engineering achievement—going down the first time it sailed. But it went beyond that. If this supreme achievement was so terribly fragile, what about everything else?

Indeed, it *was* a terrible blow to those who had put so much of their physical and emotional energy in readying this marvel for its place on World Ocean. On the other hand, it was *not* the greatest engineering achievement of humankind, nor was it terribly fragile. Because of the immense popularity of *A Night To Remember*, these ideas nonetheless have wide currency. Whatever the merits of Lord and his book, he was no engineer, he was a journalist whose words just quoted rankle those who *are* engineers. He was no more a judge of ships than those journalists of 1912 who thoughtlessly hung the "unsinkable" albatross 'round the necks of those who designed, built, and operated these great ships. He might have given credit to the sister ship, *Olympic*, that served in war and peace—equally a "supreme achievement"—and thereby gave the lie to that "terribly fragile" canard. But he did not, and hence deserves the criticism he receives here.

Notes for Chapter 19

1. The comment is quoted from E. T. Chamberlain, 1914, "International Conference on Safety of Life at Sea, with special reference to ship construction," *Transactions*, Society of Naval Architects and Marine Engineers, Volume 22.
2. Conrad's letter to J. B. Pinker, April 22, 1912.
3. Conrad's two articles on the *Titanic* can be found in Joseph Conrad, 1921, *Notes on Life and Letters*, Doubleday, Page & Company, New York.
4. *Report on the Loss of the Titanic*, 1912, His Majesty's Stationery Office, London.
5. The paper by Professor Purvis was printed in full in *Engineering*, 1912, Volume 94.
6. Paul Heyer, 1995, *Titanic Legacy,* Praeger, Westport CT, London, (tells of the biscuit-tin analogy.)
7. A secondary source for *Empress of Ireland* is James Croall, 1980, *Fourteen Minutes: The Last Voyage of The Empress of Ireland*, Sphere Books, London.
8. Conrad's *Illustrated London News* article is reproduced in *Notes on Life and Letters,* Note 3.
9. Contributions by Shaw and Conan Doyle appeared in the *Daily News and Leader* over several issues in May 1912.
10. Walter Lord, *A Night to Remember*, Henry Holt and Company, New York, 1955.

CHAPTER 20

Unsinkability—The Technical Problem

During the 19th century the advance from wooden sailing ships to steel steamships brought about the thought of building an unsinkable ship. Of course, these thoughts were not based on the comparative densities of wood and/or iron and its alloyed forms. Woods (unless grossly waterlogged) used in ship construction float, have densities less than that of water. Metals used in ship construction sink, having densities greater than water. But a hull built of wood is often burdened with dense cargo, and machinery made of iron and other metals, so that the low density of its structure is no protection. But a hull built of iron or steel—although burdened by the same cargo and machinery—may not sink because its structural material has made possible watertight internal structure that can limit flooding to a volume that will not sink the vessel.

With slightly more detail, we point out what the metal construction does provide is the feasibility of inner bottoms and internal watertight bulkheads that will limit any flooding to a small portion of the hull. This is an important safety feature that was recognized early in the age of iron. It was put to practical use in that greatest ship of the 19th century, the *Great Eastern* (1850s). It was recognized in the technical literature that began to appear after the Institution of Naval Architects was founded in 1860. Several example papers are cited in the Chapters of Part I.

The principle is simple. Subdivide a hull into X number of cells so that if the hull envelope is ruptured by a collision of likely magnitude, only x of those cells will be flooded. Of course, x must be less than X, and enough less that the ship is kept afloat by the X–x unflooded cells. But how much less is that? A general answer is "x can be only a small fraction of X." The naval architect should be able to answer more specifically in any actual case.

Another general answer might be "the larger the X, the safer the ship," another way of stating the generally correct notion that the finer the subdivision of a hull, the more difficult it is to sink. The ultimate in large Xs is a hull filled with plastic foam—or any low-density solid that is impermeable to water. The x would be large also, but would be a comparatively small fraction of X.

But these are maunderings. Long before a ship became unsinkable by filling it with foam, its subdivision would preclude any commercially attractive application for it. Of course, use of foam fillings for emergency flotation is common in the world of small boats. For such craft, filling even a small fraction of the internal volume with foam can keep the entire vessel afloat when its unfilled volume is flooded. Such an expedient is thus practical and common, and may even be required by law, for small recreational boats. It is a practicable expedient because the foam usually usurps only a small fraction of internal volume. On the other hand, a ship requires internal life support necessities for a large number of passengers and crew. It also must have a large fraction of its hull volume available to carry cargo. The economic loss from excessively fine subdivision far outweighs the marginal increase in safety gained by such extreme measures.

Here is a what-if question to ponder. Suppose a replica of the *Titanic* were to be built.[1] The builders would have great difficulty in reproducing her propulsion machinery. Why not substitute diesel (or diesel-electric) propulsion instead? This sounds like a good idea, but then you will not need those six boiler rooms that occupied so much space in the original ship. Since you do need that space for machinery, just fill all that superfluous volume with foam, making the ship truly unsinkable. The owners might even be bold enough to trumpet that now they have a *really* unsinkable *Titanic*. So the question is this: would six foam-filled (former) boiler rooms be enough for absolute unsinkability?

Let us look at warships. They have a subdivision that features many watertight cells of which a fraction may be flooded without sinking the ship. Obviously, though, the foam-filled ultimate is similarly impractical. Several special factors apply, however. Not being cargo carriers, large unobstructed spaces low in the hull are not required. Openings between watertight compartments that can be closed are much more feasible when there are no passengers whose comfort and convenience are to be considered. When the vessel is used for its intended purposes, the probability of hull damage is much higher than for the cruise ship, say, conferring much higher priority on means of limiting flooding. In short, the ship can come much closer than a commercial ship to the ultimate of a foam-filled hull.

If one continues to look beyond the realm of canoes, john boats, outboard runabouts, and the like, then the great dreadnought battleships of the World War II era come closest to being unsinkable, largely because of their intricate subdivision. Consider the German *Bismarck* and *Tirpitz*, and the Japanese *Musashi* and *Yamato*. They arguably might be labeled the ships that came closest to being truly unsinkable, a title that they earned, ironically indeed, by sinking. Or more correctly, earned by the number of weapons that their enemies were required to expend in sinking them. The two Japanese ships were both sunk by air attack at sea: the *Yamato* sank after 14 torpedoes and seven heavy bombs; the *Musashi* went down only after 20 torpedoes and 17 heavy bombs. The *Tirpitz* was also sunk by aircraft, the immediate instrument being three 12,000-pound "earthquake bombs" that hit the ship[2] while she was moored in a Norwegian fjord.

The *Bismarck* was overwhelmed by an overwhelming assemblage of British ships that came to the spot where their opponent was unmaneuverable after an aerial torpedo had crippled its rudders. After a 90-minute gun battle, the ship was rendered defenseless, but defiantly she stayed afloat. The gunfire had riddled the superstructure, but had not caused damage that would cause internal flooding. Following routine procedures, the crew set scuttling charges as they began to leave the ship. This plus some final torpedo hits caused the vessel to sink. Kurt Illies, a German marine engineer who had participated in the design of *Bismarck,* boasted in later years that although the ship was torn apart above the waterline, and several shells did pierce the armor belt, the British shells never penetrated the innermost watertight envelope. Therein lay the propulsion machinery and other vital parts of the ship.[3] From these experiences, the *Bismarck*-class battleships can claim to have at least come close to deserving the label of unsinkable.[4,5]

We may rightfully conclude that no amount of subdivision will confer the elusive crown labeled "unsinkable" on any ship. This is especially true for a passenger liner that must handle a large number of passengers. This becomes even truer when ships of this type are pressed into war service and are subjected to hazards of war. These ships are designed for peacetime where only collisions and groundings can lead to sinking.[6] Groundings will not be discussed since none of the three ships that were lost were involved in a grounding incident that led to their loss. The universal application of inner bottoms, beginning long before the first of the five were built, has proven to be a near infallible protection against serious flooding as a consequence of grounding. A supporting example: in the late 20th century, the cruise ship *Sovereign of the Seas* sustained damage to seven watertight compartments in her inner bottom from a grounding on a coral reef off St. Marten in the Caribbean. Although seriously damaged, the ship remained afloat. All the flooding was confined to the inner bottom.

Collisions in peacetime are the most probable events that affect passenger vessels. The *Olympic*, for example, was involved in three notable collisions in its quarter century of service, and although two of her partners involved sank, the German submarine *U-103* and the *Nantucket* lightship, the *Olympic* survived to be scrapped.

The classic collision that sinks one of the participants has the bow of one ship striking the side of the other at a large angle, 45 degrees or greater. Although the bow of the striking ship is badly damaged, an internal "collision bulkhead" close to its bow limits flooding to a trivial volume. The side of the other ship, however, is sliced open for a width that may be as great as the beam of its tormentor, and may have a vertical extent from turn of the bilge to a point well above the waterline. Water will pour into the victim, and if transverse watertight bulkheads do not limit the flooded volume, that unfortunate ship will sink. The classic example of this type collision was the *Stockholm* and *Andrea Doria* in July 1956 off Nantucket in a dense fog. The *Stockholm*'s bow penetrated half way into the *Andrea Doria* causing fatal consequences for the Italian liner. The *Andrea Doria* quickly listed to 23 degrees and stayed that way for 11 hours before progressive flooding silenced the electrical plant and the pumps. With pumping no longer possible, flooding reached ventilation trunks, and in consequence the ship capsized and sank.

Watertight longitudinal bulkheads and watertight decks can limit flooding, but also may be hazards. They will be discussed in the next chapter. This chapter assumes all watertight boundaries, except for the inner bottom, to be transverse bulkheads.

To attain a reasonable approach to that goal of unsinkability, the designer must space the bulkheads so that the water is confined to such a volume that the ship does not sink. Calculations have to be made, applying the esoteric formulae of the naval architect's trade. Actually, these formulas are far from esoteric, being only the application of Archimedes' Principle to determine how much the weights and moments added by the flooding water will cause the hull to sink, trim, and list. True, practical calculations are complicated by the curved boundaries of the hull, and by the nettling circumstance that a certain weight of water in the hull will cause sinkage, which in turn will lead to more water and sinkage. Until late in the 20th century, these calculations were daunting in their magnitude. The computer rendered these easy to handle and what took a naval architect days of computation was reduced to minutes. Even the most complicated calculations now could be done in the earliest stages of design to improve the ship's ability to survive damage.

A computer program produces the *floodable length curve*, one of the standardized hazard-analysis tools of the naval architect. To understand the implications of this curve, one must first be familiar with certain terminology that is essential to its making and use. The non-specialist might gain that familiarity by reading the three definitions that follow.

Margin Line. When the space between two transverse watertight bulkheads is flooded, the buoyancy represented by the compartment is lost, and hence the draft of the ship must increase. It is assumed that the lost buoyancy will cause an increase in draft. A trim by bow or stern is very likely. If the ship is not to sink, then the "bulkhead deck" (the deck to which the bulkheads are made watertight) and any non-tight opening in the hull must remain out of the water. Put another way, the waterline that is established by the flooding must not lie above the top of any watertight boundary, external or internal. To allow for uncertainties, the naval architect designs the ship so that this waterline will not come within three vertical inches (an arbitrary standard value) of the top of any watertight boundary. This imaginary line is termed the "margin line" that is evident on floodable length curves. Regulations or cautious designers often have required greater margins. For example, Board of Trade rules applied to the *Olympic* class required the margin line to be 3/100th of a hull's depth below the bulkhead deck. That stipulation required a margin line of at least 2 feet, 7¼ inches below the bulkhead deck, which was "E" Deck on these ships.

Bulkhead Deck. This is the highest deck to which bulkheads are made watertight.

Permeability. This is a fraction of volume that an internal space can fill with water. If that space is a ballast tank already filled with seawater, then its permeability is zero; if the tank is empty, the permeability is 1.00. The likely presence of internal piping and structure reduces that value to the 0.98 used by naval architects. A machinery room, though it is crammed with engines, pumps, heat exchangers, pipes, cables, and their appurtenance, is nonetheless largely air, and might have a permeability of 0.80 in steam ships, or 0.85 in a latter-day cruise ship equipped with diesel engines or gas turbines. Spaces for human occupancy are usually assigned a permeability of 0.95. Spaces of other types will have values appropriate to type and extent of things that may be found in them. In general, standard practice as well as formal rules guide a designer in choosing these values.

Floodable Length. At any point in the hull of a ship there is a "floodable length." This is the length of hull, with the point in question placed at its midlength, that can be flooded without causing immersion of the margin line *anywhere*. Permeability is a strong factor in determining this length, since the amount of water that can enter over a given length is inversely proportional to the value of that factor.

With a floodable length curve in hand, the designer can see immediately how much flooding, centered at any point of interest, will sink the ship below the margin line. If, say, this point is 250 feet aft of the bow, and the curve reads 120 feet at that point, then flooding that extends from 190 feet to 310 feet will sink the ship to the margin line, i.e., place it at the boundary between just sinking and just surviving. Locating the bulkheads at 191 feet and 309 feet reduces the flooding length to 118 feet, and in the ideal world of calculations the ship cannot sink if the hull is penetrated at the example point 250 feet aft of the bow.

But suppose the penetration is centered at 309 feet (or at 191), i.e., right at a bulkhead. The 118-foot compartment floods, but the adjacent one also floods. It is indeed quite likely that the misfortune that brings a collision will center it at a spot where two adjacent compartments are ruptured. To produce a high probability of survival, those transverse bulkheads have to be placed close enough together that any two adjacent compartments can be flooded without sinking the ship to the margin line. Such was the objective of the *Titanic* designers, and such is the requirement of subsequent passenger ship safety regulations.

The x of the discussions earlier in this chapter thus usually has the value 2; the X (total number of compartments defined by watertight boundaries) is determined by the necessity of keeping x at or above 2 at every point along the hull. For *Titanic* and *Olympic*, X was 16. (In this discussion we are not including inner bottom compartments, since they are not included in the standard floodable length calculation.)

Figure 20-1 shows a floodable length curve for the *Titanic*.[7] Note that there is actually more than one curve, dictated by the use of more than one permeability. The effect of different permeabilities is thereby illustrated, and the reader should thereby realize that lack of precise data

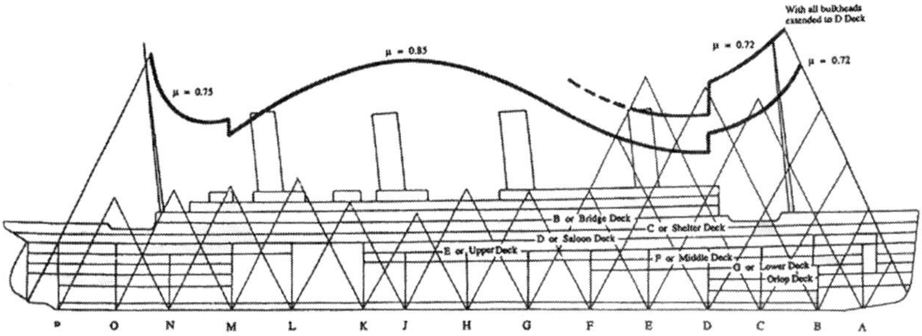

FIGURE 20-1 Floodable length curve for *Titanic*. It shows that any two adjacent spaces between transverse bulkheads can be flooded without having the length of flooding equal the floodable length. The ship thus meets "two compartment standard." The upper partial curve forward shows how extending the *Olympic*'s watertight bulkheads to "D" Deck increased that ship's floodable length as a consequence of her 1913 upgrading. (*Curve constructed by the authors*)

on permeability values causes uncertainty among the users of these curves. This uncertainty is kept at a reasonably low value by use of conservative standard values of this important parameter.

The curves in the illustration are fitted to a profile of the ship, featuring the locations of the ship's transverse watertight bulkheads. The pyramidal figures fitted to these boundaries are simply triangles whose altitude equals their bases, i.e., the bulkhead spacing along the hull. They show at a glance that flooding of any single compartment produces a flooded length less than the floodable length. Similarly they show that flooding of any pair of adjacent bulkheads produces the same result. Along most of the length, however, flooding of three or more adjacent compartments has the opposite result: the ship would sink.

A floodable length curve for the transatlantic liner *Normandie*, built to the rules of SOLAS 1929 (Safety of Life at Sea conference held in 1929) appears in Chapter 23. This international conference on passenger ship safety had the benefit from the many ship losses in peace and war that had occurred since the night of the *Titanic*. Nonetheless, the "two-compartment standard" that had earned the White Star liner its unsinkability label remained the standard, as can be seen from the *Normandie* curve. After 1912, however, and without the need for authority to speak in either 1914 or later, the word "unsinkable" disappeared from a ship designer's lexicon.

From the simple circumstance that the two-compartment survivability standard applied to the *Titanic* was not changed afterward, one may conclude that the famous victim was not deficient in its watertight subdivision. A refutation of that assertion lies in the alterations that White Star ordered for the *Olympic* soon after her sister's loss: a inner skin was added to the hull over the entire length of the machinery spaces, and watertightness of its bulkheads was extended one deck higher. The third sister, *Britannic*, received these modifications during her construction. The forbidden word, unsinkable, was no longer spoken in connection with these improvements, and they were not to be recommended by SOLAS 1914, the international conference that met in 1913–1914 to determine what lessons were learned from the loss of the *Titanic*. The *Olympic/Britannic* changes protected those ships against repetitions of the *Titanic* event, but little else. They were of little help, maybe no help at all, when a German mine exploded against the *Britannic*.

The White Star Line's attempts to increase the safety of its ships may have been mere window dressing, but to some degree the inner skins and higher watertight bulkheads must have increased the safety of those ships vis-à-vis collisions at sea. Judging by the comment of W. T. Chamberlain, quoted in the preceding chapter, there appear to have been many other improvements that *might* have been adopted. Relying on his application of the word "foolish" to those that had apparently been showered upon both American and British authorities, we shall not delve deeply into this topic. However, we shall look at two improvements that were discussed at the Mersey[8] hearings: longitudinal watertight bulkheads and watertight decks.

The iceberg impinged on the *Titanic* hull over the extent of six compartments, rupturing the shell over all six by a series of scrapes and blows. The ice apparently did not penetrate more than a few inches into the hull. The add-on inner skin of the *Olympic* machinery spaces was set 30 inches inboard from the shell, and hence would, almost certainly, have prevented flooding of Boiler Rooms Nos. 5 and 6, thereby reducing the flooded length to less than the floodable length in the forward part of the ship. The ship would not have sunk. In other word, longitudinal subdivision as exemplified by this barrier only 30 inches inboard might have been of incalculable benefit.

When the water flooding the forward end of the *Titanic* reached the tops of the bulkheads at "E" Deck level and the bow compartments filled with water increasing the bow trim, it began to flow aft along this deck and began to fill other compartments in the ship. This action eventually caused the foundering. If only "E" Deck had been watertight! Another incalculable benefit! So it apparently seemed to some marine practitioners of the *Titanic* era.

If watertightness in the transverse plane is good, then the *Titanic* experience suggests that watertightness in the longitudinal and deck planes would be even better. Nonetheless, Lord Mersey stopped short of recommending these improvements; he merely suggested that they be studied. The conferees of SOLAS 1914, with a much greater breadth and depth of expertise than those assembled at the Mersey hearings, did not endorse them either. Here, then, is a more-than-meets-the-eye situation that we shall explore in later chapters.

Notes for Chapter 20

1. Replica of the *Titanic*? Even using the name has been anathema since 1912, but at century's end there was indeed talk of building a cruise ship that would be a replica, and bear the name.
2. Two sources used here are William Garzke, 1985, *Axis and Neutral Battleships in WWII*, U.S. Naval Institute, Annapolis, and Richard Hough, *Dreadnought: A History of the Modern Battleship*, Macmillan, New York, 1964.
3. The Illies message came via personal communication to John Woodward.
4. Robert Ballard, *The Discovery of The Bismarck*, Warner Books, New York, 1990.
5. Did any navy ever claim that its ships were unsinkable? Probably not, but we will reproduce part of an article that appeared in Volume 35 (1913) of *The Engineer*:

   THE UNSINKABLE SHIP

   > The new battleship *Pennsylvania* and her two immediate predecessors, now under construction, the *Nevada* and *Oklahoma*, of the United States Navy, are to be equipped with a novel installation for localising and minimising the effects of underwater injuries. The scheme broadly means nothing short of turning all of these watertight divisions into so many separate caissons, from which the water can be expelled or held within limits through the medium of compressed air.

   The article proceeds at some length to explain the origin of the scheme in underwater salvage or construction caissons that are evacuated of water by compressed air. It also diagrams its fitting to the ships in question. If this was a "good" idea, it is certainly one that apparently never saved a sinking battleship. In fact, one of the ships named, the *Oklahoma* was sunk at Pearl Harbor. It was raised, but sank again, this time under tow to the West Coast for permanent repair, almost taking the towing vessel with her.
6. Fire may be the deadliest hazard of all, and there are cases of ships sinking as a consequence of a major fire. Since none of our five ships was lost by fire nor even experienced a major fire, we omit discussion of fire as a pertinent topic in this chapter.
7. This floodable length curve was not made by the *Titanic* designers, but by the authors from information available in the 1990s. The margin line is taken to be three inches below "E" Deck, but at the scale used for the illustration, deck and margin line are indistiguishable.
8. *Report on the Loss of the Titanic*, His Majesty's Stationery Office, London, 1912.

CHAPTER 21

Unsinkability III

A naval architect, armed with the hull offsets (dimensions) of a ship under design, prompts a computer to deliver the floodable length curve. With the curve in hand, it is a fairly simple task for the naval architect to sketch in transverse watertight bulkheads to produce the desired degree of watertight integrity, while leaving compartment lengths suitable for machinery, cargo, and other items needing large unobstructed spaces. But watertight barriers in horizontal planes (decks) and in vertical fore-and-aft planes (longitudinal bulkheads) surely limit flooding as well, and hence seem to be alternatives to transverse bulkheads. They also might seem to offer valuable supplements to those bulkheads. This notion gains some notice in the preceding chapter. We even suggest there that had the *Titanic*'s "E" Deck been watertight, or had that ship been fitted with watertight barriers similar to those of *Lusitania/Mauretania*, she would have survived the iceberg.

Such notions were bruited in 1912. Lord Mersey,[1] in his report on the events of that April night, suggested that the naval architectural profession study these two forms of longitudinal barrier to flooding. And they were. Before plunging ahead to the conclusions, let's review what was being said by seemingly competent practitioners in the interim between the *Titanic* disaster and the convening of the SOLAS 1914 conference in November 1913.

One of these practitioners was Joseph Conrad,[2] whose stature as a designer was virtually nil, but was nonetheless regarded as a voice of authority in maritime matters. He has been quoted and criticized already in Chapter 19, hence only a key sentence from his ideas about longitudinal bulkheads appear here (repeated): "it is obvious to the simplest intelligence that the more watertight spaces you provide in a ship the nearer you approach safety."

Now let's introduce J. Bernard Walker, editor in 1912 of the *Scientific American,* a magazine that is still to be found in the living rooms and studies of thoughtful persons in the 21st century. Mr. Walker was not a ship designer, but an exploration of the magazine as it was during his tenure shows that he had a professional interest in maritime technology. *Scientific American* had published an article on the *Olympic* and *Titanic* in 1911. That widely distributed picture of the *Olympic* standing on end to show her length equaling the height of one of the tallest towers (the Woolworth building) in Manhattan comes from that article, and is a graphic artifact that can be credited to J. Bernard Walker. The magazine also published articles on warship design, and on the technical controversies that were current in the decade before World War I, such as the merits of steam turbine propulsion. These did give him certain bona fides in the technical marine community, but as will be seen, his principal recommendation was stunningly flawed.

Walker also apparently enjoyed considerable support from his staff at the *Scientific American*, for his judgments were rendered in an "instant book"—it was published in 1912, and the date at which he signed his preface was June, a mere two months after the Great Ship went down. (What wonders could he have wrought had he possessed the resources for desktop publishing?)

The book was *An Unsinkable Titanic,*[3] subtitled *every ship its own lifeboat.* A quotation from the preface tells of the theme to be expounded.

> Since her day, the shipbuilder has eliminated all but one of the safety devices that made the *Great Eastern* a ship so difficult to sink. Nobody, not even the shipbuilders themselves, seemed to realize what was being done, until, suddenly, the world's finest vessel, in the pride of her maiden voyage, struck an iceberg and went to the bottom in something over two and a half hours time!
>
> If we learn the lesson of this tragedy, we shall lose no time in getting back to first principles. We shall introduce in all future passenger ships those simple and effective elements of safety—the double skin, the longitudinal bulkhead, and the watertight deck—which were conspicuous in the *Great Eastern*, and which alone can render such a ship as the *Titanic* unsinkable.

Yes, his cynosure among ships was the *Great Eastern.*[4] That she was a remarkable ship in several ways cannot be questioned. The benefit of her double-walled hull was demonstrated convincingly when the outer envelope was holed by a rock in Long Island Sound, all without any penetration of the inner hull. From that incident one can conclude that she also could have scraped against an iceberg without internal flooding. Given the low ship speeds of its day, and the much smaller size of any ship that might collide with the *Great Eastern*, we might conclude that it was nearly invulnerable to the ship/ship collision hazard. It also may well have been a tough sink in naval warfare, given the absence of armor-piercing artillery and automobile torpedoes among the navies of the 1850s.

But the *Great Eastern* was both an experiment and a ship that would, especially on its intended service to the Far East, face more serious hazards than the ships of later days, and she would pass through lonely seas with no means of communicating any distress. To employ a high degree of overkill in its construction was a commendable decision on the part of its builder, I. K. Brunel, but uncritically projecting its experience into the much different marine milieu of 1912 does not seem creditable for an editor of a high-level technical publication.

That last statement is not intended to deny Walker's implications that ship designers misapplied the experience gained from the success or failure of earlier designs, and that shipowners pressured their designers to sacrifice safety for commercial expediency. In the half-century or so between the *Great Eastern* and the *Titanic*, the latitude for slighting safety in favor of economic return was much greater than it was to become, post-*Titanic*. No international safety standards were extant before 1914, and national standards, such as those of the British Board of Trade, had their defects. The Board of Trade's failure to have lifeboat rules suitable for ships the size of the *Titanic* is an obvious example.

But Walker was shooting from the hip, and would have had difficulty in proving his implication of malfeasance. No matter—he was a seemingly qualified spokesman for popular post-*Titanic* views. Our purpose is to use him as an example of wrong-headedness.

Walker criticized the Board of Trade, but his argument was not with inadequate rules. Maybe he was aware how long it may take rule-makers to act, and how many competing pressures are likely to influence them. He appeared to want the designers to return voluntarily to the super-safe features of the *Great Eastern* in post-*Titanic* construction. He put it simply: his recommended return "can be made without any burdensome increase in cost." That indeed might have been true if the level at which costs are "burdensome" is favorably defined. But true or not, tradeoff between level of safety and level of cost was, and is, a legitimate aspect of engineering design. Fortunately for later-day travelers, in most instances safety standards are applied equally to all players by government regulation that usually is based on international consensus.

In the early 20th century, the double skin, the longitudinal bulkhead, the watertight deck admired by Walker had become a feature of *warship* design. He praises them in his book, but concedes that the degree of protection suitable for a warship is neither suitable nor necessary for a commercial ship. In his conclusions he nonetheless advocates that passenger-ship designers re-adopt those three essentials of unsinkability pioneered by the *Great Eastern*.

However, without explicitly saying so, he seems to realize that commercial pressures will not allow his favorite ship to be resurrected—that the watertight deck and double skin of the *Great Eastern* will not return to favor. (At the time he was writing, Walker would not have known that the *Olympic* and *Britannic*, surviving sisters of the *Titanic*, were to be given double skins as part of White Star's reaction to the sinking.) In an admitted compromise, he seizes upon an art-of-the-possible design; see the illustration in which he shows cross-sections of the *Titanic*, *Mauretania*, and *Great Eastern* (Figure 21-1). The *Mauretania* lacks the double skin and the watertight deck, but she does have longitudinal bulkheads (the bulkheads separating the coal bunkers from the boiler rooms).

Another illustration in the Walker book is labeled, "Comparison of Subdivision in Two Famous Ships—*Titanic* and *Mauretania*" (see Figure 21-2). The former is shown in the act of sinking, water egregiously slopping over the tops of its transverse bulkheads. Since the *Mauretania* never sank, it is not shown sinking; the actual comparison between the ships is only in their two deck plans. The *Titanic* plan shows her single skin and her 16 compartments provided by her transverse bulkheads (heavy lines). The *Mauretania* plan shows her transverse bulkheads and her longitudinal bulkheads dividing her hull into 34 compartments. (The "double skin" in the *Mauretania* label apparently refers to those longitudinal bulkheads, which is not the type of double skin provided by the *Great Eastern* construction. An illustration that compares the two ships in more detail should make this clear.)

No, the *Mauretania* never sank; she went to the scrap yard in 1935, mostly because of the ship's obsolescence and the effects of the Great Depression on passenger traffic. Oh, but her sister, the near identical *Lusitania,* sank in 1915 following a submarine attack—an event described in other chapters.

Walker spoke of the *Mauretania*, but apparently could not find a suitable photograph of that ship in his files, probably because she was the second ship of a pair, and hence had received little media attention. He

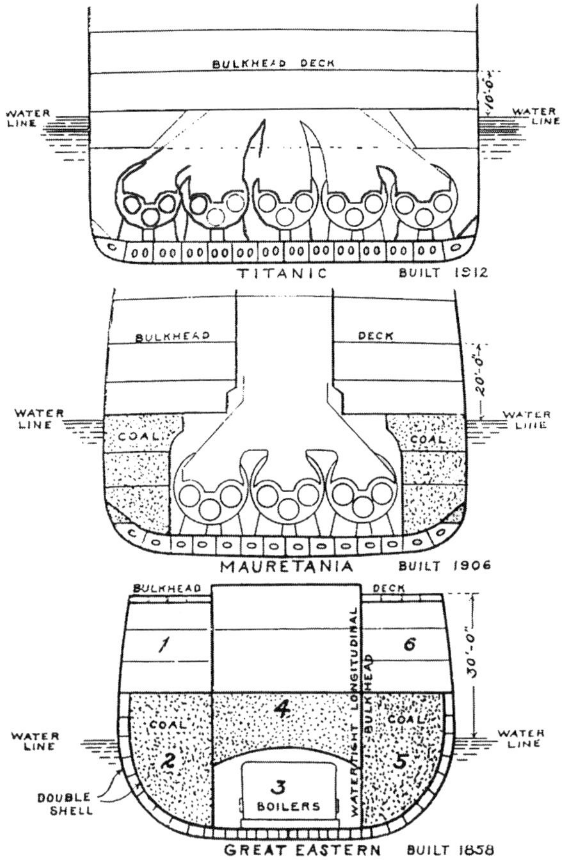

FIGURE 21-1 Cross-sectional sketches of three ships, taken from Walker's book. He advocated the *Great Eastern* protective scheme, but offered the *Mauretania* as a compromise for designers who would find a watertight deck too difficult to apply. Note that the longitudinal bulkhead in the *Great Eastern* is further inboard than that of the *Mauretania*.

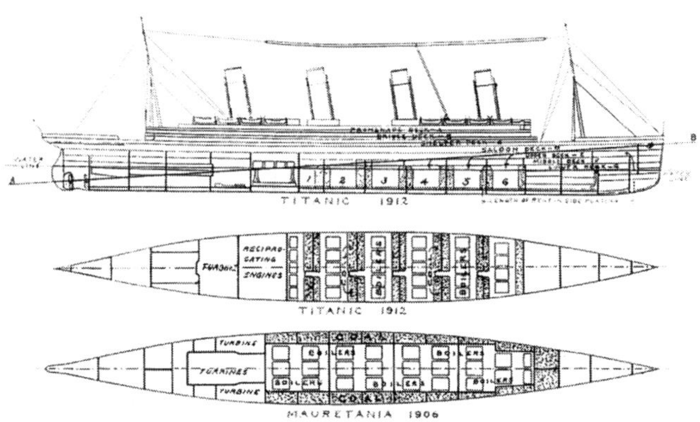

FIGURE 21-2 Walker's comparison of the *Titanic* and the *Mauretania*. (*From his book*)

used instead a photo of the *Lusitania,* boldly labeling it as the ship that "can serve as its own lifeboat" (a euphemism for "unsinkable"). We hope Mr. Walker was suitably embarrassed when he heard how poor a lifeboat it really was. In his defense, he might have retorted that an honorable person such as he should not have expected that a hidden enemy would fire a torpedo into a passenger liner. And we should allow him that defense, for we have already noted that the objective of unsinkability seems to have been abandoned as hopeless once war had been widely waged against commercial shipping.

The *Lusitania* incident was, of course, the result of military action, and hence scarcely a fair test of the safety under ordinary circumstances of transatlantic liners. However, the lesser known sinking of *Empress of Ireland*[5] in early (pre-war) 1914 was the result of a ship/ship collision. The victim was fitted with longitudinal coal bunker bulkheads in the manner of the *Lusitania* and with an outcome quite similar to that of the more famous ship: immediate heavy list, quick sinking, large loss of life. Even the *Andrea Doria* case, to be cited later in this chapter, speaks against the use of longitudinal subdivision. Although that ship did not sink as a direct consequence of one-sided flooding, the 23-degree list to starboard that developed soon after the collision had serious consequences for the deployment of its lifeboats. (The port-side boats could not be launched, and some went down with the ship while others broke free.)

Obviously, it is unfair to review J. Bernard Walker's book after hindsight has provided so much evidence to dump upon his ideas, but one can voice a few negative thoughts that remain after setting aside all question of the book's technical merit. The work is patently an "instant book," as shown by the date of signature on the preface. There must have been no time for developing ideas with input from professional ship designers, nor is there evidence of consultation with anyone. In tone it is assertive, with scant pause to develop argument supporting the assertions. The tone suggests the text of a speech to a nontechnical audience. The book is short, with its textural length padded by small pages, large type and numerous illustrations, some of which don't contribute anything to the message.

The longitudinal watertight bulkhead was a bad idea. Although in a later chapter we quote a seemingly professional opinion, post-*Titanic,* that supports it, this form of subdivision was never required. On the other hand, it was never condemned. The whys and hows of its continued existence appear in later chapters.

A truly professional critic who published his ideas on unsinkability in response to *Titanic* was George Dickie,[6] an American naval architect.

He wrote a paper for presentation to the Society of Naval Architects and Marine Engineers (of which he was vice-president). The paper appeared in the 1913 *Transactions* of the Society; it carried the impressive title, "On the Possibility of Building a Large Passenger Liner That Would Not Under Any of The Known Mishaps at Sea Lose Her Buoyancy or Stability and Sink." Mr. Dickie was a designer of passenger ships, and mentions in his paper that he was writing while travelling on one of his ships, the *Congress*. That ship was intended for freight and passenger service along the West Coast of North America, and he was aboard as it made its delivery voyage to San Francisco from an East Coast shipyard. As a preface to his prescriptions for the *Titanic*, he remarks that the more stringent American lifeboat regulations that followed the *Titanic* experience—regulations that resulted in an additional 25 tons of weight high in his ship—came into force while he was designing the *Congress*. Compensation for the loss in stability required a 21-inch increase in ship's beam. Well-meaning law required him to make these changes, but he scoffed at the notion that any number of boats could be an effective life-saving agency in the rough seas so often encountered along the northern Pacific coast. With regard to his newly designed *Congress*, he said:

> With passenger list full, this vessel carries 850 people and, should it ever be necessary that they leave the ship during that part of the voyage north of San Francisco, I can hardly conceive of its being accomplished without serious loss of life.

For quite logical reasons such as this, Dickie preferred the Walker concept of "every ship its own lifeboat" i.e., the never-never-land goal of "an unsinkable ship," to quote words he was bold enough to use. His paper describes such a ship, shown in Figure 21-3. In size, power, and general appearance, it is much like the *Titanic*.

The principal innovation of the Dickie design is a watertight deck within the hull; indeed, a double watertight deck that the cross-sectional view in the illustration shows quite clearly (Figures 21-3 and 21-4). The space between these decks, only 5½ feet high over the mid-portion of the hull, he describes as "a sub-basement for the hotel part of the ship above," through which piping, wiring, and ventilation would be distributed from the machinery spaces to the passenger accommodations and public rooms. The convenience of this feature apparently was intended to compensate for the lost cargo and storeroom volume, and for the difficulty of access to any such spaces below the watertight deck.

Large openings in the watertight deck over the boiler rooms obviously could not be avoided; the exhaust gases must go up, and the air for

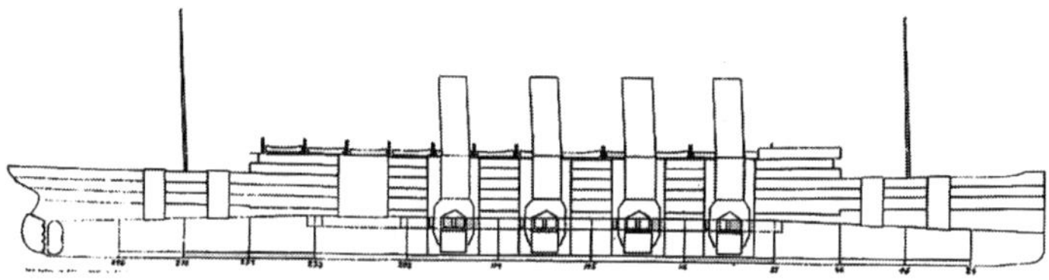

FIGURE 21-3 Inboard profile view of the unsinkable ship sketched by George Dickie. His watertight deck is the one that slopes downward at both ends. (*From his paper in vol. 21,* Transactions, *Society of Naval Architects and Marine Engineers*)

FIGURE 21-4 Cross-sectional view of the unsinkable ship sketched by George Dickie. His watertight deck is the second one above the boilers. His longitudinal passage appears in the space between the boiler uptakes. (*From his paper in vol. 21,* Transactions, *Society of Naval Architects and Marine Engineers*)

combustion and for ventilation of the machinery spaces must come down. Dickie's design, therefore, requires four watertight trunks rising from the watertight deck to a point well above any conceivable flooding level. This necessity he does point out, though regards it as no particular handicap.

Dickie does offer some quantitative discourse on the possible extent of flooding in the event of collision, but the paper is short, and is actually little more than a vehicle for introducing a concept to a technical audience. It closes with the words "along the lines I have indicated, a ship can be designed . . . as practically an unsinkable ship."

Several of the professionals in the audience offered comments after the presentation. They were all polite and temperate whether supporting or criticizing. Perhaps the most critical words were those of Professor Herbert Sadler,[7] pointing out that if safety features made a ship commercially impractical—as was frequently said about watertight decks—then that ship would not be built. Naval Constructor David Taylor[8] countered in favor of watertight decks by saying that if such a feature would make a ship commercially infeasible, then a designer of Dickie's stature would not have proposed it.

Would a ship with a watertight deck such as George Dickie advocated be unsinkable? The reflex answer should be "no, if bombs and torpedoes are the cause," but we already have conceded the impossibility of thwarting serious military assault. How about unsinkable in the face of the more natural hazards of the sea? A reasoned reply, based on what has been said in the preceding chapter, might be, "If the *Titanic* had been fitted with such a deck forward and over its boiler rooms it probably would have survived its brush with the iceberg, but. . . ."

But what? Recall from two chapters previous that an intemperate Joseph Conrad scorned "the high priests of the modern cult of perfected material and of mechanical appliances" for expressing doubts about the efficacy of longitudinal coal bunker bulkheads. Your present-day high priests offered a defense of their spiritual predecessors with the words "they know that a design choice that is so obviously suitable in *one* situation also must be suitable in many others." Right—what is effective against the iceberg below the waterline also must be effective against the more likely bow of another ship slicing into the hull. The damage in that kind of collision is likely to extend over the full depth of that hull. A contemporary of Conrad, Professor Fred Purvis,[9] wrote of that possibility with the words that follow.

> In collisions with other steamers the area of damage is not at all restricted to the underwater portion of the ship; the impact which might be expected to tear open the ship's side would also tear away a large portion of any deck near the waterline; the stringer plate certainly would be carried away for practically the same length as the damage of the outside skin; for that length, therefore, the transverse bulkheads would be called upon to play the same part whether the watertight deck were fitted or not.

The damage to the stringer plate (the deck plate closest to the side of the ship) would destroy the watertightness of the deck in way of the damage. Further, in a point that Purvis does not mention, listing or sinkage might dip the exposed edge of the deck below the waterline; if so, flooding water might flow *along* that barrier (still watertight except at the collision point), producing a "free surface effect" high in the ship. The baneful consequences of that were seen in the high-fatality capsizings of several European vehicle ferries (e.g., *Herald of Free Enterprise*) in the 1990s. No collision was required: faulty sealing around forward vehicle doors allowed water to enter and run along the vehicle deck. The deck was not watertight, but was close enough to that state that water could not drain below fast enough to preclude its collecting in fatal amounts on one side of the deck.

Of course, the fleets of seagoing vehicle ferries that serve the European coasts were beyond the imaginations of 1912's naval architects as well those of sea-wise novelists. However, the problem of water unleashed on decks above the waterline serves to illustrate the occurrence—found in many venues—of downside consequences of reasonable sounding safety measures. The putative high priests must be aware of the ever-lurking law of unintended consequences.

Mr. Dickie's watertight deck was a bad idea. The centerline passage that was a prominent feature of his unsinkable design was a bad idea also. This passage is evident in the illustration that shows his double watertight deck. It apparently was not essential to his concept of a watertight deck, but was rather a feature that would facilitate access throughout the lower reaches of the hull, and thereby at least partially deflect the criticisms of watertight decks as interrupters of access.

Recall Dickie's statement, quoted previously, that "under no conceivable circumstances . . ." This boast was unchallenged for 40 years, but in July 1956 the Italian liner *Andrea Doria* and the Swedish liner *Stockholm* collided near Nantucket Island.[10] The ice-strengthened bow of the *Stockholm* sliced through empty fuel tanks in the forward part of the Italian ship. Flooding of the tanks caused a severe starboard list of about 23 degrees to afflict the *Andrea Doria* almost immediately. The list made launching of the port-side lifeboats impossible, and made it difficult for passengers to reach the starboard-side boats. But no matter, since the ship took 11 hours to sink, giving time for a flotilla of rescue ships, that included such ships as the *Ile de France*, USNS *Private William F. Thomas*, *Cape Ann*, *Hornbeam*, *Hopkins*, and even the *Stockholm*, took on all but 51 of *Andrea Doria*'s passengers and crew. This situation was an example of J. Bernard Walker's concept that a ship was the best lifeboat. The collision had occurred in a very busy sea lane that passed south of Nantucket Island. Everyone who was not killed by the collision itself was saved.

The *Andrea Doria* never recovered from the original starboard list, but that embarrassment was not the cause of its foundering. Penetration of a centerline passage by the bow of the *Stockholm* was the wound from which there was no recovery. Because it was inconceivable that such a well-protected space could be opened to the sea (paraphrasing Dickie), no watertight door was fitted where the passage ended at the Generator Room just aft of the fuel tanks. The inrush of seawater must have been slow, given those 12 hours in which it continued, but it could not be stanched, nor could the bilge pumps overcome it because the generators were among the first casualties of the flooding.

The Dickie paper appears to be the only professional work, either pre- or post-*Titanic*, that advocated watertight decks for passenger ships. No remembered technical paper condemns them, and they were not to be outlawed by safety regulations. Even without such objections as that raised by Professor Purvis, the severe handicap they would place on internal arrangement made them an unlikely feature of a passenger ship. Ventilation ducts can run up and over a watertight bulkhead, but how do you get them past a watertight deck? Watertight doors can provide access through watertight bulkheads, but what contrivance could have closed the Grand Staircase of the *Olympic* ships in an emergency?

Walker and Dickie did little more than provide convenient targets for criticism of the popular ideas that they espoused. The designs for watertight decks that they advocated were recognized as vaporware by their contemporaries and by their successors. A similar remark might be made about longitudinal watertight bulkheads—but no, that idea was not so easily dismissed. Indeed, it might be said to have thrived in spite of the lessons from the *Lusitania* and *Empress of Ireland* disasters. Tanks for petroleum and its products have given it enduring life. For example, the cargo space of large petroleum carriers ("tankers") is typically divided by bulkheads into many cargo tanks; among the bulkheads are two running longitudinally to make three tanks abreast. More appropriate to this discussion is the common use of wing tanks for fuel oil by the many transatlantic liners that were built after World War I, and by the cruise liners that followed them.

An illustration—a machinery deck plan of the liner *Normandie*—shows the typical fuel tankage of a petroleum-era passenger ship built under the strictures of SOLAS 1929 (Figure 21-5).

So in spite of *Lusitania* . . . *Empress of Ireland* . . . and lesser known disasters, no SOLAS conferences (and the resulting national laws) have

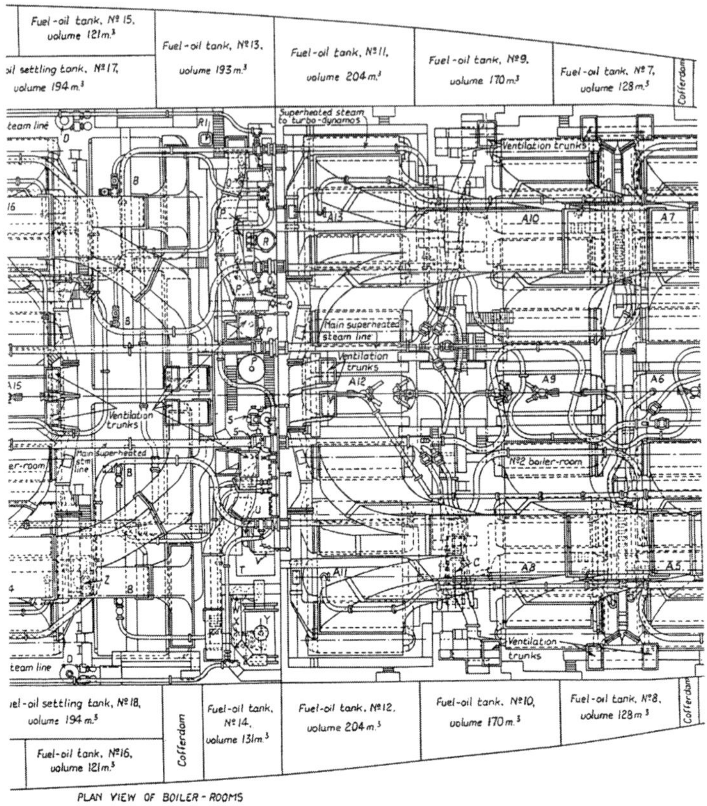

FIGURE 21-5 Plan of a portion of the *Normandie* boiler rooms, showing the fuel oil wing tanks. (*From a special edition of* The Shipbuilder, *heralding the new ship in 1936*)

ruled against longitudinal watertight subdivision. Has convenience of tankage drowned out the cries of the victims? No, what *is* outlawed is the consequence of the asymmetrical flooding: a list that endangers the ship and that makes launching of lifeboats difficult. Showing survivability by means of a floodable length curve is, therefore, not sufficient since a margin line shown to be above water by that designer's artifice may be dipped below the waterline by listing of the damaged ship. If there are wing tank boundaries or other impediments to free cross-flooding, damage stability calculations are a necessary supplement to the floodable length.

The Board of Trade rules that were existent in 1912 do not use the terms "floodable length" and "damage stability," but do specify that listing shall not dip the margin line below the water surface. So far as post-

Titanic regulation was concerned, SOLAS 1929[11] said this in Regulation V of its Convention:

> Where it is proposed to fit watertight decks, inner skins or longitudinal bulkheads, watertight or non-watertight, the Administration shall be satisfied that the safety of the ship will not be diminished in any respect, *particularly having in view the possible listing effect of flooding in the way of such structural arrangements*" (emphasis added).

This stipulation became more specific when translated in national maritime law. For example, U.S. Coast Guard rules require that a list does not exceed 15 degrees, nor dip the margin line below the waterline.

The damage stability calculations are not a favorite exercise among traditional naval architects because of the essential feedback loop that complicates the procedure: flooding produces sinkage and listing, which in turn allow more flooding, which in turn produces more sinkage and listing. However, the typical ship hydrostatics software that was developed late in the 20th century includes the ability to make the necessary iterative calculations, and from the same input that leads to the floodable length curve.

Lacking complete data on the ship, on its state of loading and weight distribution as well as the amount of coal burned on the voyage at the moment the torpedo struck, it is difficult to determine the extent of the flooding with the most sophisticated computer software to reproduce the *Lusitania*'s stability history in the short-term aftermath of the torpedo explosion. However, *estimates* can be made, based on any number of possible circumstances in which the ship would be holed along its midbody. If, say, the coal bunkers amidships, flanking Boiler Room No. 3, were flooded while partially filled with coal, and with the ship's GM at 2.0 feet, a list in the 20-degree range could follow.

Among the complicating factor in a damage stability analysis is the presence of a matching space on the undamaged side of the ship. If we are speaking of fuel spaces, then is it likely that both are either filled or empty—or both filled with ballast? Both empty is probably the most dangerous condition because no weight on the undamaged side will balance the weight of the flooding water. As stated, such was the case assumed in the *Lusitania* estimate of the preceding paragraph.

In the *Andrea Doria*'s case, the both empty condition is known to have been present: the tanks in way of the collision were fuel tanks that were empty, as well as the matching tanks on the far side of the centerline. Operational doctrine called for empty fuel tanks to be refilled with ballast water, but the ship was close to New York. Pumping oily ballast to a barge in harbor would be expensive, and would delay refueling. When the collision opened all of these tanks on one side, a list of over 20 degrees developed very quickly—too quickly for counterflooding to be effective. Launching and boarding of lifeboats was very difficult on one side of the ship, impossible on the other.

Notes for Chapter 21

1. Lord Mersey, 1912, *Report on the Loss of the Titanic*, 1912, His Majesty's Stationery Office, London.
2. The source used here for the opinions of Joseph Conrad is his *Notes on Life and Letters*, 1921, Doubleday, Page & Company, Garden City, NJ and Toronto.
3. J. Bernard Walker, 1912, *An Unsinkable Titanic,* Scientific American, New York.
4. A source for the *Great Eastern* is James Dugan, 1953, *The Great Iron Ship,* Harper, New York.
5. A source for the *Empress of Ireland* is James Croall, 1980, *Fourteen Minutes: The Last Voyage of The Empress of Ireland*, Sphere Books, London.
6. Dickie's paper outlining the design features of an unsinkable ship was published in the 1913 *Transactions*, Volume 21, Society of Naval Architects and Marine Engineers.
7. Professor Herbert Sadler was a founding member of the University of Michigan faculty of naval architecture and marine engineering. He was lured to Ann Arbor from Glasgow, then a world center both for shipbuilding and the teaching of ship design.
8. Naval Constructor David W. Taylor was then rear admiral, and already famed for his work in developing model basin techniques for the propulsion power prediction of naval vessels.
9. Professor Purvis' paper appeared in *Engineering*, Volume 97, 1913.
10. A source for the story of the *Andrea Doria* sinking is Alvin Moscow, *Collision Course: The Andrea Doria and the Stockholm,* G. P. Putnam's Sons, New York, 1959.
11. The results of the SOLAS 1929 Conference are reported by George H. Rock, 1929, "The International Conference on Safety of Life at Sea, with Special Reference to Ship Construction," *Transactions*, Society of Naval Architects and Marine Engineers, Volume 37.

CHAPTER 22

SOLAS 1914

If that iceberg collision had been just a little bit less ardent, the *Titanic*'s entire passenger complement might have reached New York by train from Halifax. Captain Smith of the designers in Belfast still would have been grilled by the British Board of Trade. However, given the lack of drama and loss of life, the public probably would have ignored the proceedings. The important finding may have been the type of damage that the collision caused and the possibility in change of the rules in how ships were constructed.

The deaths of 1,522 could have been prevented by a few changes in outfit and operating practice. The enormity of the disaster caught the public's attention and generated books, songs, poems, and sermons as well as ill-informed analyses such as those of J. Bernard Walker and Joseph Conrad, cited in earlier chapters. Criticisms fell upon the designers, builders, operators, and owners of the *Titanic* like the showers of April.

Following the criticism came the advocacy of remedies that would attempt to make ships unsinkable or at least provide a certain and hazard-free salvation for the innocent. Most of the ideas were worthless as a consequence of their authors' ignorance of ship technology. Other remedies, advocated by semi-professionals such as J. Bernard Walker, involved changes in the design of ships that might be feasible, such as addition of longitudinal bulkheads in the style of the *Lusitania*. Professional ship designers were more reticent[1] in that they did not rush to get out in front with innovations in that murky area of unsinkability. They preferred to wait, with good reason, until they and their peers had met to dissect and to propose rules that should remedy any defects to be found in the *Titanic*'s design.

Marine professionals were greatly affected by the magnitude of the loss of life and stunned by the ship's sinking. The beneficial consequence of the clamor from the inquiry of Senator William Smith in the United States and the public outrage put pressure upon the professionals to act quickly and to devise effective remedies. The burden was especially heavy on the British Board of Trade, which was, after all, responsible in that it was the rule-making body under which the *Titanic* had been built. It did act with commendable speed; by late May of 1912 it had its renowned wreck commissioner, Lord Mersey, to chair the first investigation in which the builders, designers, and operators testified for the record. However, its speed did not match that of U.S. Senator William Smith, clutching subpoenas in one hand, the collars (figuratively, of course) of the surviving passengers and crew in the other, as they disembarked from the *Carpathia*.

The Smith Investigation, 1912.[2] Senator William Alden Smith of Michigan, a member of the Senate's Commerce Committee, received quick authorization from both the Senate and President William Taft to proceed with an investigation. He and members of his committee rushed to New York on the day of the *Carpathia*'s docking to confront Bruce Ismay before he could leave the ship. Then, armed with his impressive authority, the senator conducted hearings in New York and Washington, April 19 through May 25, 1912, well ahead of Britain's slower moving Board of Trade. He had acted quickly to subpoena witnesses, especially the British subjects that Bruce Ismay wanted to rush from the *Carpathia* to an outbound ship. Ismay failed in that attempt and he himself became the Senator's prize witness. However, Ismay could not (or would not) provide any significant revelations. There were, nonetheless, recommendations distilled from the many testimonies, and predictably, they dealt extensively with the number of lifeboats and how the wireless was used aboard ships. The senator had no access to witnesses who could speak of the design and building of the ship; they were beyond the reach of his subpoenas. The closest he came to expert input on such matters was a memorandum from Admiral H. I. Cone of the U.S. Navy, and it was the inspiration for the Smith recommendations that relate directly to ship design. They are brief enough to quote in full here.

"All steel ocean and coastwise seagoing ships carrying 100 or more passengers should have a watertight skin inboard of the outside plating, extending not less than 10 percent of the load draft above the full-load waterline, either in the form of an inner bottom or of longitudinal watertight bulkheads, and this construction should extend from the forward collision bulkhead over not less than two thirds of the length of the ship."

These American hearings resulted in a number of recommendations. All steel ocean and coastwise seagoing ships carrying 100 or more passengers should have bulkheads so spaced that any two adjacent compartments of the ship could be flooded without destroying the floatability or stability of ships. Watertight transverse bulkheads should extend from side to side of the ship, attaching to the outside shell. The transverse bulkheads forward and abaft the machinery spaces should be continued watertight vertically to the uppermost continuous structural deck. The uppermost continuous structural deck should be fitted watertight. Bulkheads within the limits of the machinery spaces should extend not less than 25% of the draft of the ship above the load waterline and should end at a watertight deck. All watertight bulkheads and decks should be proportioned to withstand, without material permanent deflection, a water pressure equal to 5 feet more than the full height of the bulkhead. Bulkheads of novel dimensions or scantlings should be tested by being subjected to actual water pressure.

These were, again, only recommendations, since they had progressed no further than a senate committee. The two-compartment flooding standard was already law for those ships built under the authority of the British Board of Trade, and has continued to be the regulatory standard since then. The other recommendations were obvious responses to the particular circumstances of the *Titanic* sinking, but may not have been suitable for general application. The experts who were to meet in London, 1913–1914, to develop regulations that would become law did not specify either double skins or watertight decks. Admiral Cone's contribution in 1912 was solely that of translating the suggestions of persons like the Walkers and the Conrads into the language of the maritime professionals. He gave Senator Smith a certain imprimatur, but the notions of passenger ship safety that he packaged for the Senator were rejected by more thoughtful contemporaries when they met as SOLAS 1914.

Senator Smith's report was given to the United States Senate, dated 28 May 1912. Those of his conclusions that survived to be considered by SOLAS 1914 are paraphrased in the following.

- The absence of suction or disturbance of the sea surface as the ship went down allowed those in the water near the ship to swim away. A few of the survivors owe their lives to this phenomenon.
- Lifeboat capacity sufficient to accommodate every person on board should be provided. Every boat should be manned by at least four crewmen who have been drilled in the lowering and handling of the boats.

- Wireless should be manned 24 hours a day. Auxiliary power should be provided for the wireless set.
- Firing of rockets or candles at sea should be prohibited.
- All vessels carrying 100 passengers or more should be fitted with an inner skin that extends above the full-load waterline by 10% of the draft. This construction should extend from the Forepeak bulkhead over not less than two thirds of the ship's length.
- All vessels carrying 100 or more passengers should be equipped with two-compartment subdivision with bulkheads that extend side to side, and extending above the full-load waterline to a distance 25% of the draft above that waterline. In addition, the bulkheads forward and aft of the propulsion plant should be extended to the uppermost continuous deck. All watertight bulkheads and decks should be designed to withstand, without permanent deflection, a water pressure equal to at least five feet above the bulkhead. These bulkheads should be tested under actual water pressure.

The last two of these recommendations came directly from Admiral Cone's report. They have received comment in an earlier paragraph.

Senator Smith concluded that the ship went down gradually, by the bow, and assumed an almost vertical position before sinking at 0047, New York time, on 15 April 1912. He noted the many conflicting statements bearing on the question of whether the ship broke in two, but concluded that the ship sank intact. That conclusion remained part of the *Titanic* canon until the wreck was discovered in 1985.

The Board of Trade (Mersey) Investigation, 1912.[3] Senator Smith's hearings were greeted with both scorn and resentment in the United Kingdom. There was anger and concern because some of his questions appeared to reveal ignorance of maritime matters, resentment because he seemed to be treading on British maritime jurisdiction. It was also British feeling that they had a tradition in maritime practice and America did not. In contrast, Lord Mersey was an experienced wreck commissioner, and was one of England's own. On the other hand, some of those who write about the *Titanic* affair are bold enough to say that Lord Mersey may have used a façade of quiet professionalism to conduct a whitewash of British maritime practice. Maybe yes, maybe no, but unlike Senator Smith, Mersey had access to the designers/builders, especially in the person of Edward Wilding of Harland & Wolff. Wilding, who, after the death of Thomas Andrews in the sinking appears to have been the most authoritative person on the design of the *Titanic*, may not have been without vested interest, of course. However, we see no air of defensiveness in his testimony; he appears to have testified candidly and without need to justify.[4] In consequence, the hearing was able to delve with apparent disinterest into the construction of the ship, particularly with regard to its watertight subdivision. The following is quoted from Lord Mersey's concluding recommendations on this vital matter.

1. The newly appointed Bulkhead Committee [of the Board of Trade] should enquire and report, among other matters, on the desirability and practicability of providing ships with (a) a double skin carried up above the waterline: or, as an alternative, with (b) a longitudinal, vertical, watertight bulkhead on each side of the ship, extending as far forward as convenient: or (c) with a combination of (a) and (b). Any one of the three (a), (b) and (c) to be in addition to watertight transverse bulkheads.
2. That the Committee should also enquire and report as to the desirability and practicability of fitting ships with (a) a deck or decks at a convenient distance or distances above the waterline which shall be watertight throughout a part or the whole of the ship's length: and should in this connection report upon (b) the means by which the necessary openings in such decks should be made watertight, whether by watertight doors or watertight trunks or by [something missing in original].

These measures had been discussed by Edward Wilding, and recognized as possible saviors of the *Titanic*, but note that Mersey's recommendation was not that they be *mandated* henceforth, but that competent authority study them. Was this a matter of covering up? No, it was a professional's recognition that the problems of watertight integrity were not as simple as the Conrads and Walkers were advancing, nor as easily fixed as Admiral Cone, speaking through the voice of an American senator, seemed to think. In similar fashion, the Mersey professionals did not *reject* the concepts of longitudinal bulkheads and watertight decks, in spite of being "high priests of the modern cult of perfected material and of mechanical appliances" in Conrad's analysis.[5]

The 1914 Safety of Life at Sea Conference.[6] Another factor in safety regulation was the need for uniform standards among all maritime nations. That necessity was (and remained so a century later) difficult to satisfy, given the diverse attitudes likely to prevail among the many nations that can be called "maritime," and the circumstance that even after the many reach agreement, the individual nations must enact laws that make the agreement enforceable. Before the *Titanic*, such a task had never been considered, but after the disaster it became essential. The British government took the lead and called for a conference among nations to establish regulations for the building and classification of passenger ships. On 12 November 1913 the first International Conference on Safety of Life at Sea (a SOLAS Conference, as such meetings were soon to be called), convened in London.

SOLAS 1914 convened approximately 19 months after the *Titanic* sinking. Some of that interval was inevitably occupied in justifiable waiting for the Mersey conclusions. Some was needed for the complicated task of organizing an international conference. The largest part of all was probably necessary for the engineering studies that the individual delegates had to undertake before they could present and argue their positions at the conference. Given the immense labor involved in analyzing matters of ship buoyancy and stability before the age of the digital computer, a lengthy interval for the technical aspects of preparation is not surprising.

The SOLAS Conference labored through 20 January 1914. Delegates represented the United States and most of the maritime nations of Europe. Japan was not officially present, but had observers on hand. The conference produced agreements on many matters, though the necessary progress into law did not follow. That step involved ratification by the governments concerned, and within a few months World War I became an overwhelming distraction. Regardless of the legal problems, the agreements are instructive, since they, in effect, constitute the experts' judgments on the *Titanic*, as well as on the other liners of our interest.

The documents produced by the Conference were a Convention. They were a statement of principles covering the many facets of ship safety, a set of Regulations, which stated how those principles were to be implemented, and a Final Protocol, that, inter alia, contained some recommendations that had not been resolved into items of the Convention and its Regulations.

The Convention was to apply to powered vessels that carried more than 12 passengers on international voyages between the nations participating.

The documents themselves were long and tedious so we shall only scan them for the important points that seem to have been generated by the *Titanic* experience (and hence reflect on the ships under discussion). In some cases, quotation (identified by contrasting type face) seems appropriate; in others, paraphrasing and condensation seem better for the purposes of this book. The present authors' comments are identified by brackets [ ].

To begin the review, let us quote words from the closing address by Lord Mersey, president of the conference.

It is important to point out that, even after the most careful attention to all practicable details of design which increase the safety of a vessel at sea, there still remains the possibility of a serious and totally destructive accident. Therefore, it is imperative that those charged with the management of vessels should never relax their vigilance on the supposition that any vessel is unsinkable. On the contrary, they should strive to add to the safety provided by the vessel itself that very great increase in safety which results from prudent and skillful management and navigation.

Convention, Chapter III—Safety of Navigation

Articles 6 and 7. Establish an international ice patrol in the North Atlantic, and instruct the United States to undertake the patrol, with the other nations together to contribute 85% of the expense. [The United States had already established an ice patrol before the conference and has been recognized since as a major factor in eliminating collisions between ships and icebergs.]

Articles 8 and 9. Required the master of every vessel sighting dangerous ice to disseminate the information.

Article 10. Required that when ice is reported, on or near his course, the master of every vessel was bound to proceed at night at a moderate speed, or to alter his course so as to go well clear of the danger zone. [Everyone who knows the story of the *Titanic* will recognize the justification for Article 10.]

Convention, Chapter IV—Construction

This section of the Convention treats subdivision by watertight bulkheads, their construction, openings in these bulkheads, and openings in a ship's side. We'll bypass this important subject in favor of referring to its details as they appear in the Regulations.

Convention, Chapter V—Radiotelegraphy

Article 31. Required all vessels carrying 50 or more persons (even if none of them are passengers) to be fitted with a radiotelegraph installation. There are some exceptions, principally involving length of voyage, distance from shore, etc., but no transatlantic voyage qualifies for exception.

Article 33. Divided ships into three classes. The first, which encompasses the largest and fastest ships, are required to maintain a continuous radiotelegraph watch.

Article 34. Specifically states that ships required to have radiotelegraph, and engaged in the transatlantic trade, must maintain a continuous watch. [The *Californian* incident surely prompted this regulation.]

Convention, Chapter VI—Lifesaving Appliances

Article 40. Fundamental Principle: At no moment of its voyage may a vessel have on board a total number of persons greater than that for whom accommodation is provided in the lifeboats and the pontoon life rafts on board. [This was international recognition that everyone on board should have a seat on a lifeboat.]

Article 42. Strength of Boats: Each boat must be of sufficient strength to enable it to be safely loaded with its full complement of persons and equipment.

Article 49. Strength and Operation of The Davits: The davits shall be of such strength that the boats can be lowered with their full complement of persons and equipment, the vessel being assumed to have a list of 15 degrees. [Many *Titanic* lifeboats were launched when only partially full. In sum, they were launched with about 400 empty seats. Second Officer Charles Lightoller,[7] who supervised the launching of the port-side boats, defended his part in this distressing omission by saying that he feared that the boats were not strong enough to support their rated capacity while suspended from the davits. It is not clear if he was present when one of the lifeboats was tested at the Harland & Wolff Shipyard. This boat was lowered with 65 shipyard workers aboard.]

Regulations, Construction

Articles V, VI, VII. Definitions relating to floodable length. Since these have been cited in a preceding chapter, they are not repeated. However, in addition to defining the word *permeability*, numerical values for it are specified by the Regulations, and are stated below:

- For steam vessels, the machinery space, including double bottom: 80%
- Cargo spaces, bunkers, storerooms, baggage and mail rooms, chain lockers, watertight shaft or pipe tunnels, fresh-water tanks above the double bottom: 60%
- Passenger and crew spaces, peaks, trimming tanks, double bottoms, and all other spaces not listed above: 95%

[These percentages appear to be higher than those used by the *Titanic* designers. Since higher permeability implies more flooding water to be allowed for in calculating floodable length, increased percentages indicate a conservative outlook.]

Articles VIII and IX. Permissible Length of Compartments. All parts of these articles that appear pertinent to our discussion are quoted following.

(1) The maximum permissible length of one compartment having its center at any point in the vessel's length is obtained from the floodable length by multiplying that length by an appropriate factor, called the *factor of subdivision.*

(2) This factor of subdivision depends on the length of the vessel, and, for a given length, varies according to the nature of the service for which the vessel is intended. This factor decreases in a regular and continuous manner—

(a) As the length of the vessel increases; and

(b) As, for a given length, the vessel departs from the type of vessel engaged in a mixed cargo and passenger service, and approaches the type of vessel primarily engaged in the transportation of passengers.

(3) For each of the two types of vessels referred to in the previous paragraph (2)(b) the variation of the factor of subdivision may be expressed by a curve, of which the coordinates represent the length of the vessel and the value of the factor. The following table gives certain points on two curves the higher of which corresponds to the minimum requirements for the "mixed" type, and the lower the minimum requirements for the "passenger" type.

A	B	C
1.00	295	259
0.90	374	285
0.84	404	305
0.65	489	380
0.50	571	489
0.39	699	685
0.34	899	899

Column A gives the maximum permissible values of the factor of subdivision for the length of vessels given in columns B and C. Column B is applicable to vessels engaged in mixed cargo and passenger service. Column C is applicable to vessels primarily engaged in the transportation of passengers. [The original table gives lengths in meters also.]

(4) For a given length, the value of the factor of subdivision appropriate to a vessel between two extreme limits will be between the values of the factors determined by the two curves mentioned before, and will be

automatically fixed by a "criterion of service" which is to form the subject of further study.

Article X continues the above.

(1) When the factor of subdivision is equal to or less than 0.5, it may be doubled in order to give, at any point of the vessel's length, the total length of two adjacent compartments; but, in that case, the length of the shorter compartment of any pair shall not be less than one quarter of the total length so obtained. If one of the two adjacent compartments is situated inside the machinery space and the second is situated outside the machinery space, and average permeability of the portion of the ship in which the second is situated differs from 80%, the length of the pair of compartments shall be adjusted to the proper value by applying a suitable correction.

(2) In no case whatever shall the length of any watertight compartment exceed 28 meters (92 feet).

(3) When the factor of subdivision applicable to any vessel is less than 0.84, but more than 0.5, the combined length of the two foremost compartments shall not exceed the floodable length at the extreme forward end, provided also that the length of the second compartment is not greater than that permissible by Article VIII above, and not less than 3 meters (10 feet).

(4) When the length of the vessel is more than 213 meters (699 feet) but less than 251 meters (823 feet) the floodable length at the forward end of the vessel shall be at least 20 percent of the vessel's length; and the vessel; forward of a bulkhead placed either at the distance of the actual floodable length abaft the stem or not nearer to the stem than 20% of the vessel's length, shall be divided into at least three compartments.

(5) When the length of the vessel is equal to or greater than 251 meters (823 feet) the same method shall be adopted, but the floodable length shall be at least 28% and the number of compartments at least four.

Article XV—Openings in Watertight Bulkheads

(5) In the case of watertight bunker doors, satisfactory arrangements shall be made by means of screens or otherwise, to prevent the coal from interfering with the closing of the doors. [Since the *Titanic* coal bunkers did not have openings in their watertight boundaries, Article XV does not appear to come from experience with that ship. It is of inter-

est nonetheless, because the boundaries of the *Lusitania* bunkers were supposedly watertight, and the flooding through doors that could not be closed because of interference from coal was said to be a significant factor in its rapid plunge to the bottom. The presence of this article, promulgated a year before the *Lusitania* torpedoing, indicates that professionals were aware of the problem. Evidence that at least some of them were aware can be seen in excerpts in Chapter 19 from a 1913 paper by Professor F. P. Purvis. Naval Architect Edward Wilding made the same point in his testimony before the Mersey Inquiry and again at the New York liability hearing on the *Titanic* in 1915.]

Article XXI—Double Bottoms

This article details the requirements for double bottoms. Item (5) is of particular interest, and is quoted verbatim.

(5) In vessels over 213 meters (699 feet) in length, the double bottom, for at least half the vessel's length amidships and forward to the Forepeak bulkhead, shall extend up the vessel's sides to a height above the top of the keel not less than 10 percent of the vessel's molded breadth. [The White Star Line, in protecting the *Olympic* and *Britannic* against raking damage following the loss of the *Titanic*, fitted an inner skin up to the load waterline (Chapter 24), even though the resulting space between inner and outer skins was only 30 inches, which was only about one half the vertical spacing of the inner (double) bottom. Now 10 percent of molded breadth would be just a little over nine feet, whereas the load waterline was 34½ feet above the bottom. This article, therefore, does not seem related to the *Olympic* alteration. It might be interpreted as an attempt to increase the protection against grounding, but it also might have come from Mersey's finding that a double bottom of the extent specified here would have prevented flooding of the *Titanic*'s forward-most boiler rooms.]

Article XXII—Going Astern

Vessels shall have sufficient power for going astern to secure proper control of the vessel in all circumstances. [Does "in all circumstances" imply imminent collision? Apparently not, since the steam turbine ships that dominated the transatlantic passenger traffic for the next 60 years had *less* astern power than the *Titanic*. Of course, the relevant criticism of that ship was its supposed inability to turn, not its inability to stop. Incidentally, the convention did not discuss maneuvering characteristics.]

Regulations, Life-Saving Appliances and Fire Protection

Lifeboats, their number, size, cubic capacity, construction, outfit, etc., had to be specified in detail. The principle that there must always be boat capacity for all persons aboard already has been quoted from a statement in the Convention.

Final Protocol. The Conference documents ended with several recommendations on matters that apparently could not be resolved into regulations. Two that seem worthy of note are quoted here.

1. "Binoculars should not be provided for lookout men." The *Titanic* lookouts were not provided with binoculars, and those responsible for this apparent neglect have been excoriated by the critics. However, the severe limits that binoculars place on a user's field of vision is a reasonable justification for the recommendation.

2. "The governments . . . should consider . . . securing that vessels crossing the North Atlantic shall not pass over the banks of Newfoundland during the fishing season." The *Titanic* never reached the banks of Newfoundland, but the recommendation is worthy of discussion because it seems to confirm the allegations that transatlantic liners steamed with more regard for schedule than for safety.

After the Experts. The 1914 Conference said nothing about maneuvering (the ability to avoid a looming iceberg) or about structural strength (the ability to resist the blows of an iceberg with more success).

Second, it said nothing about requiring longitudinal watertight bulkheads, like those of *Lusitania* and *Mauretania*. It also did not cover regulations that would require watertight decks. Senator Smith's principal recommendations relative to ship design thus did not attract the support of the SOLAS experts.

Third, it *did* make explicit regulations concerning the number and placement of transverse watertight bulkheads, but said nothing explicit about their height. The White Star line already had increased the height of the *Olympic* watertight bulkheads, which could be taken as a tacit admission that the *Titanic* was deficient in this regard. In the next chapter we'll see how the *Titanic* did (or did not) meet these after-the-fact regulations.

As noted earlier, the 1914 SOLAS was never ratified by enough nations to become international law. World War I delayed the process long enough, and provided enough additional experience with ship sinkings that the alternative to continued pursuit of ratification was a second SOLAS Conference, held in 1929. We'll have to look at a few aspects of that in a later chapter. Meanwhile, we need explicit discussion of how the ships of both White Star and Cunard may have been ex post facto defective because of ways in which they failed to meet the regulations of 1914. That's also the next chapter.

Notes for Chapter 22

1. An exception to the reticence of the professionals is found in the alterations to the *Titanic*'s sister *Olympic*, quickly put in train to ensure that she would not be sunk by a similar accident. These are treated in Chapter 24.
2. Hearings before a Subcommittee of the Committee on Commerce, United States Senate, 1912. See longer note on the senator himself in Appendix 2.
3. Mersey. *Report on the Loss of The Titanic*, 1912, His Majesty's Stationery Office, London.
4. Wilding provided the hearing with calculations that showed from the rate of flooding that the area opened to the sea was about 12 square feet. Since the damage was known to extend over 300 feet in length, the *average* vertical extent must be about ½ inch. Those figures suggest the narrow gash concept that was to persist until images of the damaged area could be made in 1996.
5. Joseph Conrad's intemperate comments on the Mersey hearings are quoted more fully in Chapter 19.
6. The complete text of the SOLAS Conference was published in the 1914 *Transactions* of The Society of Naval Architects and Marine Engineers, Volume 22.
7. Second Officer Charles Lightoller was the highest ranking survivor from the ship's crew. He testified before Senator Smith, before Lord Mersey, and much later wrote his autobiography, 1935, *Titanic and Other Ships*, Ivor Nicholson and Watson, London.

CHAPTER 23

SOLAS 1914 Reviewed

SOLAS 1914[1]—the first-ever international conference on safety of life at sea—assembled in London, November 1913, in response to the *Titanic* tragedy. That it decreed lifeboats for everyone scarcely needs to be stated. But what else did this conference settle? Many voices, including a few with some professional standing such as J. Bernard Walker and George Dickie, had spoken of deficiencies in the *Titanic*. Here, then, was a forum in which such ideas would be debated, and in which any remedial measures would be proposed. These pronouncements can be regarded as either an indictment of the *Titanic*'s safety features or a dismissal of charges. If indictment is the proper label, then obviously deficiencies in the surviving sisters *Olympic* and *Britannic* were illuminated as well. The illumination would be most glaring upon those ships, but *all* passenger ships were within the circle of light, by no means excepting the two Cunard liners of our interest here.

Before proceeding, let's reiterate a remark on SOLAS Conferences that has appeared earlier: their pronouncements do not become law until individually enacted into national laws by a requisite number of maritime nations. In the case at hand, World War I interrupted the ratification process. Only after passing through the filter that was SOLAS 1929 did any pronouncements originating in 1914 become law.[2] Nonetheless, SOLAS 1914 serves admirably as a landmark of judgment on passenger ship design.

In the immediate pre-war context, the 1914 judgments appear to be the best possible at the time, i.e., those delivered by the most expert opinion possible, and free of self-serving bias. The individuals attending the conference were experts in their profession. They did not wholeheartedly adopt British maritime practice because most of those individuals represented national interests that were wary of British maritime policy. That they were free of any tilt toward profit at the expense of safety—that's more difficult to assert, but a "very likely" judgment seems in order. That judgment is based on 80-plus years of hindsight, but close acquaintance with that valuable resource shows no significant passenger ship accident, post-1914, that can be charged to the refusal of the SOLAS conference to act on the *Titanic* lessons.

The conference statements on an ice patrol, wireless operation, and the provision of lifeboats need no comment, since their absence or insufficiency in the *Titanic* case is incontrovertible. However, the question of watertight integrity through subdivision, expressed in Articles V through X of the regulations (given verbatim in the preceding chapter), impacts the design of ships, and is the material of principal concern in this book. Probably the implications of these regulations are not clear to a nontechnical reader, since even those of us who claim technical expertise must give them study in order to relate the regulations to the design of *Titanic* or any other ship. But the results of that effort can be summarized with decent brevity.

Study by the naval architect and its subsequent evolution into exegesis for the nontechnical reader must be based on a floodable length curve: the one presented and explained in Chapter 20 is repeated in this chapter.

The question of *permeability* is obviously important in floodable length. Values specified by SOLAS 1914 are given in Chapter 22, but the regulations quoted there appear to leave too much latitude for interpretation, especially when one has to make a choice of permeability for compartments that have mixed use. The values used in the illustration are the more explicit formulas offered 15 years later by SOLAS 1929.

The floodable length curve confirms the statements relative to *Titanic* made in the Mersey report,[3] namely:

> Any three of the four forward compartments could have been flooded without sinking the ship to the top of her bulkheads.
>
> Even if the four forward compartments had been flooded . . . the ship would have remained afloat.
>
> The ship as actually constructed was doomed as soon as the water in Boiler Room No. 6 and all compartments forward of it entered in the quantities that it actually did.

As constructed, the *Titanic* sank from the flooding it experienced. However, did it nonetheless conform to the regulations that SOLAS 1914 devised in response to the sinking?

The first step in answering that question is to select a value for the ship's "criterion of service," which is a fraction that indicates where it lies in the spectrum that stretches from a pure (i.e., 100%) freight carrier to a pure passenger carrier. Having a value for the criterion is the first step in determining a "factor of subdivision," an essential element in evaluating floodable length. The SOLAS 1914 document does not specify a criterion of service, but merely calls for further study of the topic. In the absence of the vital number, one must take a conservative approach and treat the *Titanic* as a passenger carrier. The appropriate value for factor of subdivision must, therefore, come from column A and the last row of the table reproduced in Chapter 22; that value is 0.34.

Following the regulation, the 0.34 is doubled to 0.68 as a basis for calculating the combined length that two adjacent compartments may have. The *Titanic* compartments between bulkheads C and E have a total length of 105 feet; dividing this by 0.68 produces 154, which just obeys the rule. If the 1914 value of machinery space permeability is used, or if an appropriate average of the 0.85 and 0.72 permeability values is used in this neighborhood, then the rule is easily met.

The *Titanic* length[4] was 850 feet. Twenty-eight percent of that length is 238; floodable length is 234 feet at the forward end. The length of the first four compartments is 193 feet, which meets the requirement of four compartments within the forward floodable length.

The partial curve (see Figure 23-1) at the forward end of the floodable length curve demonstrates the benefit of raising bulkheads to the "D" Deck level, an improvement that was explored in the Mersey re-

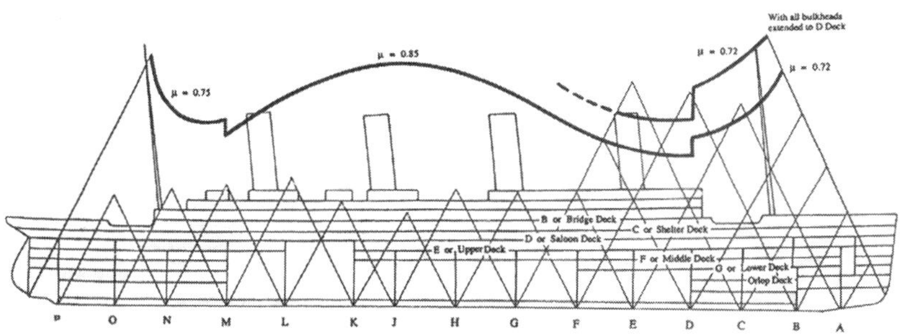

FIGURE 23-1 Floodable length curve for *Titanic*. It shows that any two adjacent spaces between transverse bulkheads can be flooded without having the length of flooding equal the floodable length. The ship thus meets "two compartment standard." The upper partial curve forward shows how extending the *Olympic*'s watertight bulkheads to "D" Deck increased that ship's floodable length as a consequence of her 1913 upgrading. (*Curve constructed by the authors*)

port, and was a retrofit applied to the *Olympic* before the 1914 regulations were published. There will be a comment later in this chapter on the *Olympic* alterations.

A tentative conclusion is that the watertight subdivision of the *Titanic* would have been no different if the ship had been built to the 1914 regulations. However, in 1914 there was that unfinished business of the criterion of service. The study needed in 1914 was performed during the 1920s. That business was finished at the 1929 SOLAS. The latter conference produced the curves shown in Figure 23-2. Their use produces a value of 0.39 for the *Titanic* factor of subdivision. If that value replaces the 0.34 used in the previous discussion, the change is *favorable* to the *Titanic* design.

As a matter of interest, the floodable length curves for the transatlantic liner *Normandie*[5] which was built under the regulations of SOLAS 1929, appear in Figure 23-3. Her permeabilities and factor of subdivision are noted on the figure. With that in hand, a question is asked: Would an opening in the hull to the sea from the bow through Boiler Room No. 1, or from her bow through Boiler Room No. 2, have sunk the *Normandie*? The answer should be "probably not," though the curves seem to indicate otherwise. This contradiction lies in something the figure doesn't show. The *Normandie*, like all transatlantic liners of its era (i.e., the 1930s), was oil-fueled, and was fitted with fuel "wing tanks" on the sides of the boiler rooms (see Figure 23-4). These tanks provided some measure of the longitudinal subdivision so fervently advocated by Walker and Conrad and other critics of 1912. Had these tanks been empty before the sea came pouring in, they might have caused a problem with list. How-

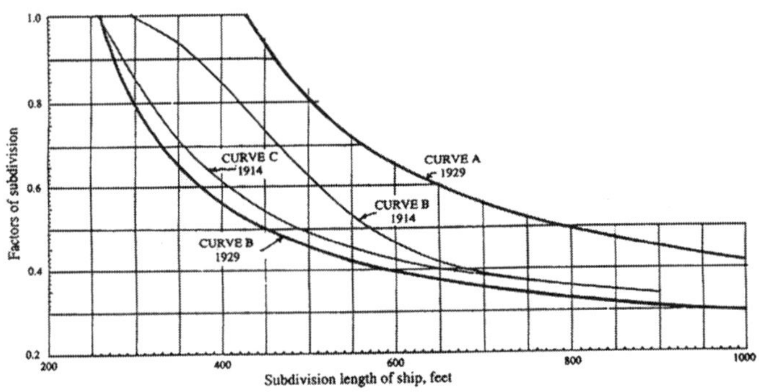

FIGURE 23-2 Curves for "factor of subdivision" developed by SOLAS 1929. Curve B is from SOLAS 1914, given as a table in Chapter 22.

ever, absent some other complicating circumstance, the *Normandie* would not have sunk.[6]

Do the *Normandie* floodable length curves tell us that the ship was designed to a higher standard of watertight integrity than the *Titanic*? The answer is no.

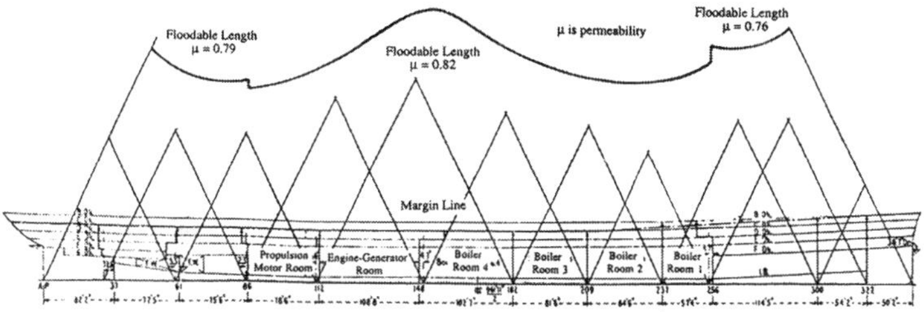

Criterion of service = 93.1 —CURVES OF FLOODABLE LENGTHS AND SUBDIVISION, S.S. "NORMANDIE" Factor of subdivision = 0.339

SHIP CHARACTERISTICS

Length between perpendiculars	961′ 11″
Length at subdivision load line	961′ 11″
Beam, mean shell	117′ 9″
Depth to bulkhead deck amidships	54′ 1½″
Depth to margin line amidships	53′ 10½″
Subdivision draft, molded	36′ 7″
Freeboard to margin line amidships	17′ 3½″
Displacement (tons S.W.) at subdivision draft	67,934

VOLUMES BELOW MARGIN LINE

Volume of passenger and crew spaces forward	(e)	185,720
Volume of whole portion of ship forward	(v)	502,760
Volume of passenger and crew spaces in machinery spaces	(o)	487,276
Volume of cargo, stores or fuel within machinery spaces	(r)	219,035
Volume of whole portion of ship within machinery spaces	(v)	2,918,380
Volume of passenger and crew spaces aft	(a)	295,916
Volume of whole portion of ship aft	(v)	643,694

SUBDIVISION—PER INTERNATIONAL CONVENTION

Total volume of passenger and crew spaces below margin line	(P)	1,168,912
Total volume of ship below margin line	(V)	4,064,834
Volume of machinery space + volume of fuel-oil deep tanks forward or aft	(M)	2,918,380
Number of certified passengers	(N)	1,972
$P_1 = KN$ ($K = 0.6L = 677.15$)		1,138,140
$P > P_1$		
Criterion of service $= C_s = 72 \times \frac{M + 2P}{V}$		93.10

$A = \frac{190}{L - 198} + 0.18$	0.429
$B = \frac{100}{L - 138} + 0.18$	0.301
Factor of subdivision $= F = A - \frac{(A - B)(C_s - 23)}{100}$	0.339

PERMEABILITIES

Forward spaces	$\mu = 63 + 35\frac{a}{v}$	75.93
Machinery spaces	$\mu = 80 + 12.5\frac{a - c}{v}$	82.00
After spaces	$\mu = 63 + 35\frac{a}{v}$	79.09

FIGURE 23-3 A floodable length curve for the *Normandie*. (*Adapted from a paper by James L. Bates and Ivan J. Wanless, 1946, "Some Aspects of Large Passenger Liner Design,"* Transactions, *Society of Naval Architects and Marine Engineers, vol. 46*)

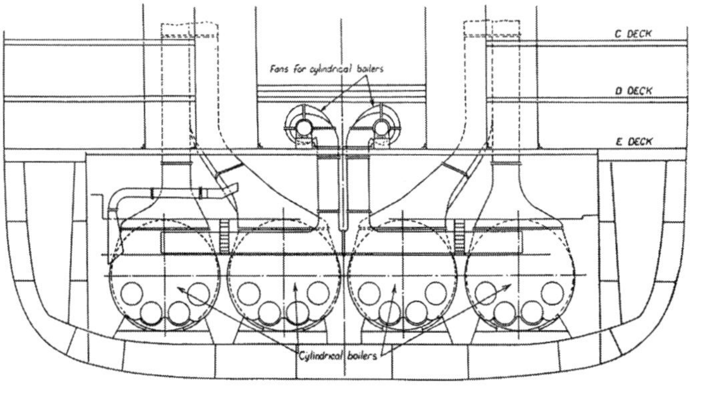

FIGURE 23-4 Cross section of the *Normandie*'s auxiliary boiler room, showing how the innerbottom, which formed a boundary of the fuel tanks, was carried up the sides to "E" Deck. (*From a special edition of* The Shipbuilder, *heralding the new ship in 1936*)

Should it be concluded that the wing tanks flanking the machinery spaces of the *Normandie* and other passenger ships offer a degree of safety far beyond that of the *Titanic*? If the only hazard out there is an iceberg, then they surely do. Sinkings of ships from a collision with an iceberg rarely happened in a post-*Titanic* world. The greatest hazard in peacetime is a collision with another ship. With that in mind, one can say that those tanks offer little to a nervous passenger wary about a ship's watertight subdivision. If war is to be considered, with bombs, shells, mines, and torpedoes, then wing tanks are also a doubtful protection.

The 1914 convention does specify (Article XXI) that the inner bottom of ships over 699 feet in length shall be carried up the sides for a height of at least 10% of the hull's width. In the case of the *Titanic*, that height would have been 9.2 feet. The inner bottom was approximately five feet above the outer, extending all the way to the side (but not up it), so that by the 1914 rules there was a deficiency of about four feet. From what we know of the flooding of the two boiler rooms that were holed, a 1914 inner bottom *might* have protected these rooms, and if it did, then the ship would not have sunk. So, on this point the *Titanic* was judged to be deficient. Forensic evidence from the 1998 *Titanic* Expedition shows that the damage in question was actually about 10.5 feet above the keel, so that an extended inner bottom built strictly to the 1914 rules would *not* have been sufficient to save the ship.

Then there is the matter of lifeboats. There can be no quarrel with the "lifeboats for all" rule that was formalized by SOLAS 1914, but there can be concern in the realm of lifeboat davits. The 1914 rules would have required 24 pairs of davits, whereas the actual ship had only 14. Compliance would, therefore, have required more than simply adding boats for the existing davits to handle. Indeed, the many circumstances of an emergency severe enough for evacuation of a ship (e.g., listing, lack of time, difficulty in recovering boat falls that had been used) can make multiple launchings per davit-pair impossible. However, at this point we are speaking of ship design judged against ex post facto rules. With that in mind, we note that davits are devices that are added easily, so that one can argue that the *Titanic* was not deficient in design, but merely in outfit.[7] In the chapter concerning the later sister *Britannic*, note how different the boat-handling gear is from that used in the *Titanic*.

The *Titanic* boats could be lowered without electrical power; the four electrically driven boat winches on the Boat Deck were intended for use in lifting. Since compliance would require almost doubling the number of davits, and since multiple launchings per davit would be required, more winches would seem to be in order.

Is there anything else? Article XXII requirement that "vessels shall have sufficient power for going astern to secure proper control . . ." might be construed as stemming from the inability of the *Titanic* to stop short of the iceberg, but that cannot be the case. Requiring a ship, even one much smaller than the *Titanic*, to stop within the distance it would travel when at a normal sea speed, and in the half minute or so that had been available to the *Titanic*, would be specifying the impossible. The ability to *turn* to avoid is another matter, and later-day critics have raised the question of *Titanic*'s ability to make high-speed turns.[8] That matter is discussed in a later chapter.

Beyond this matter of control while going astern, SOLAS 1914 says nothing about maneuvering ability.

The 1914 convention did not touch on ship strength. Joseph Conrad provided a source of controversy with his biscuit-tin analogy, but authority was unmoved. On the other hand, the structural failure of the hull in the last moments of the sinking was not recognized until the wreck was discovered 73 years later. Even had it been known in 1912, the circumstance is unlikely to have generated any rule-making, since the ship did not sink *because* of the hull failure. Nonetheless, the hull strength as revealed by the now known failure during sinking is of interest to latter-day naval architects, and is discussed elsewhere in this book.

Final judgments must be deferred until a concluding chapter, but at least we can now offer conclusions relative to the *Titanic*'s "unsinkability." They are:

- *Calling* the ship unsinkable was folly, but had nothing to do with its merits (or lack of) as a safe ship.
- The sinking was the consequence of operational error coinciding with unusual circumstances. The fact that those circumstances occurred had nothing to do with its merits (or lack of) as a safe ship.
- The watertight subdivision met the test for adequacy by meeting the requirements of ex post facto subdivision rules.
- Later transatlantic liners (typified by the *Normandie*) exhibited no greater degree of subdivision.
- An inner bottom carried upward a few feet past the turn of the bilge might have saved the ship. The regulation requiring that was also ex post facto, and hence one might argue that it shouldn't be used in supporting any negative judgments concerning the *Titanic*. We'll leave that argument to other authorities.

With regard to the extended inner bottom, it is interesting to note that it was not an innovation that was born by afterthought following the

Titanic sinking. Figure 23-5 shows an inner bottom turned upward for a few feet at its outboard extremities in the German liner *Kaiser Wilhelm II* of 1905, and Figure 23-6 shows a less extreme instance in the Cunard liners *Lusitania* and *Mauretania*.

With that said, it now may be time to offer some concluding words about the general notion of unsinkability. Except in the pejorative term "unsinkable *Titanic*," not much has been heard about unsinkability since the *Titanic* and *Lusitania* sank, but it was a term of common currency among ship people before that. It continued for a while, as, for example, seen in words of J. Bernard Walker and George Dickie that were quoted in previous chapters. It even appeared in a surprising place—in a report by *The Marine Engineer and Naval Architect*[9] describing the refitting of the *Olympic* following the loss of her sister. The anonymous author proclaimed that the latter ship had become "practically unsinkable." Perhaps it was the many ship losses of World War I that finally disabused the world of hope for that unattainable ideal, the unsinkable ship.

Farewell to Watertight Decks and Longitudinal Subdivision. These putative safety features, so obviously beneficial according to critics of the Walker-Conrad school, never gained the imprimatur of regulation. Can we say that those features were never built into a post-*Titanic* passenger ship? That would be a risky statement, since there are just too

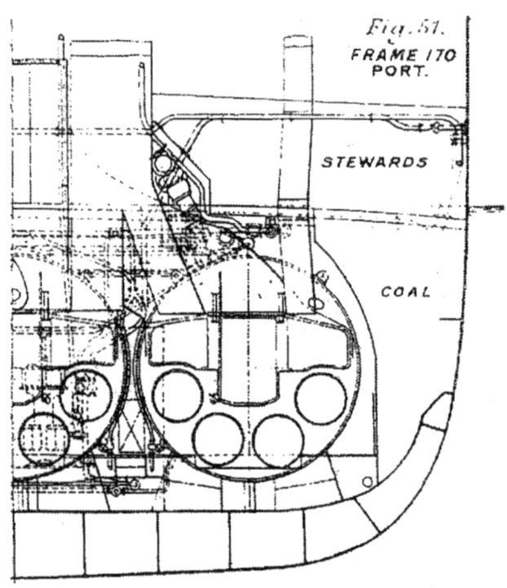

FIGURE 23-5 Cross section in way of a boiler room of the German transatlantic liner *Kaiser Wilhelm II*, showing the innerbottom carried up the sides of the ship. (*From* Engineering, *vol. 84, 1902*)

many passenger ships that would have to be exhumed before the word *never* would be justified. It *can* be confidently said that watertight decks and longitudinal subdivision were never required by safety regulations, and that they never became an accepted feature in commercial ship design. Let's note a few high points in brief history of their demise.

A good beginning is an excerpt from an article by an anonymous author that appeared in the English technical magazine *Engineering* a few weeks after the conclusion of the Mersey hearings in July 1912. Recall that Mersey had done nothing more than to recommend a *study* of the merits of watertight decks and longitudinal bulkheads. The author of that *Engineering* article may have assumed that the ship design profession would commence such a study based on Mersey's recommendation and

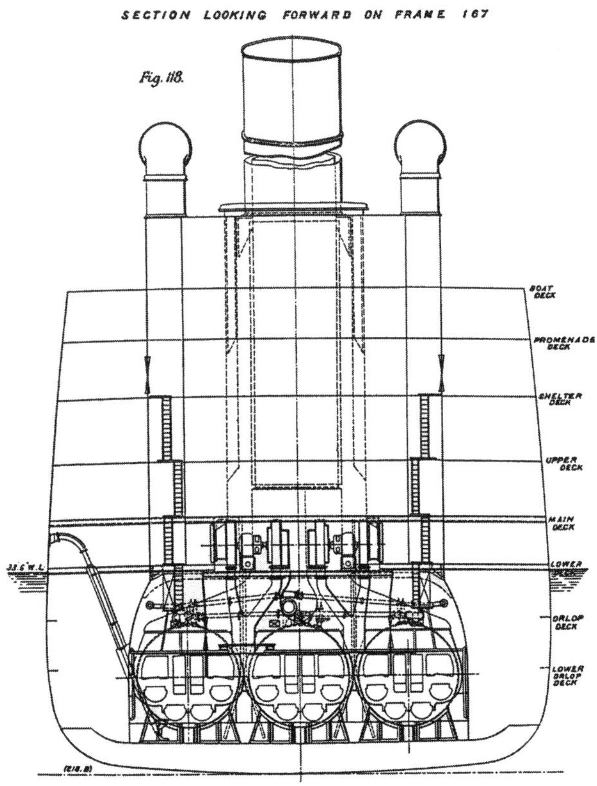

FIGURE 23-6 Cross-section in way of a boiler room of the *Lusitania* and *Mauretania*, showing a partial turn of the inner bottom up the sides of these ships. (*From* Engineering, *vol. 94, 1907*)

endorse these features that were omitted from the *Olympic*-class design. It's presented here for its interest as a historical curiosity.[9]

> In a first consideration of the question of subdivision, it appears indisputable that the ship which has the additional protection of longitudinal as well as transverse bulkheads has the advantage both as regards strength and safety; and, in so far as the former is concerned, the benefit in the case of a very large ship is undeniable. But the question of the greatest safety is complicated by the fact that, if a ship fitted with longitudinal bulkheads is damaged and flooded on one side, the vessel will take a list. To counteract this list it is necessary to fill with water the compartments on the opposite side of the ship, and the general trend of opinion has been that the flooding valves should be arranged whereby counterflooding could be effected. No evidence or correspondence brought to light in the *Titanic* inquiry has shown, however, that such a system has been adopted on a ship with a combination of longitudinal and transverse bulkheads, and the reason for this will be seen by all practical men. In fact, such an arrangement would be far from foolproof, and might be seen as such a source of inconvenience and danger that other alternatives have been preferred.
>
> In the case of the *Lusitania* and *Mauretania*, for example, practical provision has been made for counterbalancing the flooding effects, and they are for the most part instantly available. Should misfortune befall either of the ships referred to and damage be received to the extent of flooding two wing compartments, this would give the ship a considerable list. This, however, could be counterbalanced by blowing down of wing boilers, giving a righting moment more than equivalent to the upsetting moment caused by flooding two wing compartments. Moreover, the ship would rise bodily, due to the expulsion of a greater quantity of water from the boilers than flowed into the ship through the damaged area. There are, moreover, other additional ways of countering any great list caused by damage to the hull that could be effected, for instance, by flooding wing engine rooms, filling the double bottom on the undamaged side, and emptying those tanks on the damaged side that best suited the particular case; or by flooding the wing bunkers, etc.
>
> Generally, recent investigations have shown that subdivision by a combination of longitudinal and transverse watertight bulkheads, particularly when used in conjunction with a system of watertight decks, is a very efficient method of keeping a vessel afloat in case of accident, and the general appreciation of this has become obvious from the latest developments in the design of the highest class of transatlantic liners and large warships.

> Regarding the carrying up of the double bottom above the turn of the bilge, there is no doubt that this precaution may be in some cases advisable. But if the double bottom is also subdivided longitudinally into four watertight compartments and the longitudinal bulkheads are well stayed and made watertight, there is no great gain to be secured by carrying the double bottom any further than a little above turn of the bilge.

The arguments quoted can be judged "well reasoned" with respect to longitudinal subdivision, but just as in the case of the Conradian arguments to the same end, the word "naïve" comes to mind also. In 1914 and 1915 the catastrophically rapid sinkings of the *Empress of Ireland* and *Lusitania* demonstrated that safety based on calmly measured actions by men in the lowest levels of the ship was no safety at all. Those two violent confirmations of that viewpoint can be regarded as the death knell of the longitudinal idea.

Recall from Chapter 20 that Professor Fred Purvis,[10] in a 1913 paper, speaks of carrying the question of closing coal bunker doors to the men who would have to close them in emergency. He concluded from their responses that a time on the order of 10 minutes might be required. In the face of rapid flooding, that was equivalent to leaving them open (and scurrying up the ladders to possible safety). In the face of flooding slow enough to allow a boiler room crew to carry out the procedure, bilge pumps could probably take care of the situation.

Since he was writing before the *Empress of Ireland* sinking, the professor was unaware of the circumstances of that event, and the *Lusitania* sinking was a year in the future. He, therefore, had no examples to offer against the dangers of "unclosable" watertight doors, but does go on to state in strong terms that something should be done to make such doors closable surely and rapidly. This necessity was recognized by the SOLAS delegates; their Article XV states "In the case of watertight bunker doors, satisfactory arrangements shall be made by means of screens or otherwise to prevent the coal from interfering with the closure of the doors." However, events at sea during World War I took the ship designers' attention to more important matters, and soon afterward the near universal change to oil fuel obviated the need to tinker with bunker-door design.

The ultimate fate of longitudinal subdivision was not total rejection, in spite of the experiences of the *Lusitania* and *Empress of Ireland* cases. When oil fuel became universal for the transatlantic ships following World War I (the surviving liners, such as the *Olympic* and *Mauretania* were converted), a favored place for stowage of oil was the "wing tank." This concept is mentioned earlier in this chapter in connection with the floodable length of the *Normandie*, and a plan of the *Normandie* machinery level shows those tanks, lying between the outer shell and the boilers. This arrangement does present the problem of a severe list from flooding on one side following an accident. Post-*Lusitania* ship designers were most thoroughly aware of that lingering problem, and either kept the tank volumes small enough that the unavoidable list would not exceed a tolerable level, or furnished instructions for fuel and ballast management (i.e., measures to be taken *before* an accident) that would forestall possible list problems.

The safety rules that eventually became law after SOLAS 1929 were based on recognition that while longitudinal subdivision is, in general, a hazard, it does have some benefits. Internal tanks for fuel, water, lubricants, and cargo are not only beneficial, but also virtually essential. The rule makers thus avoided draconian prohibitions in favor of specifying the maximum allowable consequences of longitudinal subdivision (and watertight decks, as well). The essence of the matter appears in Regulation V of its Convention, SOLAS 1929. That item has been quoted in an earlier chapter, but is nonetheless worth repeating here:[11]

> Where it is proposed to fit watertight decks, inner skins or longitudinal bulkheads, watertight or non-watertight, the Administration shall be satisfied that the safety of the ship will not be diminished in any respect, particularly having in view the possible listing effect of flooding in the way of such structural arrangements.

The supposedly watertight longitudinal bulkheads were reincarnated as boundaries of fuel tanks, and are regarded as a safety feature against minor penetrations even though careful attention must be paid to the listing problem. Indeed, the presence of these tanks has the benefit of obviating the need for extending the inner bottom up the sides, the lack of which was a weakness in the case of the *Titanic*. The tank boundaries, in effect, provide an inner bottom that typically extends well above the load waterline.

How about the watertight deck that once seemed such an obvious necessity to both intemperate critics and professionals? In spite of an occasional favorable mention, such as those of Walker and Dickie quoted in an earlier chapter, watertight decks were regarded with disfavor throughout the commercial ship design field. They complicated internal arrangements and communications. Their dangers were recognized, as exemplified by a contemporary quotation in Chapter 21, attributed to Professor Fred Purvis. It should be recalled that he alluded to the danger of water running along such decks after a collision, and to the likelihood that their watertightness would be destroyed by the collision that supposedly justified that quality.

With regard to the once advocated deck and bulkhead, the regulations in 1929 stated that if a ship is to have these safety features—regarded with such favor by some authorities of the 1912 era—then its designers must demonstrate that they are *not* a hazard. Consider the watertight deck concept in the catastrophic sinkings of several European vehicle ferries, principally the *Estonia*[12] in 1994 and the *Herald of Free Enterprise* in 1987. These passenger ferries were not fitted with watertight interior decks, but their vehicle decks did play a role suggested by the prescient Professor Purvis. Water entered their hulls *above* the vehicle deck, where it ran along the deck until it collected into enough topside weight and free surface to bring about a rapid capsizing. The capsizing of the *Herald of Free Enterprise* resulted in the loss of 193 persons out of 539 on board, while in the *Estonia* 852 persons perished. In the *Herald of Free Enterprise* the bow wave and trim by the bow combined to bring the open bow door below the water surface while the ship was in Zeebrugge harbor. Water entered the vehicle deck, causing the ship to lose its way and suddenly while still in the supposedly hazard-free waters of the harbor.

A paper presented by John Bedford and Chris Hackett[13] summarized their findings on comparing the subdivision of the *Titanic* with modern SOLAS requirements. The basis for the analysis was that:

1. The vessel must have positive residual righting arms with a list of 15 degrees.
2. The area under the right arm curve be at least 0.015 meter-radians (0.5 foot-radians) measured to the lesser of:
   a. The angle that progressive flooding occurs.
   b. 27 degrees measured from the upright condition in the case of flooding of two or more adjacent compartments.
3. The residual righting arm is not less than 0.33 feet (0.10 meters)
4. With unsymmetrical flooding the angle of list for the simultaneous flooding of two or more compartments shall not exceed 12 degrees.
5. With symmetrical flooding there will be a positive metacentric height (GM) of at least 0.17 feet (50 mm).

Based on these criteria, Bedford and Hackett concluded from their stability analyses:

> It is our view that these results must be considered good for a design nearly 90 years old. We would suggest that the ship would comply with

modern requirements without many detail modifications but only that the fundamental configuration is a sound basis for compliance and this stands as a vindication, if ever one was needed, of the design produced by Pirrie, Andrews, Carlisle, and Wilding on behalf of Harland & Wolff for what was a big step forward in the development of the transatlantic ocean liner. The good service given by the surviving sister vessel, the *Olympic* endorses this view.

Notes for Chapter 23

1. The complete text of SOLAS 1914 was published in the 1914 *Transactions* of The Society of Naval Architects and Marine Engineers, Volume 22.
2. The results of the SOLAS 1929 Conference are reported by George H. Rock, 1929, "The International Conference on Safety of Life at Sea, with Special Reference to Ship Construction," *Transactions*, Society of Naval Architects and Marine Engineers, Volume 37.
3. *Report on the Loss of the Titanic*, 1912, His Majesty's Stationery Office, and London.
4. This is the length at the waterline.
5. *Normandie*: maiden voyage 1935. Destroyed by fire in New York harbor, 1942. Overall length 1029 feet, 66,400 tons displacement, 80,000 gross tonnage, 1972 passengers, 1345 crew, and 160,000 shaft horsepower on four shafts driven by steam turbine-electric engines. Dimensions, etc. from the *Normandie* Souvenir Number, *The Shipbuilder and Marine Engine Builder*, June 1935.
6. Wing tanks are a potential source of list as so forcefully demonstrated in the sinking of the *Andrea Doria* in 1953. Empty fuel tanks, flooded following collision, gave that ship such a list to starboard that none of its port-side boats could be lowered. Empty tanks refilled with seawater as the designers' instructions required, would not have been such a potentially devastating factor. Among other facets of the "prudent and skillful management and navigation" advocated by Lord Mersey at SOLAS 1914 (see the full quotation in Chapter 22) is attention to the operating instructions given by the designers.
7. About adding boats and davits: adding boats and davits high in a ship raises its center of gravity ("decreases its GM," a naval architect would say), and thereby degrades its stability. The boats presumably could have been added safely to the *Titanic,* since its design called for a full complement, but additional davits is another matter that we lack sufficient information to analyze. However, recall from Chapter 21 the mention of a ship under design in April 1912 whose beam had to be increased to counteract the effect of the post-*Titanic* requirement for more boats (and presumably more davits). We also have noted in Chapter 18 that the *Britannic*—construction barely underway when her sister *Titanic* sank—was fitted with a heavier complement of boats and davits, but had a beam two feet greater than the first two *Olympics.*
8. Criticism of turning ability is a feature of Colin Carmichael, "Was *Titanic* Unsafe at Any Speed?" *Steamboat Bill,* Steamboat Historical Society of America, Spring 1972.

9. The anonymous article appeared in the August 9, 1912 issue of *Engineering*.
10. The paper by Professor Purvis was printed in full in *Engineering*, 1913, Volume 97.
11. From SOLAS 1929. See Note 2.
12. We know of no book that has been published that details the sinkings of the *Estonia* and *Herald of Free Enterprise*. A technical paper, "The Safety of RO/RO Passenger Ferries by J. R. Sprouge, presented to the Royal Institution of Naval Architects in April 1988, is the best source of information on the loss of the *Herald of Free Enterprise*.
13. Hackett, C. and Bedford, John G., "*The Sinking of SS Titanic, Investigated by Modern Techniques,*" Northern Ireland Branch of the Institute of Marine Engineers and The Royal Institution of Naval Architects, 26 March 1996.

PART V

Design Criticisms

The lessons learned by ship designers from the *Titanic* disaster were exemplified by the findings of the international safety conference known as SOLAS 1914. We have explored those findings in Part IV, in many instances by verbatim quotations. So far as deficiencies revealed by the *Titanic* are concerned, the number of lifeboats and their launching gear were the most obvious fault, and principal item of concern to the conferees. In the first chapter of this part we comment on that very thing, in particular with regard to the *Britannic*, the only ship of the five that was built after the *Titanic* disaster.

Added to the lifeboat problem that was addressed by those attending the SOLAS 1914 conference are the varied other criticisms leveled in haste by professionals and amateurs alike. The advocacy of longitudinal subdivision and watertight decks is a prominent example, and one that has been explored in Part IV. Criticisms that go beyond those seemingly reasonable concepts were promulgated by an array of non-professionals. Their works received some credence because of the popular standing of some of the critics. Joseph Conrad and Walter Lord are two prominent *Titanic* authors that have been cited earlier. Conrad's criticisms were centered on the concept of unsinkability; Lord's were critical, but less specific. The points they raised have received their answer in Part IV.

Other authors have raised specific issues. As one example, *Titanic* maneuverability has been questioned because the ship could not turn fast enough to miss that iceberg, and its structural adequacy has been questioned because a mere chunk of ice could inflict a fatal damage. Part V has a chapter devoted to each of these issues.

To naval architects there were some important questions to be answered on stability. In following their peaceful pursuits, none of the ships in question demonstrated a lack of stability. Of course, with their hulls ruptured by means not envisioned by their designers, both *Lusitania* and *Britannic* capsized. In the case of the former ship, the pernicious effect of longitudinal subdivision and non-tight subdivision were discussed. The *Mauretania*, had it been damaged in its peaceful pursuits (in the manner of the *Empress of Ireland*, perhaps), might have shared the same fate. However, longitudinal subdivision was discussed earlier, and hence is not discussed further in Part V.

The *Britannic*'s capsizing may reflect on its design since its "inner skin" was a form of longitudinal subdivision. However, watertight doors left open or sustaining shock damage from the mine explosion, portlights open in contravention to orders, and the effects of non-tight subdivision were much more significant factors.

The *Titanic* sank without capsizing, even though there were noticeable lists at the commencement of the flooding, and near the end when the great amount of flooding significantly reduced her transverse stability. Her designers, criticized rightly or wrongly for other things, should be praised for her survivability over a period of two hours and forty minutes. Contributing to this expanse of time was the fact the naval architect who helped design was aboard to give Captain Smith and his crew invaluable insights to countermand the flooding. However, what author has taken the trouble or even realized these factors? These questions will be discussed in a chapter on stability in Part V.

CHAPTER 24

The White Star Response to the Loss of the Titanic

Bruce Ismay was extremely depressed over the loss of his new ship and the 1,522 persons who perished with the *Titanic.* The well-known story of his hiding behind the code word "Yamsi" was a clear indication of his state of mind. By the time the *Carpathia* had reached New York he had at least fashioned a bold front with which to confront the local reporters: he announced that henceforth every White Star ship would carry enough lifeboats for everyone aboard. Was this a hint of guilt, a hint that he was the one who had said NO to a plan (see Chapter 27) to carry two, or perhaps three times the number of lifeboats actually fitted? At the British Inquiry he denied that he had, claiming he never saw such a plan advanced by Andrew Carlisle of Welin Davit Company (former managing director of Harland & Wolff). However, the Ismay pronouncement in New York was a tangible response to a terrible tragedy, and a response that would be carried out by his shipping line. The *Olympic* was on her westward passage when her new sister sank. Unable to render assistance she continued her voyage arriving in Southampton. Her next westbound voyage was scheduled for 24 April 1912 and 24 collapsibles (Bertron Boats) had been placed on board to supplement her lifeboat complement. However, her engine room complement of greasers, trimmers, and firemen would not report aboard until wooden lifeboats were substituted instead. Unable to sail and reach agreement with the striking crew, White Star canceled the westbound voyage two days later. When the ship was finally ready to resume transatlantic service on 15 May 1912, wooden lifeboats replaced the Bertron Boats.

The Ismay response was such an obvious one, and one so easy for the travelling public to observe, that it was hastily repeated at many passenger terminals. In effect, it also was repeated in many maritime design offices, and often with more than superficial impact on designs on the drawing board at the time. Lifeboats and their gear have weight, and placing that weight high in a ship inevitably raises its center of gravity. Stability is thus degraded, perhaps enough to require major remedial measures, such as adding solid ballast. George Dickey, at the time designing the coastwise liner *Congress*, has been quoted in Chapter 21 as saying that his ship required an additional 20 inches in beam to compensate for the change in center of gravity caused by additional boats mounted high. A small change, indeed, but one that surely required many additional hours at the drawing boards by Dickey's firm.

Some existing ships must have faced the same problem. Resolution in their cases almost certainly came through the addition of fixed (i.e., nonliquid) ballast low in their hulls. The loss of carrying capacity (i.e., "cargo deadweight") is unlikely to have been a significant penalty. However, increase in weight from any cause inevitably means higher fuel consumption, so that there would be some penalty in operating costs.

Then, there were those ships that were given the additional lifeboat outfit without adding a stability compensation. One ship that may have fallen in this category was the Great Lakes passenger steamer *Eastland*, infamous for its capsize in the Chicago River while loading passengers one Saturday morning in 1915. In his 1995 book *Eastland: Legacy of The Titanic*, transportation historian George Hilton[1] pursued this thesis. He had found no record of any stability compensation when lifeboats were added in 1912, or of any professional-level knowledge of stability principles among its designers and builders. He, therefore, reached the conclusion that the deaths of at least 1000 passengers just a few feet from the bank of the Chicago river is also part of the *Titanic* story.

In any case, the forever-after requirement of lifeboats for everyone was an imperative need that no shipping line of that era would ignore. It also would be mandated by the SOLAS Conference of 1914, but had already become the law of many maritime nations by the time that conference closed in January 1914.

No rescue ship arrived in time to save any of the *Titanic* victims in the water. The event was nonetheless a dramatic demonstration of the virtues of wireless to summon rescue ships. It also revealed flaws in this miraculous new communication device. The most obvious one was shown by the failure of the *Californian* to reach a still-floating *Titanic*. This was the ship of a sleeping captain and a sleeping wireless operator. No word came to its crew via wireless set because only one wireless operator was aboard, and that one was sleep.[2] As a result of the *Titanic* experience, ships that were required to carry wireless were required to man them at all times and, obviously, to carry the personnel necessary for that new imperative.

Now, if the wireless was no better than its manning, it was also no better than its power supply, which was the ship's electrical service system. For the *Titanic*, as well as for all ships of its era, that system was dependent on steam that was generated by the same boilers that served the propulsion engines. The electric plant of the *Titanic* performed admirably in the face of flooding boiler rooms. Nonetheless, the vulnerability of the boiler rooms, a ship's sole source of energy for any purpose, was obvious. Indeed, the late-sinking dimming of the *Titanic* lights and the weakening of the wireless signals indicated that the plant was running on a failing steam supply as the sinking event approached its climax. In later years, ships were to be required to be equipped with diesel-driven generators, placed high, that would automatically pick up emergency service loads in response to falling voltage in the main system. In 1912 the internal combustion engine did not seem to be a serious challenger to the established order in the realm of ship propulsion, but it was serving reliably in many other ways. Its use as an emergency source of power for lighting was a near-obvious application, and White Star immediately began to install emergency generators driven by internal combustion engines in its ships, placing them on the highest deck, and arranging for them to furnish emergency lighting, and power to the wireless sets. The *Megantic* received the first one in mid-May 1912, the *Olympic* emergency generator followed soon after.[3]

An international patrol of the iceberg area of the North Atlantic was another near-obvious response that was soon put in place. It was first patrolled by the U.S. Navy and then by the U.S. Coast Guard since 1912, first with ships and then aircraft.

Adding lifeboats . . . enhancing the distress capability of wireless . . . making wireless independent of a ship's boilers . . . patrolling the region frequented by bergs . . . these were things that *had* to be done, and could be done quickly. But deeper changes, perhaps touching on that magical quality *unsinkability* in post-*Titanic* ships might be required as well. Conrads, Walkers, and Dickies offered their ideas, which we have seen and criticized earlier. The changes that were to come from consensus opinions of professionals appeared in the pronouncements of SOLAS

1914. However, as we have seen in Chapter 22, ship design as it was manifested in the *Titanic* was criticized only by the requirement that inner bottoms be carried up the side of passenger ships for a distance equal to at least ten percent of the molded beam. On the other hand, no one could object if the owner of a ship were to go beyond Authority's recommendations. White Star, still planning to operate two sisters of the *Titanic*, apparently felt the need to distance them, safety-wise, as far from that ship as possible.

Alterations to the Olympic. The action of SOLAS 1914 was foreshadowed by one recommendation in the Mersey report:

> Extending double bottom up the sides. It was shown in evidence that there would be increased protection in carrying the double bottom higher up the side than was done in the *Titanic*, and that some of the boiler rooms would probably not then have flooded, as water could not have entered the ship except in the double bottom.

The White Star Line was under no compulsion by the Mersey report to take any action to implement this recommendation. The same can be said with respect to SOLAS 1914, since it had no authority until national laws were made in conformity to its findings. Adding an inner skin, or even enhancing the inner bottom to turn up the sides of the hull, would be a near impossible task. The White Star Line, wanting to gain the confidence of the traveling public, decided to add the inner skin anyway.

The *Olympic* was taken from service during the 1912–1913 winter for alteration at her builder, Harland & Wolff.[4] At a cost of $1,215,000 (and considerable lost revenue) the ship was fitted with an inner skin that extended fore-and-aft from the forward bulkhead of Boiler Room No. 6 to the after end of the Generator Room, and vertically from the inner bottom to "D" Deck. The space between the hull and the new inner surface was 2.5 feet in way of the boiler rooms, as much as four feet at points aft of those rooms. The 2.5-foot space was less than half of the inner bottom depth, and hence did not constitute "carrying the double bottom higher up the side," but was the best that could be done with the existing hull, and probably would have prevented the boiler-room flooding experienced by the *Titanic*.

This alteration was indeed a major task since working space within the boiler rooms required temporary removal of all boilers that were located next to the shell. That in turn required major derangement of piping and uptakes, and indeed, removal of the funnels themselves. The improvement, whatever the time, trouble, and cost, was surely a wise measure for any ship that might scrape against an iceberg or other steel-rending object, but would be of no value in a more conventional ship-against-ship collision or against the explosion of a torpedo or mine. Almost certainly the major benefit was in reassuring the potential passenger that the *Titanic*'s sister never suffer the same fate.

The space between inner and outer skins was subdivided into tanks for the stowage of oil. The pumps and piping for an oil-firing system were not to be fitted until after World War I, when the boilers were converted to oil firing as part of the conversion of the ship back to passenger service. However, the provision of oil tanks in 1913 indicates that White Star must have realized the inevitability of petroleum fuel even at a time when the third *Olympic* sister was still under construction. The thought stirs speculation: what if White Star had boldly opted for oil in 1910? And what if the *Titanic* had then been fitted with the wing fuel tanks that became common in later transatlantic liners?

On a more mundane note, let's add that the *Olympic*'s "skin tanks" were able to provide 1,760 tons of the 5,500-ton oil-stowage capacity given the ship.[5]

During the addition of the inner skin, several of the watertight bulkheads in the forward part of the ship were extended upward. After this improvement, the bulkhead heights forward of Boiler Room No. 4 were the following:

Forepeak/Cargo Hold No. 1:	"B" Deck
Cargo Hold No. 1/Cargo Hold No. 2:	"D" Deck
Cargo Hold No. 2/Cargo Hold No. 3:	"E" Deck
Cargo Hold No. 3/Boiler Room No. 6:	"B" Deck
Boiler Room No. 6/Boiler Room No. 5:	"E" Deck
Boiler Room No. 5/Boiler Room No. 4:	"B" Deck

Compare these values to those in Table 8-2. Some reductions had to be made in the first-class accommodations to allow these enhancements of bulkheads. It is probable that the loss in revenue potential was more than offset by the increase in public confidence that it was safe to venture across the Atlantic on a sister ship of the *Titanic*.

Was the *Olympic* then unsinkable? Obviously not, but one might think that this misleading word would no longer be used in reference to a passenger ship, especially with respect to the only surviving sister of the *Titanic*. That was not the case because an anonymous author of an article that has been cited (Note 4) as a source for the refitting used it. He wrote that the *Olympic* had become "practically unsinkable."

Alterations to the Britannic. The *Britannic* was in an early state of construction in April 1912. The innerskin change desired by the White Star Line was, therefore, more nearly one of design than extensive alterations to an existing ship. These changes have been discussed in Chapter 18. Only a summary follows here.

Britannic was laid down at Belfast on 30 November 1911. Since this date was only 4½ months before *Titanic*'s fatal encounter, alterations in response to that sinking were carried out more readily than with *Olympic*. The provision of an inner skin could be accomplished without the expensive and laborious removal of boilers. Fortuitously, as well, the maximum beam of the new hull was to be 18 inches greater than that of the first two ships, making the inner skin less intrusive on working space in the boiler rooms. (Beam was 94'-0" instead of 92'6".)

The greater beam was truly fortuitous, since this change had nothing to do with collision protection. Rather, it was made as a precaution in the face of an alteration from the *Olympic* design that had already been included in the *Titanic*. The essence of the matter was the possibility that the alteration (an alteration to the accommodations area high in the ship) might be found to require a compensating addition of weight low in the ship to restore lost stability. The compensation preferred, and possible for the still-nascent *Britannic* (which was to have the same alteration as the *Titanic*), was the increase in beam. The inner skin was added to the design when its desirability became evident post-*Titanic*.

The increased vertical extent of watertight bulkheads, accomplished in *Olympic* during her 1912–1913 refit, was included in *Britannic* during her construction. In addition, the bulkhead between the turbine engine room and the generator room (just aft of the engine room) was made watertight.

The most radical change evident in *Britannic* is in her lifeboat outfit, especially in the means of launching the boats. Davits were replaced by cranes which were much more versatile in that they could reach across the decks to transfer boats from one side of the ship to the other, and could adjust more readily for problems caused by list and trim of a sinking ship. In the sinking of the ship this outfit performed well. The most obvious criticism can be based on the deaths caused by the drifting of a boat into one of the still-turning propellers as it rose from the water; one pair of cranes was located so far aft that its boats had to be launched near the wing propellers.

A more extensive description of the *Britannic* lifeboat outfit is found in Chapter 27.

Notes for Chapter 24

1. George W. Hilton,1995, *Eastland: Legacy of the Titanic*, Stanford University Press, Stanford, CA.
2. There is some question of whether the *Californian* could have reached the site of the sinking. The last reported position of the *Titanic* was off by 14 miles from her actual position. If Captain Walter Lord had steered for that position, rather than toward the lighted steamer visible from his bridge, his incorrect course, plus his likely troubles in evading the ice, might well have made his arrival at the scene too late to save any of the *Titanic*'s people who had not found places in the lifeboats.
3. The placing of an emergency generator aboard the White Star liner *Megantic* as a source of power for the wireless was reported by the *New York Times*, 17 May 1912.
4. The alterations to *Olympic* is described in the periodical *Marine Engineer and Naval Architect*, Volume 35, 1913. The author is anonymous.
5. The alterations to *Olympic* following World War I are described in *Engineering*, Volume 113, 1920.

CHAPTER 25

Boiler Explosions—Lusitania

When discussing the explosions of marine boilers we must make some important distinctions. Is the underlying topic watertube or firetube boilers? Is the phenomenon called "explosion" truly that, comparable to the explosion of a bomb or torpedo, or is it merely a violent escape of steam from its normal confines (a ruptured steam pipe, perhaps)?

Both the Cunard and the White Star ships were fitted with firetube boilers—the cylindrical or "Scotch" boilers common in their era—and hence the discussion that follows applies strictly to that type. We delay the principal discussions of this chapter for only a few words to observe several reasons why the firetube type is regarded as a much more dangerous alternative than the watertube type.

One reason for the greater hazard attending a cylindrical boiler is the much greater amount of potentially explosive energy stored in its water. An example comparison cited in Chapter 5 shows weight of water in a cylindrical boiler to be approximately 3½ times that in a watertube boiler of comparable rating. In both cases, the water is under pressure and at a saturation temperature for that pressure. In consequence, a significant amount of thermal energy per unit mass is present. If a rupture in the pressure envelope of a boiler allows a sudden release of that energy, then the potential for destruction, possibly explosive, is present in either case, but is much greater for the firetube boiler family.

As a matter of passing interest, one authority,[1] in writing about boiler explosions, estimated that a typical firetube unit contained enough thermal energy to lift the boiler two miles above its foundations.

Recall from our illustrations that the furnaces of a cylindrical boiler are themselves cylinders fitted into the lower half of the boiler's outer envelope. Therein lies a second reason for that greater hazard: external pressure bears upon the furnace walls of the cylindrical boiler, ever tending to collapse the furnace inward. The furnace interior is exposed to flame, its exterior surface is exposed to the inwardly acting pressure of the boiler water. Let the water level fall to the point where the top of a furnace is no longer cooled by that water and the metal will weaken to the point where it yields. If it yields to the point of failure, the immediate consequence is surely sudden death for the firing crew, and perhaps an explosive destruction of the boiler. In contrast, only the tubes (unlikely to be greater than four inches in diameter) are exposed to flame in a watertube boiler, and no pressure acts inward against these tubes, nor against any other pressure-containing surface.

The watertube boiler, with its lighter weight, its potential for almost unlimited pressure, and its relative freedom from explosive hazards, was to supercede the cylindrical boiler almost completely by mid-century. Elsewhere we have discussed the possibility that watertube boilers might have been chosen for the Cunard and White Star ships. But they were not. All five ships of interest were powered by steam from cylindrical boilers, and the three that sank faced the possibility of boiler explosion as they went down.

But to the distinction between a true explosion—a bomb-like release of a boiler's thermal energy—and an eruptive (but comparatively slow) release of steam. The non-explosive release is unlikely to destroy the boiler nor damage the hull of the ship. The explosive release almost certainly causes severe damage to its surroundings, including the watertight envelope that normally holds the sea at bay. To illustrate the bomb extreme, let's look at two such events. The first occurred aboard the USS *Bennington* (a gunboat) at anchor at San Diego in 1905. The following is a brief contemporary account, quoted in full.[2]

> On the morning of July 21st, while getting up steam, preparatory to proceeding to sea, the collapse of a furnace in one of the *Bennington*'s boilers resulted in the death of one officer and sixty-one enlisted men, the injury, more or less seriously, of forty others, and much damage to the ship.
>
> The lower furnace of boiler B (the port forward one) collapsed, causing this boiler to be driven aft against boiler D, liberating the steam contained in both boilers, wrecking the steam piping and, through breaking of sea valve connections, permitting entrance of water to the hull.
>
> The vessel began to settle with heavy list to starboard, making it necessary to beach her, with the assistance of tugs, to prevent sinking in deep water.

> The heavy loss of life was caused principally by scalds and burns due to escaping steam and water. No one in the firerooms escaped.
>
> The accident is supposed to have been due to the accidental closing of the valve or pipe leading from boiler B to its steam gauge and the failure of the safety valve to lift upon the formation of excessive steam pressure. The steam gauge, being disconnected, registered no pressure, and the water tender in charge was evidently not aware of the fact that pressure had formed until a few moments before the explosion, when a slight leak in the boiler was detected.
>
> An examination and test of the material of the collapsed furnace showed it to be of good quality.
>
> The boilers of the *Bennington* are about fourteen years old, having been in the ship since the time of her building in 1891; they are four in number, with three furnaces each.

A post-accident photograph of the ship shows the upper works in the after two thirds of the ship rendered into twisted metal. A suicide plane, crashing on the deck from astern, might have done no greater damage.

The second case of true explosion to be cited occurred aboard the ferry *Omar D Conger*. Until a fatal moment in 1922, the *Conger* was a small steamboat (one boiler) that linked Port Huron, Michigan, with Sarnia, Ontario, across the St. Clair River. The vessel was at her dock, being made ready for the day's service, when the boiler exploded as steam was being raised. Only four men, all members of the crew, were aboard. All were killed instantly. A shore-side witness, quoted in the Detroit *Free Press*,[3] said:

> I heard the explosion and looked up and saw a cloud of smoke where the *Conger* had been. I was deafened and sort of dazed for a few seconds and the next thing I saw her hull was sinking and debris was scattered all around. The boat sank almost in an instant. It looked as though the explosion blew the boiler right through the side of the hull.

Lesser accidents associated with boilers and their connecting piping could cause multiple deaths among the boiler crews, and were usually called explosions even though damage to the boiler in question might be trivial. A common scenario for such "minor" explosions began as waterhammer in a steam pipe, a consequence of carryover of liquid into the pipe from the boiler. The shock of the waterhammer might fracture a cast-iron valve body, releasing a flood of steam into the boiler room. The ship might suffer no major damage to its hull or machinery, but fatalities in the neighborhood of the rupture were almost certain.

What was the frequency of occurrence of boiler explosions? Table 25-1, taken from Rimmer,[4] lists the number of boiler mishaps reported to the Board of Trade in Great Britain from 1882, when legal requirements for reporting went into effect, through 1909, and then separately for the year 1910. Considering the many thousands of steam vessels under the British flag, the table seems to tell us that boiler explosions afloat had become rare during the first decade of the century, even if we observe that collapses and "sundry mishaps" were reported as explosions.

Water hammer incidents, as well as a variety of other fatal events, are included in the sundry category. One of the latter occurred[5] in 1903 aboard the cruiser HMS *Blake* while that ship was on maneuvers, and apparently was being driven hard. As described in contemporary news, "the crown of a combustion chamber came down," an event that did not cause damage beyond the boiler itself, but the eruption of steam through the furnaces into the firing isle of the boiler room killed three men, and injured several others. (A collapse is the failure of a furnace from being ruptured inward by the boiler pressure acting upon its walls. Figure 25-1[6] shows what could happen.)

Another fatal "sundry" occurred[7] aboard a British merchant vessel in 1906. A manhole cover in the outer boiler shell blew off while the unit was being tested for leaks under operating pressure and temperature. No great material damage was inflicted on the surroundings, but two crewmen died, and two others were injured. The cause was assigned to faulty closing up of the door following recent maintenance.

These examples illustrate the point that accidental releases of pres-

TABLE 25–1 *Summary of Boiler Explosions*

	1882–1909		1910	
Description	Number	Deaths	Number	Deaths
Explosion of land boilers	257	184	8	0
Collapse of furnaces, land	308	151	5	1
Explosions afloat	21	17	0	0
Collapses afloat	125	38	2	2
Sundry mishaps ashore	615	10	58	4
Sundry mishaps afloat	624	112	30	5
Explosions of shore heating boilers	61	7	2	0

sure from a Scotch boiler ran a gamut from minor damage to a boiler to its explosive destruction with severe damage to the surroundings.

Among the foregoing examples of marine boiler accidents, none occurred in consequence of a ship's sinking. Their stories do offer useful background, but something of great interest here is missing: boiler explosions that accompanied sinkings, especially those that were *caused* by the sinking. In those cases, what survivors believe to be a boiler explosion may be some other violent event. Physical evidence, that should be obvious on both the remains of the boilers and the humans who served them, becomes inaccessible. Reduced to speculation on what might happen, one might think that, absent a triggering explosion from a hostile missile, a boiler would *not* explode. After all, the immersion of a boiler surely douses its fires, and chills the boiler shell and associated piping. A great bubble of steam might erupt from the seawater as it pours into a furnace (and perhaps cause witnesses to report "explosion"?). The immersion of exterior surfaces of a boiler might condense the steam within, not only reducing the pressure that would cause an explosion, but even reducing it to the point where the boiler shell would collapse inward.

The foregoing suggests that boilers in a flooding boiler room should not explode. Water quenches fire. Water cools hot metal. Water can prevent explosions. However, there are three missing factors in this simple analysis. One is thermal stress, a consequence of uneven heating or cooling of massive metal objects. With heavy-walled pressure parts subjected

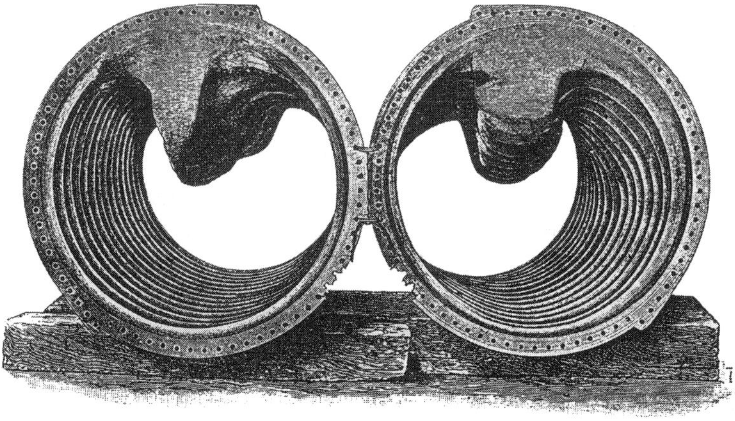

FIGURE 25-1 Two Scotch boiler furnaces that have suffered collapse from overheating. (*From Robert H. Munro, 1903,* Steam Boilers: Their Defects, *C. Griffin & Company, London*)

to flame temperatures, the cylindrical boiler was especially vulnerable to this phenomenon, and in consequence its normal handling required very careful heat-up and cool-down procedures. Twelve hours! Cautious marine engineers in the firetube boiler age might take that much time to bring a cold boiler to the point of service. And to bring it safely back to the cold state when repairs were to be undertaken.

Half a day for a planned cool-down, but now a collision or torpedo explosion has suddenly breached the hull. There *may* be time for remedial measures, as there was in the case in four or five of *Titanic*'s boiler rooms. On the other hand, the breach may be in way of the boiler room itself, and survival for the boiler crew comes only from abandoning all duty for immediate flight. That was *Lusitania*'s story, and nearly that of the *Titanic*'s Boiler Room No. 6. In both instances, boilers under steam met cold seawater. We know that no boilers exploded in the *Titanic* case. Aboard the *Lusitania*, fire and pressure may have been squelched by the inundation, or instead thermal shock stress may have triggered an explosive rupture of the steam piping. Although there is an absence of incontrovertible evidence for explosions of munitions or coal dust, there is evidence that either the firetube boilers *did* explode or the associated steam piping had some failures during some sinking process. "It is known that in the engine room there was some damage to steam piping, and the steam supply to the steering gear and turbo-generators was cut off around the time the second explosion occurred. Did this happen before the water immersed the boilers in the two forward firerooms? If so, then the lower steam pressure would diminish the possibility of a boiler explosion in those two spaces.

The second missing factor is "water hammer" in steam piping: a slug of liquid in a pipe is slammed against an obstruction—a closed valve, or even a short-radius elbow—causing the cast-iron (likely) body of the obstruction to fracture. The steam erupts with a violence that justifies the term "explosion." As will be seen later in this chapter, we state a case for this type of event occurring in the *Lusitania* machinery spaces. Such an event could have contributed to the second explosion, and hence been yet another factor in the ship's sinking.

Remember that two of the *Lusitania* survivors spoke of being swept into the ship's funnels by inpouring water during the late moments of its sinkings, only to be saved by sudden upwellings of warm water. Since these events were something that these survivors must have actually felt (as opposed to things they thought they saw or heard), we accept them as factual, and assign their cause to sudden ruptures in the boiler room far below—but not to boilers or steam piping, since these sources had surely been immersed in cold seawater for some minutes.[8] However, the source of the second explosion is not affected by these recountings, since witnesses place that violent event as much as 30 seconds after the torpedo hit, while the ship could not have listed to the point of having her stacks lie in the sea in less than 10 minutes. The significance of the upwellings through the funnels is their confirmation that some type of explosive event did occur in the *Lusitania*'s forward boiler rooms. But observe the careful use of "explosive events" rather than "boiler explosions." Could something other than a boiler or steam pipe under pressure have ruptured? Advocates of the coal dust or munitions scenarios might respond with a positive rejoinder to such a question, but we are speaking here of something that happened deep in a flooded hull, and at a time when sources of ignition for explosive combustion certainly were gone.

The ship was almost lying on its beam ends and sinking rapidly. Surely there were pockets of air trapped here and there, and under increasing hydrostatic pressure. In a boiler room the relatively thin-walled boundary of an uptake collapses, allowing a burp of air through the funnel. Such events seem likely, as do the violent openings of steam pipe joints, and possibly boiler seams, under thermal shock of cold seawater. But the *Lusitania*'s second explosion appears to have been a much more violent event than those, and to have occurred very soon after the torpedo explosion. There appears to be good reason to assume that boiler explosions often occurred when a ship was hit by a torpedo in way of her boilers. We might quote the words of Wesley Frost,[9] an American consular officer in Queenstown (Cobh) who interviewed survivors of torpedoed ships who landed in that port during World War I.

> The dangers attendant upon warningless torpedoing are by no means confined to actual drowning from the sudden engulfment of the ship. In at least one case out of every three, to begin with, there occur boiler explosions. If the submarine makes good shooting the torpedo strikes amidships and induces an engine explosion with the fatalities more horrifying than those from shellfire. I saw the magnificent Scotch chief engineer of the *Salmo* frightfully scalded about head and torso; and when I left Queenstown he was said to be dying in agony at the Civil Hospital there. In a score or more of cases I listened to the most sickening stokehold stories of how brave men, frequently Americans, were blown into ribbons or scalded to death in live steam from bursting pipes. In the *Gafsa* case seven men were killed in the engine room, and in the case of the *Camara* nine men were annihilated by the explosions.
>
> These boiler explosions, contrary to the fallacy current among sea-

> men, are not caused by the sudden immersion of hot boilers in the cold water of the sea; for this contact has only a tendency to reduce the steam pressure. They are caused either directly by the concussion of the torpedo or its explosion, or indirectly by mechanical vagaries of machinery dislocated by the torpedo explosions.
>
> The Booth liner *Crispin* . . . was torpedoed without warning while laboring in a heavy sea on a thick March night in 1917. The impact of the torpedo induced a boiler explosion in which five men were horribly macerated and killed.
>
> The Furness-Withy steamship *Swanmore* . . . was torpedoed without warning at twilight when she had reached a position two hundred miles from Bull Rock. There were eleven deaths, due in greater part to the boiler explosion which at once supervened.

Frost's evidence is second hand, and consists only of recollections recounted by traumatized witnesses, but nonetheless these should be taken as strong evidence for a linkage between torpedo explosions and boiler explosions. On the other hand, he denies the triggering of such explosions by inundation of seawater. Of course, he is really saying "no such explosions seem to have occurred on ships whose survivors were landed at Queenstown," and that is readily accepted as true, since explosions from that cause appear to have been rare. For example, we are convinced that no boiler explosion occurred in the sinking of either *Titanic* or *Britannic*.

Frost treats the *Lusitania* case in his book, but makes no mention of the second explosion. However, he contributes to our present discussion by deriding as a fallacy the notion that an inundation of cold seawater can trigger a boiler explosion. Nonetheless, it is not a conclusive contribution because he provides no support for his assertion, one that flies in the face of a widely held belief. For an instance, in February 1915, Commodore David Dow, *Lusitania*'s captain on a previous voyage, expressed the fear that if she was struck by a torpedo in way of the boilers, the steam pressure would lead to a cataclysmic explosion. For a second instance, a letter from Admiral Beresford praising the engineers of the *Titanic*, to be quoted in Chapter 26, expresses a belief in boiler explosions during sinkings. Doubters still may say "offer us hard evidence of a single bomb-like explosion of a marine boiler caused solely by flooding of a boiler room." This cannot be done because one cannot assuredly rule out a chain of events linking a torpedo warhead to a boiler event. On the other hand, it cannot be discounted because those who are witnesses to such events die instantly there in the lower depths of the ship, just as 67 (*all* 67, remember) did on the *Bennington*. We *can* assert the danger of thermal shock to a cylindrical boiler and piping ruptures caused by waterhammer. There is evidence of cylindrical boilers wreaking great havoc aboard ship.

The third[10] possibility is the rupture of steam piping expansion joints caused by the whipping response of the hull to the torpedo explosion. An examination of the machinery arrangement plans of the main steam piping supply from Boiler Room No. 1 to the steam turbines shows a piping distance of approximately 225 feet. The "take down" joints in the steam piping arrangement were spaced about every 100 feet. The specifications for the piping system called the main and large auxiliary steam pipes between the boilers and engines and branch main steam pipes from each boiler to be of iron with lap-welded seams or steel solid drawn, and riveted or screwed flanges. The pipes were to be without bends to allow expansion, but were to have the necessary expansion joints. The expansion joints (probably, the copper bellows type) were located adjacent to the watertight bulkheads and also spaced about 35 feet apart. With the boilers operating at their maximum pressure and steam at a temperature above 390°F, the piping would have expanded from its cold state about 3 to 4 inches axially for every 100 feet of length. Since the expansion joints were installed to relieve the axial thermal expansion stress in the piping and could tolerate only minimal lateral strain, the torpedo explosion and the violent whipping response that followed could have damaged the piping connections to the boilers in Boiler Room No. 1 and other branch connections.

All other steam, exhaust, bilge discharge, and donkey pipes associated with the main engines were made of strong copper, hard soldered, or solid drawn. One steam pipe, connected to selected boilers, supplied all auxiliary engines, winches, windlass, etc. The steam supply to the electric plant was taken from independent pipes at the forward engine bulkhead. There were separate valves at boilers selected for auxiliary connections. Exhaust from auxiliary units such as feed pumps was led to the auxiliary condensers, to the waste steam main, or to the feed heaters independently of any other exhaust pipe. The steam and exhaust pipes of the steering gear and refrigerating equipment were arranged to be independent in a manner similar to the electric plant, but to have cross connections to other sources of supply and exhaust. Efficient steam separators and strainers were fitted in the steam mains near the propulsion turbines and the generator engines.

It has now been seen that there are several explanations for the second explosion, even if the analysis is limited just to the boilers or to connecting piping. As stated earlier, there is likely to have been a rupture of piping caused by waterhammer. Steam condensation-induced transient fluid hammer was a common cause of explosions, especially in the Scotch boiler era. Sudden condensation, induced by much colder water, is one of the most damaging causes of fluid hammer. Its prevention is complicated by the complex nature of the underlying phenomena, so that it has occurred occasionally throughout the 20th century. The shock wave that is caused by a fluid transient from the steam condensation and its opposite, "flashing" hot steam has been actively studied by the Nuclear Regulatory Commission. One of its investigations[11] describes an event that may be similar to the *Lusitania* explosion, and hence allows us to make some quantitative estimates of what may have happened aboard that ship.

There were two possible scenarios for this type of incident on the *Lusitania*. First, a saturated water slug entered the steam piping and was accelerated through the steam lines by the shakeup motion generated by the torpedo detonation, or by a sudden flow disturbance. A second possibility was that a flood of cold seawater caused thermal shock to the boilers and created a sudden condensation of the steam in these boilers and their exposed piping. Then, when saturated water from the boiler entered the steam main, transient hydraulic waves were formed from the condensation of the steam and generated voids. The saturated water slug moved rapidly into the "void"; the generated pressure waves then traveled through the piping system. Computations based on a 200 psi differential pressure acting on a liquid slug show that this can produce a water hammer overpressure of 23 times that of the normal pressure. It is plausible that a pressure of this magnitude was capable of rupturing the expansion joint bellows and piping connections.

The thermal shock wave from a collapsing steam bubble can cause a devastating explosive effect. The event that led to the Nuclear Regulatory Commission study occurred in 1989: a buried steam line with similar steam conditions to those on the *Lusitania* exploded on 23rd Street in the Grammercy Park section of New York City. A 10-foot deep and 15-foot diameter crater was created in a sidewalk. Large blocks of concrete reportedly were projected some ten stories into the air by the force of this steam explosion and its sound was heard as far as nine miles away. Eyewitnesses to the Grammercy Park explosion reported noises and breathing difficulties similar to those that some of the *Lusitania* survivors recounted after her sinking. The cause of the Grammercy Park explosion was determined to be the sudden condensation of saturated steam caused by contact of cold ground water against the pipe.

A comparison of the Grammercy Park and *Lusitania* steam conditions is summarized in Table 25–2. The *Lusitania* half of the table is based on the assumption—which we believe to be the correct one—that the triggering event was the rupture of steam piping at a cast-iron fitting or at an expansion joint. The cause of that event was shock appearing in the form of a whipping reaction to the torpedo explosion, or was water-hammer whose cause also can be traced to that explosion. Some of the consequences were calculated be Peter Hsu,[12] based on certain assumed similarities to the Grammercy Park explosion. His findings appear in Table 25–2.

Consequences of the Explosion. The suddenness of the initial piping/boiler failure, and the rapid progression of events thereafter, meant that remedial measures such as shutting valves to isolate the site of a failure could not be taken. In consequence, steam flow to the propulsion turbines, the ship-service generators, and all other steam consumers, would fall rapidly, or be cut off completely. Boiler Room No. 1 filled with steam, as noted by the three firemen, who escaped from that room.

In the aft engine room, the pressure to the main propulsion turbines suddenly had dropped from 196 to 50 psi in a period of minutes when Third Engineer Little attempted to apply steam to the astern turbine while the engine room crew was still attempting to adjust the valves in the steam line. A catastrophic failure in one of the two main steam lines allowed a great amount of steam to billow from the ship between the third and fourth funnels. Within minutes after the torpedo struck, there was no

TABLE 25–2 *Steam Explosion Similarities*

	Lusitania	Grammercy Park New York City (23rd Street–West Side)
Date	May 7, 1915	August 20, 1989
Steam Conditions	195 psig 388°F	170–1200 psig Approx. 3800°F
Cold Water Temperature	Approx. 45°F	Approx. 60–70°F
Pipe Diameter	16–25 inches	24 inches
Steam Bubble Size	Various	Approx. 3.5 inches
Pressure Spike	Estimated at 23 times	Approx. 11 times
Explosion Centers	Expansion Joints, Piping, Boilers	Expansion Joints, Piping

steam for the ship-service generators on the steering gear. When the turbo-generators ceased to function, the entire ship was plunged into darkness. Passengers found it difficult to escape through dark and unfamiliar corridors. All auxiliaries dependent upon electric power were out of action, including the elevators, where some persons were trapped. Pumping and steering were no longer operating. The ship was able to move very slowly as long as some steam pressure remained, but she could no longer be maneuvered. With the ship in darkness, passengers and crew alike thought only of saving themselves. Discipline broke down and pandemonium ensued. Unable to maneuver, to counterflood tanks on the port side, or to slow the flooding by pumping, the *Lusitania* rapidly filled with water and capsized to starboard within 18 minutes. The question is whether this loss of steam was enough to diminish the possibility of a boiler explosion in the two forward firerooms.

Final Thoughts. The improvement in diving equipment and submersibles has allowed man to probe the ocean depths. At the time of this writing, Mr. F. Gregg Bemis is the owner of the *Lusitania* wreck. A member of the Marine Forensics Panel of the Society of Naval Architects and Marine Engineers, he has been working closely with this Panel on a plan to examine the inside of the wreck. When that investigation is completed, perhaps the question can finally be answered on what really caused that second explosion.

Notes for Chapter 25

1. The authority cited is William Henry Boehm, 1912, *Steam Boiler Explosions*, The Fidelity Casualty Company, New York. Similar material can be found in Robert Thurston, 1894, *Steam Boiler Explosions*, John Wiley & Sons, New York, and Robert H. Munro, 1903, *Steam Boilers: Their Defects*, C. Griffin & Company, London.
2. *Journal of the American Society of Naval Engineers*, 1905, Volume 17, pages 888–889.
3. Detroit *Free Press*, March 27, 1922.
4. E. J. Rimmer, 1912, *Boiler Explosions, Collapses, and Mishaps*, Constable & Company, London.
5. The HMS *Blake* incident was reported in *Engineering*, 1903, Volume 76.
6. The illustration appears in Rimmer, Note 4.
7. This "sundry" event was reported in *Engineering*, 1906, Volume 82.
8. One of the survivors of this strange ordeal was reported afterward to be covered with soot. This circumstance is evidence for an implosion of boiler uptakes under hydrostatic pressure.
9. Wesley Frost, 1918, *German Submarine Warfare*, D. Appleton and Co., New York.

10. "The Saga of the RMS *Lusitania,* A Marine Forensic Analysis," William H. Garzke, Jr., Robert O. Dulin, Jr., Peter K. Hsu, Blake Powell, and F. Gregg Bemis, New York Metropolitan Section of the Society of Naval Architects and Marine Engineers, January 1998.
11. Izenson, M. G., Rothe, P. H., and Wallis, W. G., "Diagnosis of Condensation-Induced Waterhammer," U.S. Nuclear Regulatory System, October 1988.
12. The calculations producing the numbers shown in the left column of Table 25–2 were performed by Peter K. Hsu. See Note 10.

CHAPTER 26

Titanic *Machinery in Extremis*

First Officer William Murdoch picked up the phone and heard from the Crow's Nest, "Iceberg ahead!" He looked and then shouted an order for starboard helm. He rang STOP-SLOW ASTERN on the engine room telegraphs. Then there was nothing more he could do, except explain it all to his Captain when that unfortunate man rushed into the wheelhouse.

The helm order produced a response, but one that was too late and not enough to steer the ship away from the iceberg. The engine order was equally ineffective, since neither the ships of 1912, nor those of a century later, can stop very quickly after moving at speed nor even slow significantly in the time and distance that were available to the *Titanic.*

The engineers had run the reciprocating engines in reverse for a few minutes when another order came down, SLOW AHEAD.

No engineering officer survived to testify about that half minute or so in the engine rooms. The only engine room crew in those spaces that saw the engine telegraph settings and survived was Trimmer Thomas Patrick Dillon[1] and Greaser Frederick Scott.[2] Dillon was on temporary duty in the Reciprocating Engine Room at the time of the collision and heard the alarm bells ring on the engine telegraphs. When asked how long the engines were in reverse, he answered two minutes. This testimony agrees with that of Boxhall who noted that Captain Smith changed the engine telegraph to SLOW AHEAD when he returned to the Bridge from a brief conversation on the starboard bridge wing with Officers Boxhall and Murdoch.

Scott, on the other hand, was standing at the watertight door on the starboard side of the Turbine Room near the entrance to the Reciprocating Engine Room when he heard all four engine telegraphs ring. The Bridge rang down STOP and two greasers near the engine telegraphs for the reciprocating engines rang the telegraph back to answer it. Then the Bridge rang down SLOW AHEAD. For ten minutes according to Scott the *Titanic* was going ahead. Then they rang down STOP again and when he looked at the engine telegraph about four to five minutes later, it read SLOW ASTERN. According to Scott, the ship went astern for approximately five minutes.

The evidence of the two men who saw the engine telegraphs in the engine room is conflicting. According to Lord Mersey and the attorney questioning these witnesses, each witness gave slightly different stories on what he saw. However, the evidence of Scott and Dillon also conflicts with that of Fourth Officer Joseph Boxhall who saw the telegraph settings set to FULL ASTERN. The only explanation to this conflicting testimony is that Murdoch in his haste to go astern did not push the engine telegraph to that setting. Logic, however, would be to go FULL ASTERN and that is what many author and scriptwriters for *Titanic* movies and documentaries have done. There is no question that the first order was STOP. This alerted those engine room personnel off watch that there was an emergency. The other event that the STOP order accomplished was closing the dampers in the furnaces of the 24 boilers on line at the time. According to Greaser Frederick Scott, the STOP order was necessary because they could not stop the reciprocating engines at once. It was necessary to release the steam from the cylinders first, otherwise they would blow the cylinder covers off if they tried to stop them at once.

The other conflict in these three witnesses' testimony is the time element. Trimmer Dillon's account agrees with that of Officer Boxhall. Greaser Scott, on the other hand, recounts that about 30 minutes elapsed with the engine telegraph orders that he heard and saw. Of course, minutes can be like hours in an emergency situation such as this and these men did not have wrist watches to look at.

The 1997 movie *Titanic* undertook a reconstruction of what may have happened when the alarm was sounded via the engine room telegraphs, using the engine room of the Liberty Ship *Jeremiah O'Brien*. The STOP bell activated red warning lights in the boiler rooms. The lead stokers shouted "close the dampers," a measure that cut the air flow to the furnaces, and hence reduced the rate of combustion in the coal, and the attendant steam production to a minimum. In the engine room, the amazed Chief Engineer dropped his coffee cup, and bellowing orders, raced to the throttle valve to reduce steam pressure to the engines. (Unfortunately the movie was in error because Chief Engineer Bell was in his quarters when the iceberg collision occurred.) The pointer on the engine pressure gage moved downward. When it reached a sufficiently low pressure, the Chief bellowed NOW to the man who had run to the reversing cylinder (a steam cylinder that activated the engine reverse gear). The engine valve gear moved quickly to the astern position, the engine stopped its ahead rotation in less than one turn, and then began to accelerate in reverse. The Chief frantically reopened the steam valve to increase pressure. (There were two reciprocating engines, of course, and these acts were performed almost simultaneously on the other engine by other engineers.) The stopping of the centerline turbine that operated on steam exhausting from the reciprocating engines was not shown in the movie. Indeed, it was unidirectional and could not be reversed. As a consequence, that steam was diverted to the condensers when the ASTERN order was given. A more detailed description has been given in Chapter 5.

The attempt to stop was quite in vain, of course, and the engines were ordered to STOP within ten minutes after the contact with the iceberg. They were never to turn again. The engines were not needed to drive the ship away from the berg, and no attempt was made to steam the damaged ship toward the lights of a distant ship, believed to be the *Californian*. In retrospect, an attempt to reach that ship might seem to have been an eminently sensible thing to do. But the *Titanic* officers were not to realize until the time for such a venture had slipped away that the *Californian* was effectively incommunicado. As it was, Naval Architect Thomas Andrews apparently recommended against moving the ship because he feared the effect of increased hydrostatic pressure on the watertight bulkheads that were then limiting the flooding. Andrews was particularly concerned about the watertight bulkhead between Boiler Rooms Nos. 5 and 6 for that had been where the coal had been extinguished the previous day.

The engine room phase of the drama was over, for there was nothing more that the engineers in the propulsion plant could do. However, we must comment on the likely role of the centerline propeller in the attempt to turn and stop the *Titanic*. That propeller could play no part in stopping, since it was driven by a unidirectional machine. However, in ahead operation, the propeller provided a greatly enhanced wake velocity past the rudder, and hence a greatly enhanced turning moment to the hull. The centerline propeller was still driving ahead during those fateful 30 seconds, and the ship just turned a few more degrees before impact. Maybe—if the engine-order telegraph had been left untouched. Then—maybe—slightly less damage and a consequent slower flooding rate, and the *Carpathia* arrives in time! Or maybe—slightly less damage, and the ship remains afloat, perhaps to struggle to port like the *Arizona* 33 years earlier.

However, the engines participated in the drama futilely. After some uncertainty following the impact, they were apparently put on STANDBY, never to turn again.

Then there were the boilers, their dampers shut at the first alarm. A few were needed for the steam that would keep the ship alive until her final moments, and for those unnamed few the dampers must have been reopened and their furnaces fed until an increasing trim made firing difficult, and eventually impossible. Needed or not, and unlike the engines, the boilers could not be shut down quickly because of the energy stored in hot boiler water, and in hot masses of coal. And because of their unavoidable energy storage at high pressure, the boilers were always a source of hazard until that pressure could be reduced. Eternal vigilance by instruments and by those who read them is the price of safety with any boiler, and that is especially the case with coal-fired cylindrical boilers. Among the essential measures of normal care in operation was maintenance of very slow rates of cooling when a boiler was taken out of service, for otherwise thermal stress might open a seam or crack a pressure-containing part.

Seawater immediately entered and inundated Boiler Room No. 6. It initially entered Boiler Room No. 5 at a very slow rate, Boiler Room No. 4 at an inconsequential rate, and the others not at all. However, by 0120 Boiler Room No. 5 would flood completely and Boiler Room No. 4 likewise by 0150. By 0215 all remaining boilers would be immersed in cold water. Every boiler might have been destroyed by thermal stresses, particularly those in Boiler Rooms Nos. 2 and 3. But that did not matter, since the ship was sinking. What did matter was the possibility of explosion induced by the extreme conditions that ensued during the time that the damaged ship remained afloat. What also mattered was keeping enough boilers operational to supply the generator engines and pumping systems as long as the ship was afloat.

A boiler explosion is an event of extreme noise and extreme violence. Any noise and violence accompanying the *Titanic*'s agony can be attributed to the breakup of the ship just before her climactic plunge. Positive evidence, i.e., evidence against boiler explosions, comes from events known to have occurred during the sinking, and the forensic evidence from the 1998 *Titanic* Expedition. During the sinking, the ship remained fully lighted until the fracture of the hull broke the piping between the steam plant and the generators engines. The violence of an explosion before that time probably would have destroyed the steam supply in much the same way that the hull breakage did. The continuance of the lights, the apparent continuance of pumping as long as possible, and the lack of any evidence for boiler explosions, are positive aspects of the sinking that reflect credit upon the engineering staff. The attempt to save the ship by pumping obviously had failed, but given the rate of flooding, pumping had no chance of success. The forensic evidence supports the conclusion against explosion. No damage visible on the wreck can be attributed to boiler explosion, and in August 1998, members of the Marine Forensics Panel were able to get a clear look into Boiler Room No. 2. This was surely the last boiler room where firing was maintained, and hence the one where explosions might have occurred during the sudden inundation accompanying the fracture of the hull. In 1998, all five boilers were in place in that space, and showed no exterior damage to their casings.

The bilge and ballast pumping system was described in Chapter 10. Recall that the total capacity of the pumps connected to that system was 1,700 tons per hour. The initial rate of flooding is estimated to have been about 400 tons per minute, later slowed to about 50 tons per minute, rates that obviously would overwhelm the pumping system. The condenser circulating pumps had a total capacity of 12,000 tons per hour, but this resource could not be applied to expel flooding water except in the engine rooms. There is a report from the testimony of Trimmer Thomas Dillon that the engineering staff attempted to fit a 10-inch portable leather pipe into the flooding boiler rooms via the opened watertight doors. Details are unknown, but it is unlikely that the pipe was intended for connection to the circulating pumps. Since both the bilge main and the bilge-and-ballast main were also of 10-inch pipe, the likely application of the portable pipe was to allow a bilge pump in one boiler room to be applied in an adjacent boiler room.

Even if the fitting of the portable pipe to a condenser circulating pump could have been accomplished, its use would not have affected the outcome significantly. Regardless of the combined capacity of the condenser circulating pumps, the flow rate that could have been established through a single 10-inch pipe is far short of the flooding rates that prevailed over most of the *Titanic*'s remaining time afloat. The *Carpathia* would still have arrived too late.

The iceberg opened a riveted seam in Boiler Rooms Nos. 5 and 6, placing their boilers in immediate danger. Testimony of Leading Fireman Frederick Barrett indicates that alarm lights in the boiler rooms came on, resulting in the order to close the furnace dampers in response to the major drop in steam demand that a STOP order entailed. A coal fire, deprived of air, will eventually die, of course, but not instantly. Even after combustion ceases, the mass of coal, still very hot, constitutes a significant heat source. Steam production, therefore, continues for a while after dampers are closed. Even if the rate of steam consumption drops radically, steam pressure rises until safety valves open. The discharging steam must be led safely away from the ship's interior by an "escape pipe." In most instances, the escape pipe discharge is high on the ship's funnels, often in the vicinity of the whistle. The *Titanic* safety valves opened very soon after the engines stopped, producing geysers of steam from the escape pipe openings. Some of the survivors remarked on the deafening noise of this escaping steam, and some of them spoke of this noise as coming from the ship's whistle (mistaken identification, almost certainly).

When water began to flood the boiler rooms, the fear of explosion must have come upon the stokehold crews, and they must have kept the safety valves open to lessen the danger by reducing pressure. On their own, the safety valves would have closed as the pressure fell back to its operating level, but the roar on deck of escaping steam apparently continued long after that should have happened.

A more drastic response for a threatened boiler was to "draw fires," i.e., to open the furnace doors and rake the burning coals out onto the firing aisle floor. This procedure was begun immediately in Boiler Room No. 6 (forward-most), for seawater poured in through the damaged hull plating at such a rate that the need for rapid response was obvious to the men on duty there. The necessary order was quickly given: Fireman George Beauchamp noted in his testimony that after the space began to flood an order was given to draw the fires. However, the drawing was never completed because the fast-rising water forced abandonment of the boiler room in a little over five minutes.

Hot steel and high pressure were left to confront a cold inundation, but there is no evidence that any boiler in Boiler Room No. 6 exploded. Flooding of Boiler Room No. 5 was initially at a rate a witness likened to the flow from a fire hose—significant flooding, indeed, but at a rate that could be countered by the rooms bilge suctions. After an interval of confusion and indecision, the fires were drawn in this room. No boiler exploded there.

The remaining boiler rooms were not flooded until late in the sinking, so that fires were presumably drawn and pressure reduced in a disciplined manner while sufficient time remained.

Now, we know that pressure remained in at least one boiler room until the very end because the lights continued to burn until the hull broke apart in the last moments, and the only source of the necessary power generation lay in the flooding boiler rooms. Since the aft-most room, Boiler Room No. 1, was not used on the maiden voyage, the last source of steam must have been the next forward room, Boiler Room No. 2. As noted earlier, some of the boilers of that room have been photographed by visitors to the wreck, and are obviously intact. The belief that boiler explosions occurred as the *Titanic* sank, therefore, deserves a "very unlikely" rating.

Such is the conclusion of the following century, but belief in boiler explosions had currency at the time of the sinking. Jay Henry Mobray[3] collected and published survivor stories that support the belief that boiler explosions accompanied the ship's final plunge. Some are quoted as follows.

Lady Duff-Gordon:

Then the boat's stern lifted in the air and there was a tremendous explosion. Then the *Titanic* dropped back again. Ten minutes after this there was another explosion. The whole forward part of the great liner dropped down under the waves. The stern rose a hundred feet, almost perpendicularly. The boat stood up like an enormous black finger against the sky.

Then there was another great explosion and the great stern of the *Titanic* sank as though a great hand was pushing it gently down under the waves.

Charles Stengel:

We had gone but a little way from the ship when the first boiler explosion came. It was followed in quick succession by three others, at intervals of about one second apart.

Unnamed crewman:

The end came at 2:30 on Monday morning. The lights on the ship did not go out until ten minutes before the liner sank. The inrushing seas reaching the after fires produced an explosion, which sundered the big liner. The forward half of the *Titanic* dived gently down. The after part of the ship stood on end and then disappeared.

William E. Carter:

As we left the ship the lights went out and the *Titanic* started to go down. The crash had ripped up the side and the water rushing into the boiler room caused the boilers to explode.

The author (Mobray), quoting A. H. Barkworth and W. J. Mellors:

Mellors and Barkworth both declare that there were three distinct explosions before the *Titanic* broke in two, and bow section first, and

> stern part last, settled with her human cargo into the sea. Her four whistles kept up a deafening blast until the explosions, declare the men. The death cries from the shrill throats of the blatant steam screechers beside the smokestacks so rent the air that conversation among the passengers was possible only when one yelled into the ear of a fellow unfortunate.

In a following paragraph, Barkworth is quoted "There were three distinct explosions and the ship broke in the centre."

These stories show that *some* of the survivors believed that boiler explosions accompanied the sinking. On the other hand, many other survivors said nothing of boiler explosions, yet remarked on loud noises during the interval when the stern was raised toward the sky. Some of them attributed the noise to machinery breaking loose from its foundations and crashing forward, presumably through bulkheads. Now, the ship did break in two, as some of the survivors also recognized, and that failure surely generated loud and violent noises. In short, there was the noise of ruptures from within the *Titanic* just before she sank, noise that surely came from the breaking of the hull, but also might have come from falling machinery, and also might have come from boiler explosions. The witnesses, who surely had no experience with such traumas, could easily have been mistaken with respect to the sources of noise. In short, the survivor evidence for boiler explosions is quite fragile.

Of course, the rupture of the hull, which occurred between the aftmost boiler room and the engine room, must have broken the steam pipe that was supplying steam to the generator room, aft of the engine room, and to the emergency generator room on "D" Deck aft. Since the lights were extinguished suddenly, and almost certainly as a consequence of the hull rupture, the affected pipe must have been under pressure. There must have been a sudden outpouring of steam with attending noise and a cloud of steam around the aft end of the severed bow section of the ship. Under more commonplace circumstances, the event surely would have been called an explosion, and would have had fatal consequences for any humans in the neighborhood of the break. However, it was not the explosion of a boiler, and its cause was not thermal stress from the contact of seawater with hot metal.

All of this testimony suggesting explosions of boilers comes from witnesses who surely never witnessed (or heard) a boiler explosion in any venue, nor had they experienced any other sights and sounds of a sinking ship. From their words we can, with confidence, distill only the evidence that noises indicating a break-up of some sort emanated from the *Titanic* in her last moments afloat. However, testimony of a bit more weight comes from Charles Lightoller,[4] Second Officer. Late in the sinking he had finished supervising the launching of the portside lifeboats, and was attempting to launch one of the collapsible boats that were stowed forward near the wheelhouse. When the forward funnel fell into the sea, the wave it produced washed him (and the boat) overboard. Lightoller managed to swim away towards the foremast, but then recognizing that such a place of refuge was going to sink with the ship, swam back aboard the *Titanic* in the same vicinity. Almost immediately he was swept against a boiler-room ventilator by water pouring into its opening. The grille across the opening saved him from being sucked in, though he was pressed there, doomed it seemed, by the inrushing water. A violent rush of warm air from below interrupted the inflow and released him from the grille. He eventually managed to reach the collapsible boat, overturned and drifting nearby. He credited the miraculous rush of warm air to a boiler explosion.

We question Officer Lightoller's analysis. His near immolation upon the grille of a supposed boiler-room ventilator must have taken place near the forward end of the deckhouse. If so, the ventilator must have served a boiler room no further aft than Boiler Room No. 5, and the boilers there (and in Boiler Room No. 6 as well) were icy cold beneath seawater at a time so late in the sinking. A more likely source of the saving rush of air is the forward end of first-class spaces on "E" Deck. Implosion/explosion caused a collapse of bulkheads bounding the spaces there, releasing imprisoned air. If steam is insisted upon, then one could postulate a breakage of the main steam line, still under pressure from steam generated in Boiler Room No. 2 (say), as a consequence of the same extreme hull bending that had caused the first funnel to fall a few minutes earlier.

Our conclusion must be that there was no explosive rupture of a boiler during the sinking of the *Titanic*. Ruptures of steam piping as consequences of hull rupture or of extreme hull bending are much more likely, especially since one such failure almost surely accounts for the sudden extinction of the lights.

Colonel Archibald Gracie's account of the *Titanic* tragedy reveals that Charles Lightoller had questioned a number of survivors while aboard the *Carpathia* on her way to New York. He mentioned that the hull fracture was discussed along with the possibility of a boiler explosion. Lightoller dismissed the boiler explosion when he met with Gracie and Officer Pitman to go over the accounts that he had heard. Lightoller also dismissed the hull fracture that had been seen by John Thayer, Jr. and others. This committee of three decided that the boilers and other heavy machinery had come loose and crashed through bulkheads. Gracie was so mesmerized with Lightoller that he went around on speaking engagements before his death claiming that the ship rose to the perpendicular and held that position for five minutes before beginning her plunge to the bottom. When Gracie was thrown into the water, he made the point that he was surrounded by deck planking.

The Lights. By a seeming miracle they remained burning throughout the unflooded parts of the ship until the hull failure just forward of the reciprocating engine room separated the generators (both main and emergency) from their source of steam. That word "miracle" seems appropriate because the sole source of energy for generation of electric power was the boiler plant, low in the ship, and hence vulnerable to flooding and abandonment by its crew. An essential element of the story is thus the maintenance of at least one boiler in its normal operating state for almost the entire span of the sinking. Human attendance to duty in the face of great hazard must, therefore, have been a factor.

But can one attend a hand-fired coal boiler when a sinking ship is pitched downward at the extreme angles often pictured for the *Titanic*? Shovel uphill or downhill when the ship is pitched 15 degrees? Will the fire slide downhill, either clogging the combustion chamber, or tumbling out through an opened furnace door? Answers may not be necessary because the ship probably did not trim to such an extreme angle until the final moments. Near the end, perhaps at the moment when the ship broke in two? At that point, firing of the boilers had long since ceased, but steam could have continued to be produced by the thermal inertia of the boilers, the very thing that gave them the potential for explosion even after firing had stopped. For example, a pound of water at 200 psia will produce 0.064 pounds of vapor as its pressure is reduced to 100 psia, or 0.114 pounds as its pressure is reduced to 50 psia. A generator engine running at the power required to produce 400 kW might consume about 13,000 lbs/hr at 200 psia steam pressure, or about 19,000 lbs/hr when the pressure had dropped to 100 psia. With many of the lights doused by submergence during, say, the last half hour, a single generator could probably light the ship, and one then can easily conceive of the necessary steam being supplied by one or more abandoned boilers in Boiler Room No. 2. Some survivors' tales mention a gradual dimming of lights as the end approached, a phenomenon consistent with falling steam pressure in an unfired boiler.

If iceberg damage had been limited to only two compartments (spaces between adjacent watertight bulkheads), the *Titanic* would not have sunk. If those two compartments had been boiler rooms, total steam production obviously would have been reduced, but it would not have been denied to the generator engines. Routing of steam piping to the generator room was arranged so that any two of the six boiler rooms could be flooded without isolating the remaining rooms from the generators. Further, in the case of the emergency generators on Saloon ("D") Deck, the steam supply ran vertically from the steam main below to a passage along that deck, then aft to the Generator Room. Flooding of the engine room, which might interrupt steam to the main generator rooms, therefore, would not deprive the emergency generators of their steam.

Operation of the machinery during which electric power was provided in the face of boiler room flooding and increasing trim, the engine crew was virtually invisible to the desperate passengers assembling on the Boat Deck. A few of the stokers joined them to act as lifeboat crews after flooding had driven them from their duty below. Only two of the engine room ratings survived, and they only because they had been sent on topside missions by their officers. Of the officers themselves, all perished, apparently without ever coming up on deck. This commendable circumstance was well remarked at the time. A medal was struck to honor the engineering staff, and a monument erected in England. A week after the sinking the following letter from Admiral Beresford appeared in the *Times*:[5]

> April 21, 1912
>
> Sir,—In the late appalling disaster to the *Titanic*, attention has rightly been called to the bravery, resolution, and chivalrous gallantry of Captain Smith, the officers, seamen, band, and passengers. Many comments have been justly made regarding the heroism on deck, but nothing has been said of the heroism below.
>
> Nothing can exceed the heroism of the captain, officers, and seamen of the ship, but officers and seamen are the first to offer a whole-hearted tribute of unbound admiration to those working below, as they well know how often the real grit and courage of the officers and men of these departments is called upon in moments of emergency.
>
> It is stated that the lights were burning until a few minutes before the ship took her final plunge. This proves the officers and men below remained at their posts when they must have known that a death—the most terrible and painful that it is possible to conceive—awaited them at any minute, either by the bursting of a steam pipe or water rising in a compartment. It is certain that those working below must have known the awful danger the ship was in long before anybody else, but they re-

mained at their posts, resolving to die sooner than come on deck and create a panic or attempt to save themselves.

All honour and respect to those men whose names will be recorded on the roll of fame for gallantry in a sudden and unlooked-for disaster. But I am sure the survivors of this shocking catastrophe will agree with me in placing those who worked below on "The right of the line."

I have the honour to be, Sir, your obedient servant,

Charles Beresford, Admiral
1 Great Cumberland Place, W.

Notes for Chapter 26

1. Testimony of Trimmer Thomas Patrick Dillon, British Inquiry, Questions Nos. 3715–3730.
2. Testimony of Greaser Frederick Scott, British Inquiry, Questions Nos. 5518–5631.
3. Wesley Frost, 1918, *German Submarine Warfare: The Most Appalling Ocean Horror with Graphic Descriptions of Hundreds Swept to Eternity Beneath the Waves*, memorial edition, D. Appleton and Company, New York.
4. Charles Lightoller, 1935, *Titanic and Other Ships*, Ivor Nicholson and Watson, London.
5. Admiral Lord Charles Beresford in the London *Times*, 22 April 1912. The admiral was quite the rambling writer; his letter repeated all of his thoughts several times, and in only slightly varying form. The version quoted here has been heavily edited to less than half its original length. But it says everything that he said.

   The following quotation was appended to the letter when it was reprinted a week later in *Engineering*, Volume 93.

   We give below the names of the Engineer Officers on board the *Titanic*, the great majority of whom lost their lives while attending to their duty to the very last moment:

   J. Bell, Southampton, chief engineer
   W. Farquharson, Southampton, senior second engineer
   J. H. Hesketh, Liverpool, junior second engineer
   N. Harrison, Southampton, junior second engineer
   G. F. Hosking, Itchen, senior third engineer
   E. C. Dodd, Southampton, junior third engineer
   L. Hodgkinson, Southampton, senior fourth engineer
   J. M. Smith, Itchen, junior fourth engineer
   B. Wilson, senior assistant engineer
   H. G. Harvey, junior assistant engineer
   J. Shepherd, junior assistant engineer
   C. Hodge, senior assistant third engineer
   F. E. G. Coy, senior assistant third engineer
   J. Fraser, junior assistant third engineer

H. R. Dyer, senior assistant fourth engineer
A. Ward, junior assistant fourth engineer
Renny Dodds, junior assistant fourth engineer
T. Kemp, assistant fourth engineer
F. A. Parsons, senior fifth engineer
W. D. Mackie, junior fifth engineer
R. Millar, fifth engineer
W. Moyes, senior sixth engineer
W. M. E. Reynolds, junior sixth engineer
H. Creese Woolston, assistant deck engineer
G. Chisnall, senior boiler maker
H. Fitzpatrick, junior boiler maker
A. Rous, plumber
W. Duffy, writer
P. Sloan, chief electrician
A. S. Alsopp, second electrician
H. Jupe, assistant electrician
A. Middleton, assistant electrician
A. Ervine, assistant electrician
W. Kelly, assistant electrician

CHAPTER 27

Lifeboats

Humankind has placed itself at risk upon the near waters of the habitable continents for perhaps five millennia, and upon the wide transoceanic waters for at least one. Lifeboats —small craft that save lives at hazard from the seas' many threats—seem essential to the later-day voyager, but only since the middle of the 19th century has that notion become an integral element of seagoing. Of course, large vessels commonly carried small boats aboard long before that. However, to have enough to hold all those people crowded aboard the ships of the pre-industrial age was as beyond feasibility as is provision of ejection seats for the people of the ubiquitous passenger jet liner of the 21st century. Then at sea, as now in the air, you paid your passage and trusted your life to the integrity of vessel and crew.

When ships became iron and began their rapid increase in size and speed, the possibility of their sinking from a hole consequent to collision, or being ravaged by fire, became more likely than the consequences of storm or the crunching of the hull upon the rocks of a lee shore. In a slow sinking that might follow a collision, or in a fire that might take an hour to burn from one end to the other, small boats for refuge could save lives—maybe all lives at risk. However, the "all lives" concept did not become the standard immediately. Indeed, it did not become the standard until the morning after that cold night in April 1912. The *Titanic*, with roughly two thirds of its lives lost solely (the public perception, at least) because its boats could hold only one half,[1] demonstrated the consequences of inadequate numbers. Ironically, it also called attention to an existing notion that a full complement of boats *could* have been carried.

The "public perception" caveat arises from doubts that all 2,228 persons could have been marshaled into boats in the time available. The "could have been carried" notion was publicly presented in an *Engineering* article published before either the *Olympic* or *Titanic* went to sea. Both points are discussed later in this chapter.

The *Titanic* did not have enough lifeboats. That fact is as widely and thoroughly known as its supposed unsinkability. It also is more easily understood by the lay public than the ship's ability or inability to resist sinking. There were 2,228 people aboard with 20 lifeboats that had a total capacity of 1,178 persons. The more precise tally is this: the *Titanic* carried 14 boats rated for 65 persons each, two boats for 40 each, and four Engelhardt collapsible[2] boats for 47 each, all adding to a capacity of 1,178 persons on a ship whose design capacity was 3,547 passengers and crew.

On the other hand, you may recall that the *Titanic* was quite "legal" in its safety outfit. The British Board of Trade rules—the only ones applicable since international standards did not exist—specified the number of boats as a function of a vessel's gross register tonnage. That basis is quite reasonable for most aspects of maritime rule making, and continues to be so. The rules in question were issued in 1891 and last modified in 1894 at a time when there was little experience with ships of over 10,000 gross register tons. The rules, therefore, went only as far as "over 10,000 tons." That category fit the *Olympic* ships that were greater than 46,000 gross tons, but was obviously a poor fit since they could carry many more people than could board onto a 10,000-ton ship. Obviously, this does seem like a case where hindsight should not have been needed. Ships had been growing[3] steadily in size in the two decades following 1894, surely enough time for the Board of Trade to notice, and to update their rules.

During the Mersey hearings a point was raised that shipowners had been struggling for a very long time to make a ship its own lifeboat. The intent was to make a ship so safe that lifeboats would not be necessary. There was even a provision in the Board of Trade rules that allowed a deduction in the number of lifeboats based on the degree of watertight subdivision provided. The White Star Line did not take advantage of that deduction in the provision of lifeboats. They actually provided 20 more boats; that was four more than the Board of Trade rules on lifeboat accommodation and life-saving apparatus dictated. However, the only time these rules were challenged after being established in 1894 was in 1904. Sir Walter Howell, Chief of the Marine Section of the Board of Trade consulted with his Professional Adviser, Sir Alfred Chalmers, on the matter of the number of lifeboats that should be provided in the new and larger passenger ships then being constructed and whether the rules governing them should be changed. Chalmers advised that the rules not be changed and gave this explanation of what he told Sir Walter Howell at the Mersey Inquiry:

> I found it [ships] was the safest mode of travel in, the world, and I thought it was neither right nor the duty of a State Department to impose regulations upon that mode of travel as long as the record was a clean one. Secondly, I found that, as ships grew bigger, there were such improvements made in their construction that they were stronger and better ships, both front the point of view of watertight compartments and also absolute strength, and I considered that that was the road along which the shipowners were going to travel, and that they should not be interfered with. I then went to the maximum that is down in the Table—16 boats and upwards, together with the supplementary boats, and I considered from my experience that that was the maximum number that could be rapidly dealt with at sea and that could be safely housed without encumbering the vessel's decks unduly. In the next place, I considered that the traffic was very safe on account of the routes—the definite routes being agreed upon by the different companies, which tended to lessen the risk of collision, and to avoid ice and fog. Then, again, there was the question of wireless telegraphy which had already come into force on board of these passenger ships. I was seized of the fact that in July, 1901, the "Lucania" has been fitted with wireless telegraphy, and the Cunard Line, generally, fitted it during that year to all their ships. The Allan Line fitted it in 1902, and I am not sure that in 1904 it had not become quite general on the trans-Atlantic ships. That, of course, entered into my consideration as well. Then another point was the manning. It was quite evident to me that if you went on crowding the ship with boats you would require a crew which were not required otherwise for the safe navigation of the ship, or for the proper upkeep of the ship, but you are providing a crew which would be carried uselessly across the ocean, that never would be required to man the boats. Then the last point and not the least, was this, that the voluntary action of the owners was carrying them beyond the requirements of our scale, and when voluntary action on the part of shipowners is doing that, I think that any State Department should hold its hand before it steps in to make a hard-and-fast scale for that particular type of shipping. I considered that that scale fitted all sizes of ships that were then afloat, and I did not consider it necessary to increase it, and that was my advice to Sir Walter Howell.

If this statement represents the disposition of the principal man in the Board of Trade at that time, then it is easy to understand why there were no drastic alterations made in the Board of Trade rules for life-saving appliances by the time the *Titanic* and *Olympic* were designed and constructed.

Favoring words might be said for the Board of Trade's pioneering work in establishing safety regulations, including lifeboat requirements that were adequate at the time they were made. However, the Board's blindness to change, and especially its resistance to pre-*Titanic* voices that called for updated regulations, is difficult to defend. One may marshal arguments for that defense, though all may sound specious (especially the first one to be stated here):

- At the time in question, all lives were not perceived as having the same value. If all passengers had been saved while all crew were lost, the post-*Titanic* furor would surely have been less. Or likewise if all the women and children were saved while all the men were lost.
- So few ships sank under conditions that allowed boats to be launched safely that the carriage of lifeboats in any number was a waste of resources.
- The Board was dominated by men who had financial interests in shipping, and since lifeboats are expensive,[4] it was unable to act effectively.

The third point is suggested by the makeup of the Board of Trade circa 1910, and in the failed arguments for more *Olympic*-class lifeboats at that time. The governing body of the Board consisted of decorative dignitaries—distinguished members of Parliament and the Archbishop of Canterbury—people unfamiliar with the dangers of the sea. The marine department of the Board was not staffed with naval architects and marine engineers, but with bureaucrats more skilled in policy-making than in engineering judgments. Moreover, the Board listened to an advisory committee that was dominated by shipowners who advised policies that benefited themselves rather than the public. Those who traveled by sea had no voice.

On the other hand, there *were* advocates of more lifeboats for the *Olympic*-class ships. Thomas Andrews, the principal designer of those ships, Axel Welin of the Welin Davit Company, and Alexander Carlisle, managing director of Harland & Wolff who had left that company to join Welin in June 1910, cooperated in a design that showed multiple boats per davit-set. An article and illustration in *Engineering* (quoted and shown later) give the essence of their design.

With the most glaring aspect of this tragedy being the number of lifeboats provided, evidence now has been found that a handwritten notebook of Thomas Andrews was discovered recently at Harland & Wolff. Andrews meticulously kept notes on the modifications and changes to the *Olympic* and *Titanic*. One of the entries included the subject of lifeboats. He provided the following tabulation:

Boats will accommodate	=	3,538
Passengers & Crew	=	3,473
Spare	=	65

On the opposite page, there was an entry explaining the rationale for these figures, but they were lined out by a thick red line.[16]

Meanwhile, at least one public voice spoke for more lifeboats. On 22 November 1910, Horatio Bottomley, member of Parliament, questioned the number of boats being provided for the *Olympics*, as well as the number that Cunard had given the *Mauretania* and *Lusitania*. The Board of Trade already had given him its stock reply to such criticisms: the number of boats satisfied the regulations. Rebuffed by Parliament and the somnolent Board, Bottomley turned to the media with an article on the lifeboat controversy in the magazine *John Bull*. The article inspired a significant and highly favorable commentary, published in the December 1910 issue of the same magazine; it noted that the anonymous respondent (later identified as Axel Welin of the Welin Davit Company) had in hand a design that would allow 48 or 64 lifeboats for the *Olympics*. Welin's words and illustrations were similar to those attributed to him in the *Engineering* article that is to be quoted in this chapter.

Horatio Bottomley was encouraged by Welin's reasoned approach but discouraged by that of the Board of Trade. He raised the lifeboat question again in a letter to the President of the Board, and in return was promised a review by the Board's Advisory Committee. The committee's rejection of the boats-for-everyone proposal may have been based on the cost to the shipowners who were members of the committee, but the justification was rooted in the never-to-be-refuted[5] concept that small boats would be a refuge of high hazard in the rough weather conditions that were likely to accompany a ship disaster in the North Atlantic. Alexander Carlisle of Harland & Wolff also had argued for increased boat capacity, but accepted defeat and signed the report affirming the Committee's position. Lord Mersey was to question him on the damning presence of his signature. Carlisle sheepishly admitted that he should not have signed that paper.

Regardless of rules, a proposal for carrying two or three times the number of boats actually fitted had been discussed. Further, discussion had reached the point of decision—as it transpired, only a tentative decision. A second row of boats was to be fitted inboard of the first (see Figures 27-1, 2, and 3). Additionally, Carlisle in his evidence to the Mersey Inquiry stated that a third tier of boats, nested with the second, could have been added inboard of the original boats. With this arrangement each pair of davits could handle three boats if required (the illustration shows the third boats by dotted outline). If the added boats at each pair of davits were identical to the original single boat, the addition would have brought the potential boat capacity up to the ship's complement of 3,547 persons. The post-*Titanic* arrangement of boats on the *Olympic*, as well as Welin's proposal that is illustrated here, demonstrated the reality of these attempts to give the *Titanic* "enough boats."

The original *Olympic* lifeboat outfit was the same as that of the *Titanic*. In response to her sister's sinking, the surviving ship was fitted with 28 davits, 4 of them handling 4 boats, 4 handling 3 boats, and 20 handling 2 boats, giving her a total of 68 boats that could accommodate over 3,500 persons. The davits handling 3 and 4 boats were the original 4 davits port and starboard at the aft end of the Boat Deck, suggesting an interesting point: these same davits on the *Titanic* also could have handled three or four boats.

At the time when the decision on lifeboat numbers for the *Titanic* had to be made, Naval Architect Thomas Andrews of Harland & Wolff thought there would be an imminent change in the Board of Trade rules. This is why he made the computations cited earlier and sought the help of Alexander Carlisle and the Welin Davit Company. The arrangement used on the *Titanic* had, therefore, endeavored to cover for the most severe case of boats for all being required (see Figure 27-1). Actually, the Board of Trade was thinking of *fewer* boats rather than more until the tragedy changed their minds. After this became known, White Star's plan for the *Titanic* lifeboat outfit reverted to the one-boat-per-davit-pair that was used also for the *Olympic*. Someone must have made this fateful decision. Bruce Ismay denied being that someone, but he is known to have opposed the proposal for more boats on grounds that it would have to apply to all White Star ships. Costs would be significant, appearance would suffer, deck space that was valued for first-class passenger activity would be reduced, and there would be the implication that the ships were not safe. Shipowners of that period viewed lifeboats as a necessary evil, but they did not like them because they spoiled the ship's appearance. Beyond that, the weight of boats high in the ships would degrade their stability. However, at the British Inquiry, Ismay denied that he had ever seen plans that would increase the lifeboat capacities of the two ships, *Olympic* and *Titanic*.

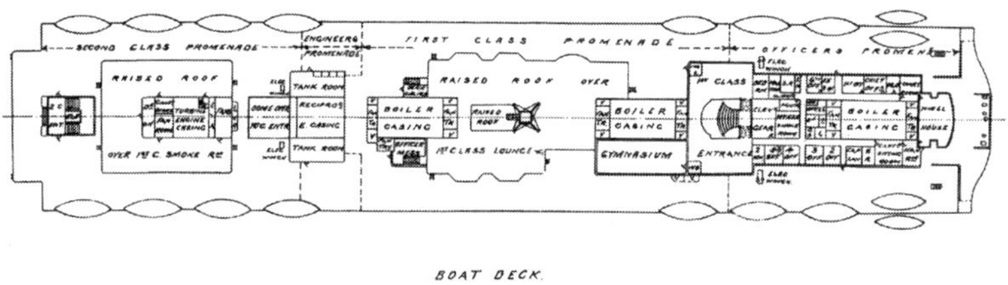

FIGURE 27-1 Plan of the *Titanic*'s Boat Deck. The first-class promenade is not obstructed by lifeboats. (*From* Engineering, *vol. 88, p. 403, 1910*)

Ismay's conduct during and after the *Titanic*'s fateful night could be taken as evidence of guilt for a decision that doomed hundreds of people to death. His actions in frantically trying to expedite the filling of lifeboats, in furtively boarding one himself, then refusing to look back as his ship sank, can be interpreted in varied ways, but a strong case can be made that his conduct aboard the *Carpathia* gave evidence of guilt. Once aboard the rescue ship he hid himself from all but the ship's doctor, having fatuously identified himself as "Yamsi." (As Mr. Ecurb Yamsi? That further fatuity has never been reported.) Then, upon landing in New York, he attempted a stroke of spinmeistership by decreeing that all White Star ships would immediately have lifeboat capacity for all aboard.

The world needed no instruction from Bruce Ismay about lifeboat capacity. The deficiency was so great and so obvious that boats-for-all-people rules immediately prevailed. Regulations promulgated by SOLAS 1914 addressed the question most directly and unambiguously in Chapter VI, Article 40 of its Convention:

> At no moment of its voyage may a vessel have on board a total number of persons greater than that for whom accommodation is provided in the lifeboats and the pontoon[6] life rafts on board.

Although the Convention was never to be ratified by the necessary number of maritime nations, national laws were quickly passed or amended to this end, and typically even before the SOLAS delegates convened in November 1913. Steamship companies who found themselves deficient rushed to rectify their deficiency even earlier.

As Andrews, Carlisle, and Welin demonstrated, the *Titanic could* have carried enough boats. It was *designed* to carry enough lifeboats, as evidenced by the quotation from *Engineering*, July 1, 1910, that follows this paragraph. The article concerns the use of Welin davits (see Figure 27-2), and by the way, accurately reproduced in the 1997 movie, *Ti-*

tanic); that constitutes the "new system" mentioned in the first sentence in the pertinent paragraph, quoted following.

> A more extensive application of the new system is being made on the White Star liners *Titanic* and *Olympic*, now in course of construction at the yards of Messrs. Harland & Wolff, Belfast. These vessels are to be fitted throughout with double-banked boats [a prediction that proved false, as noted above] operated by the new Welin gear. A plan of part of the boat deck of these new vessels is shown in Fig. 8, while Fig. 7 [reproduced here as Figure 27-3] gives a section showing the position and handling of the boats. Half the number of boats fitted on these ships will be in excess of the Board of Trade requirements.

Note the term "double banked." That means *two* boats for every pair of lifeboat davits. If applied only to the fourteen 65-person boats, there's 910 more spaces in boats. If the intentions indicated by the quotation had been carried out, there would have been space enough for every passenger, and with only modest overloading, space for all the crew as well. Ben Guggenheim . . . the Strausses . . . John Jacob Astor . . . every immigrant from steerage . . . all debark from the *Carpathia* in New York, feathers ruffled but alive and praiseful of Captain Smith. What about Captain Smith? He appears on the front page of the *Times* in the act of thanking Chief Engineer Bell for keeping the lights burning until the last soul was safely settled in a boat.

Now, a reader may have noted a discrepancy in the numbers, while the doubled boats could have saved all aboard on the maiden voyage, they still would have been insufficient for the ship's capacity of 3,547 persons. However, the davits actually fitted also could serve yet another row of boats, as Figure 27-3 shows. Soon after the loss of the *Titanic*, the *Olympic* was given that lifeboat capacity, as mentioned earlier.

The boats could have been there, but were not. Within the scope of short-term hindsight, the decision to hew to the minimum legal number was indefensible, and still is. On the other hand, with long-range hindsight now in hand, and mourning for the victims long past, the anti-lifeboat arguments that were offered in 1910 by the Board of Trade's Advisory Committee can regain an aura of validity. Those arguments have been listed earlier in this chapter; some amplification follows.

In most[7] ship disasters, at least some lifeboats become useless. In very few instances does a ship sink in a calm sea in such a slow and dignified (no significant list) manner that everyone may readily step into a lifeboat and row away, bereft only of dignity and baggage. The counter examples are many; the *Lusitania* in 1915 and the *Andrea Doria* in 1956

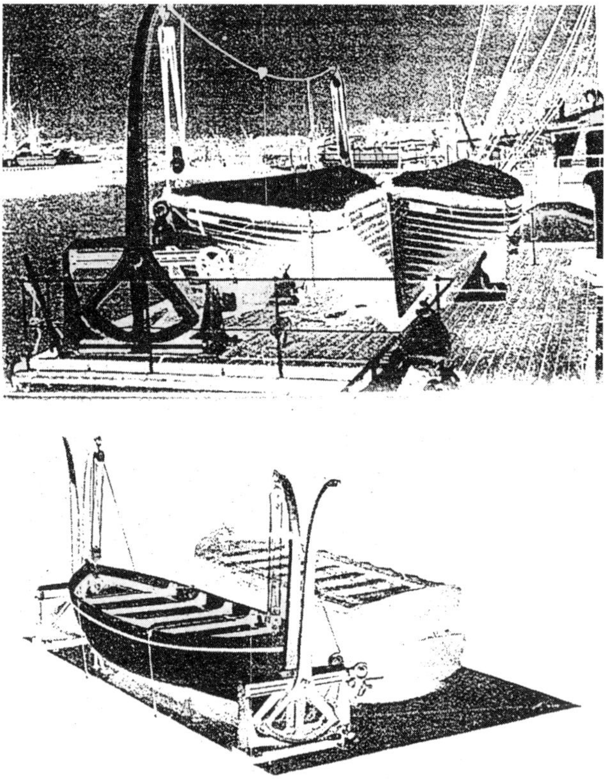

FIGURE 27-2 The Welin quadrant davits as fitted to the *Titanic*. A worm gear actuated by the hand crank swings the boat out. The boat could then be lowered manually by paying out the falls simultaneously at both ends. For lifting the lifeboat, the falls were led to powered winches (not shown) that were mounted on the Boat Deck. Stacking of two boats was possible under the same davits as shown. (*From* Engineering, *vol. 88, 1910*)

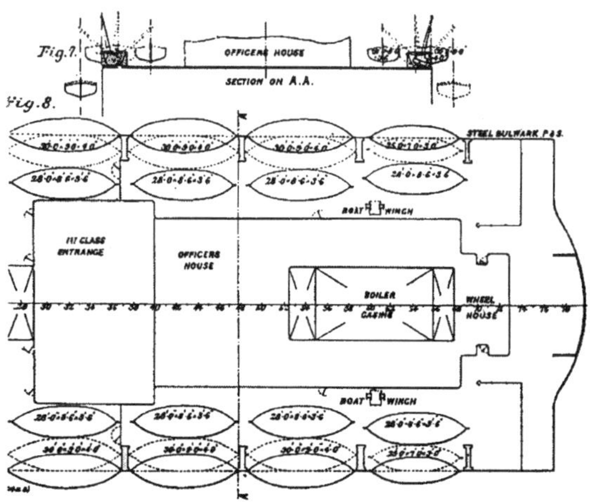

FIGURE 27-3 Partial plan of the *Titanic* Boat Deck, showing the two (or three, dotted) rows of lifeboats proposed for that ship by Axel Welin. (*From* Engineering, *vol. 88, 1910*)

are other doleful cases. Factors such as rough seas, severe lists, and rapid sinking seem to predominate. Though after the *Titanic* no one voiced opposition to a full complement of boats, the opinion that improved watertight integrity was a more viable route to safety than increased lifeboats was still supported. Edward Wilding of Harland & Wolff, in testifying before the Mersey hearings, was questioned on this point, and answered that he strongly favored emphasizing watertight integrity. His answer was probably prompted by the philosophy that the safest place to be was the ship itself. George Dickie was an American professional ship designer whose *Congress*, a passenger liner for the U.S. West Coast, was building at the time of the *Titanic* sinking. A post-*Titanic* paper[8] that he presented to the Society of Naval Architects and Marine Engineers contains the statement that he doubted the viability of lifeboats in the rough seas that the *Congress* would be likely to encounter in the Pacific north of California. His paper, therefore, pursues the idea of unsinkability. The paper has been given some attention in Chapter 21.

But neither Dickie nor Wilding nor anyone else advocated leaving off the lifeboats, regardless of the degree of "unsinkability" that could be obtained. Everyone demanded an adequate number of boats, and the boats quickly appeared.

The White Star Line was probably the leader in adding lifeboats since its *Olympic* was the most obvious target for improvement, being essentially identical to the lost ship. Fortunately, space was available to add the necessary boats as soon as they could be acquired. The number was doubled for the first post-*Titanic* voyage, then had to be tripled when the *Olympic* was taken out of service the following September. This was accomplished by stacking boats two-high under the davits, then mounting one on deck close alongside the stacked pair. The Boat Deck was wide enough to accommodate the boats on deck without unacceptable obstruction, and apparently the ship had enough stability margin that the added high weight was acceptable.

There appears to be a defect in the *Olympic* remedy. Since there is no mention of additional davits, launching of all boats would have taken three times as long as it did on *Titanic*. The actual time might be multiplied by more than three because the boat falls would have to be recovered from a launched boat before the next one could be handled. That may not appear to be a major complication, but it could be if the sinking time was that of the *Britannic*. Panic would be likely if a ship was listing, like the *Lusitania*, or burning like the *Morro Castle* in September 1934. Indeed, Edward Wilding estimated a half-hour cycle time for multiple boats per davit, i.e., the time from beginning to lower the first boat to beginning to lower the second would be a half hour. That would make one hour for the third boat to wait. Would there have been time enough to save those who gallantly stepped aside to let the women and children go first? Perhaps, if the ship sank over a period of time as did the *Titanic* and the passengers recognized the danger that the ship would sink. The first lifeboat was reportedly launched at 0045. An hour later the ship was still floating, and without trim or list sufficient to preclude boat handling. The last boat was lowered at about 0205.

If, in addition to the calm sea and the near-ideal behavior of the ship as it sank, we can posit near-ideal behavior of its people, then two (or even three) boats per davit-pair might indeed have saved everyone. As events showed, the time it took to convince the passengers that they must leave the ship, the lack of a lifeboat drill, and possible lifeboat accidents such as those of boats 11 and 13, cited in Chapter 13, contributed a significant loss of life.

An obvious response to such doubts is, "Well, you can add more boat-handling gear, and/or develop methods of launching boats more quickly," and addition of more davits to the *Olympic* might appear feasible. A glance at the first-class promenade on Boat Deck shows room for eight more davit-pairs. If the davits for the two 40-person boats are counted, the addition of eight would give 24 pairs, which is two less than what SOLAS 1914 was to require. Adding the additional davits would be accomplished when the *Olympic* was fitted for troopship duties during World War I.

The number of boats was unambiguously stated by SOLAS 1914, and many design features of the boats were specified, but a stipulation that each boat have its own launching gear was *not* included. The pertinent regulation on the gear contains the tabulation in Table 27–1.[9]

The requirements of SOLAS 1914 would have required the *Titanic*, with a registered length of 852.5 feet, to have 26 pairs of davits instead of the 16 (including the two for the 40-person boats) actually fitted.

The *Titanic* lifeboats were rated for 65 persons. Twenty-four lifeboats of 65 persons each gives a 1,560 person capacity; two boats per davit-pair gives 3,120. Multiple boats per davit-pair thus were expected to accommodate the 3,547 persons that could be aboard; four of the davit-pairs would have to handle three boats each.

The SOLAS delegates may have reasoned that lifeboats are a significant safety feature only when a ship is sinking slowly and on an even keel, as the *Titanic* did, and in that case, there will be time for repeated use of launching gear. If the ship is sinking rapidly or listing, no amount of boat and launching gear will help. Indeed, as more experience presented itself, especially from the unprecedented number of non-military ships that were sunk in two great wars, escape gear turned more toward life rafts that could simply be released to slide themselves into the water, or even to float off as the ship sank under them. By the late century, the unobtrusive inflatable life raft had become an essential lifesaving resource: in World War II often the only resource, later at least an important backup to conventional boats. Even that life raft expedient, so much more reliable in a sudden emergency than the rigid lifeboat, couldn't do it all, or even much, in some cases. Consider the 859 persons who drowned with the passenger ferry *Estonia* in September 1994.

The ability of launching gear to perform adequately through several cycles of launching might have been a matter of concern. Part of the *Titanic* lore that has appeared in many venues concerns the difficulty of one of the boats in releasing from its gear, requiring a seaman to hack at the lowering tackle with a knife.[10,11] What if another boat had been waiting on deck to be lowered by the same gear? This concern was bruited in 1912, and some remedial modifications to gear were suggested, but SOLAS 1914 offered no regulations in this regard.

Our own conclusion, looking back with a century's worth of hind-

TABLE 27–1 *Number of Boats and Davit Sets by Registered Length*

Registered Length of Ship (ft)	Minimum Number of Davit Sets	Minimum Number of Boats of the First Class*
700 and less than 730	22	15
730 and less than 760	22	15
760 and less than 790	24	17
790 and less than 820	24	17
820 and less than 855	26	18
855 and less than 890	26	18
890 and less than 925	28	19
925 and less than 960	28	19
960 and less than 995	30	20
995 and less than 1,030	30	20

*A first-class boat was defined elsewhere as one having entirely rigid sides, i.e., what we might call a "regular boat." A boat with collapsible sides, such as the four "collapsibles" of the *Titanic* sinking, were boats of the second class.

sight that was obviously unavailable to SOLAS or to the *Titanic* hearings must be the following. Even if the ship had been fitted with "enough boats for all," as proposed by Axel Welin, several hundred persons still would have perished. This would be due to confusion, inadequate training in handling the boats, no lifeboat drill, crew unfamiliar with their lifeboat assignments, and accidents in lowering or setting up subsequent lifeboats on the davits. One other factor was the passengers who would not seek the safety of lifeboats thinking that the unsinkable ship would not sink.

An independent opinion, generally supporting the conclusion just stated, was submitted to the authors by John Bedford of Harland & Wolff. It is reproduced verbatim in Appendix 5.

From the evidence, it appears that the order of women and children first was applied rigorously on the port side, but to a lesser degree on the starboard side. After the order to swing out the boats at 0020 until 0130, four boats with approximately 110 people were launched from the starboard side and three boats with 84 people from the port side. Ignoring the two collapsibles that floated off the ship, a total of five boats with 259 persons were launched in 35 minutes on the starboard side and six boats with 235 people were launched in 40 minutes to port. The order of women and children first did slow the loading of the boats on the port side and also resulted in 18% fewer people being loaded on the port side than to starboard. It is clear also that in the first 70 minutes that passengers were reluctant to board the lifeboats as the boats that were launched were only 45% filled, while over the last 40 minutes the lifeboats were 75.5% filled. The latter reflects the fact that when the forecastle started to submerge, the passengers realized that the ship was really going to sink. Even with more lifeboats there still would have been a great loss of life. If two boats had been provided at each lifeboat location, with the exception of the cutters, then perhaps 300 more lives might have been saved.

Consider the many empty spaces in *Titanic*'s lifeboats, spaces that alone might have saved 400 more people. Second Officer Lightoller was mostly in charge of launching some of the port-side lifeboats, and none of those he supervised was fully loaded. In fact, the lifeboats on the port side were on the average only 59% filled, while those to starboard were 68%. Lightoller was questioned about this at both the American and British hearings, but boldly answered the implied criticism by saying he had feared that the boats, if fully loaded, would collapse from the weight during the launching procedure. At the Mersey hearings, Edward Wilding responded, "Officers who gave evidence of danger of boats breaking were in error," and backed his comment with an account of the *Olympic* lifeboat test made before the *Titanic* sinking. A boat was loaded with the weight equivalent of the capacity of 65 persons, then lowered and raised six times without evidence of any structural distress.[12] He went on to say that he thought the *Titanic*'s officers should have known of this test. Their unfamiliarity with such information can reasonably be charged to White Star's omission of normal testing and familiarization in its haste to place the *Titanic* in service.

The Britannic *Lifeboats.* Completion of White Star's third sister *Britannic* was delayed by changes to design that appeared advisable to the company after the *Titanic* disaster. The inner skin fitted to the *Olympic* became part of the *Britannic* design, though with the ship still in the early stages of construction, it was feasible to make the hull 18 inches wider. A cross section of the hull, showing the inner skin, is given in Chapter 23. The lifeboat outfit received a much more radical upgrading, as shown in Figures 27-4, 5, and 6 in this chapter.

The planned *Britannic* outfit consisted of eight sets of gantry-type davits (see Figure 27-4), each handling six boats, but the davits were far more sophisticated than the *Titanic* outfit, a circumstance that the illustrations make abundantly clear. Among things not obvious are the ability of the davits to reach far enough out to lower boats with as much as 15 degrees list on the ship, and an adjustment between fore and after davits that allowed a boat to be lowered in a level attitude even with severe trim. Each set of davits had its own electrically powered winch, unlike the *Titanic* outfit of only four powered winches to be shared among 14 davit-pairs.[13] Even further, all boats could be lifted to the opposite side of the ship for launching, just in case a list was so severe that launching could be done only on one side.

Because of the need to place the *Britannic* into service as a hospital ship, the lifeboat outfit on the completed ship included only five sets of the gantry davits. This was supplemented by six sets of conventional Welin radial davits amidships on either side of the Boat Deck (see Figure 27-5), and two sets of these davits on the Shade Deck (see Figure 27-6). In all, the outfit of boats consisted of 45 34-foot boats, one 26-foot cutter, and two 26-foot boats.

How could the *Britannic* be equipped with only 11 crane-or-davit sets when SOLAS 1914 called for a minimum of 26 on a ship of this size? One answer is that the 1914 SOLAS regulations had not been ratified, and indeed, never would be. Further, those regulations permitted the number of davits to be reduced if it could be demonstrated that all boats could be launched within a certain minimum time. The minimum times were not specified, since each government was to decide for its ships what the time should be. Presumably the *Britannic* outfit would be able to meet that requirement, especially given White Star's drive to make its ships appear super safe.

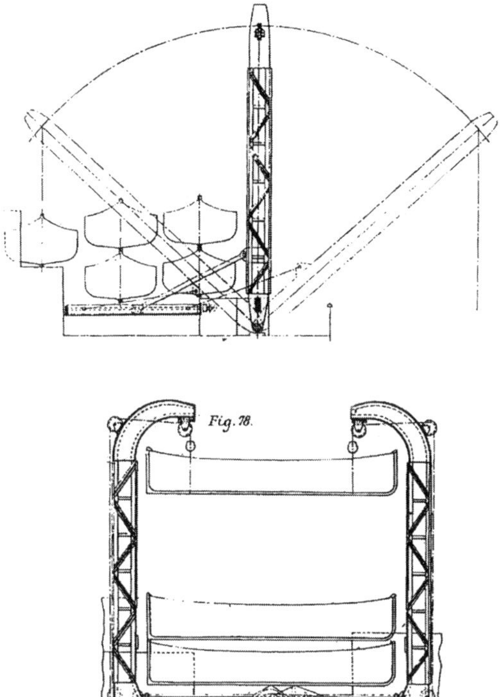

FIGURE 27-4 The Gantry Crane Lifeboat davits similar to those installed on the *Britannic*.

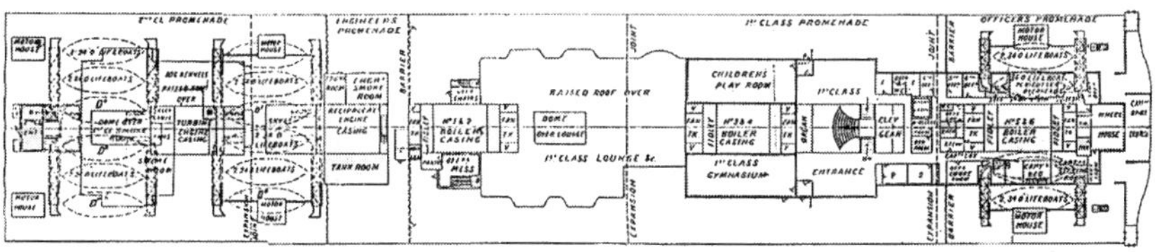

FIGURE 27-5 Plan of the *Britannic*'s Boat Deck, illustrating the radical improvement in boat outfit that White Star gave its first new liner after the *Titanic* disaster. (*From* Engineering, *vol. 94, 1910*)

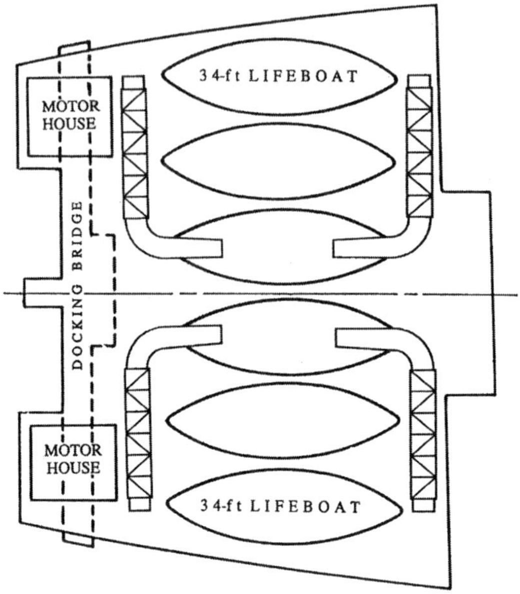

FIGURE 27-6 Sketch of the lifeboat installation on the Shade Deck (after house) of the *Britannic*. The gantries are shown schematically. (*Sketch by the authors*)

A glance at the profile and deck plans of the *Britannic* shows that its solution to the lifeboat problem did not occupy all of the deck space.

The *Britannic* sank, of course, and in circumstances that in some ways duplicated those of the *Titanic* event. The sea state and wind did not hamper launching of boats, nor endanger them when afloat. The gantry crane davits allowed a rapid evacuation of the ship, considering that the ship sank in 55 minutes. *Britannic* also took a severe starboard list, and eventually capsized, but by that time all the necessary boats had been successfully launched. Fortunately, only the crew and hospital staff were aboard; *Titanic* circumstances were different. Only one third of the persons aboard found places in lifeboats. With the *Britannic*, only one third of the lifeboat capacity was needed for the persons aboard.

The *Britannic*'s boats and handling gear did their jobs. But would they have saved everyone from a *Titanic* with all 3,547 persons aboard? Indeed, would they have saved everyone from the *Britannic* if she had her full complement aboard, even if none had been immobile hospital patients? The answer *should* be positive for either question, but one can readily argue the negative for both, based on the need for each set of gear to launch multiple boats. As it went with the *Britannic*, two boats on the port side drifted into a still-turning propeller that had been lifted from the water by the ship's list to starboard. Thirty people were killed, and 41 seriously injured. And as it went, the sinking took place in 55 minutes, roughly one third of *Titanic*'s death throes. If 3,000 patients had been aboard, would there have been time to evacuate everyone before the list became just too great?

The location of the *Titanic* lifeboats—all high and near amidships—made access for the second- and third-class passengers difficult, and hence was a feature criticized after the sinking. The *Britannic*'s arrangement of some boats on the Poop Deck may have been the answer to this criticism. If it was, this deck was the worst possible place with respect to the propeller hazard. Of course, a ship's engines should not be turning when its stern is lifting out of the water. The thought underlying these remarks is that the post-*Titanic* safety upgrades given the other two White Star ships were simply over-responses to an accident that had an extremely low probability of a repeat occurrence.

The Cunard Lifeboats. The *Lusitania* sank three years after the *Titanic*. The lifeboat lesson had been learned, and international authorities had made their declaration in 1914 for boats to save all those aboard, as well as giving details of the lifesaving outfit.[14] However, the rapid sinking of that ship, exacerbated by extreme list, made lifeboats almost useless to the unfortunate passengers and crew. Complicating matters further, there had been no lifeboat drill for passengers and crew.

Reviewing the building of the *Lusitania* (and her sister *Mauretania*), we find an outfit not significantly different from that of the as-built *Olympics*. Published deck plans show 16 boats (see Figure 27-7), and pre-*Titanic* photographs show single boats, not stacks of two or more. If these were 65-person boats like those of the *Olympics*, then the total capacity was 1,040 persons. The ship's passenger capacity alone was approximately 2,200. When the *Lusitania* sank, she quickly took a list of over 20 degrees to starboard that negated the launching of her port

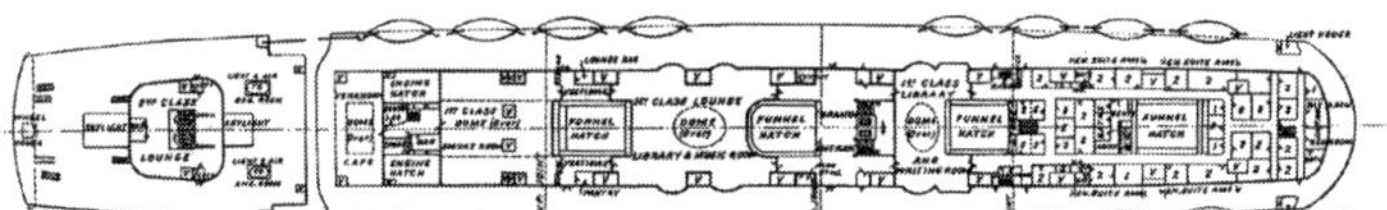

FIGURE 27-7 Plan of the *Mauretania* boat deck, showing only 16 boats. (*From* Engineering, *vol. 84, 1907*)

lifeboats. Further, she went down in 18 minutes, a time far too short for filling and launching all of her boats, even under the most favorable circumstances.

Another consideration in this tragedy was the fact that most of the able-bodied seamen familiar with the preparation of the lifeboats for lowering were in the baggage rooms when the torpedo hit. Most of those in the forward baggage room were killed from the effects of the torpedo hit or second explosion. There were snubbing chains to attached to each of the wooden lifeboats to prevent them from swinging inboard and damaging the collapsibles stowed beneath each of them. A small wooden hammer was provided in each boat to detach the chains, but most of those attempting to launch the wooden lifeboats were unfamiliar with the procedure.

Notes for Chapter 27

1. The boats saved only about one third of those aboard the *Titanic*, but if every boat had been filled to its rated capacity, the figure would be close to one half.
2. The word "collapsible" refers to the sides of the boats. The sides could be lowered or folded downward to allow more compact stowage.
3. Note: Rules for ships larger than 10,000 tons. A sampling of White Star liners shows how tonnage was increasing rapidly past 1891:

Majestic	1890	9,965 tons
Cymric	1898	13,096 tons
Oceanic	1899	17,272 tons
Adriatic	1907	24,541 tons

   Source is Roy Anderson, 1964, *White Star*, T. Stephenson & Sons, Prescot, Lancashire, England.
4. Lifeboats are expensive in ways more than just their direct cost. For example, they may require wider promenades so that passengers won't be crowded by their presence. Their weight high in the ship requires compensation for degraded stability, e.g., a wider hull.
5. Never refuted? What the *Titanic* tragedy can be said to establish is the principle that passengers and crew always be provided with the *option* of taking to boats when other means of salvation fail.
6. Pontoon life rafts. The pontoon was a watertight metal cylinder that gave buoyancy to the raft. Inflatable rafts have long since replaced this type of lifesaving gear.
7. The evidence for "most" is only anecdotal.
8. George Dickie's paper outlining the design features of an unsinkable ship was published in the 1913 *Transactions*, Volume 21, of the Society of Naval Architects and Marine Engineers.
9. The davits' regulation in which this rule is found occurs in Article XLI. The full text includes smaller ships, and also states lengths in meters.
10. Fouled boat falls. This incident appeared in the 1953 movie, *Titanic*, but had one of the male principals sliding down the boat falls to untangle the fouled lines.

11. Edward Wilding spoke of his concern for fouled boat falls at the Mersey hearings.
12. Recall from Chapter 14 that several of the *Titanic* boats were loaded beyond their rated capacity.
13. SOLAS 1914 had not been ratified (and never was) by the time of the *Lusitania* incident, but the lifeboats-for-all principle didn't need official sanction.
14. McAuley, The Liners: Motorbooks International, 1997, pp. 49–50.

CHAPTER 28

Titanic's Steering

If Captain Edward Smith had ordered a slightly different course for the night of April 14–15, his ship might have passed safely south of the area of ice hazard that several wireless messages had identified. If he had ordered a significant speed reduction that night, the collision, if it had occurred at all, might have produced damage so insignificant that his ship would have survived. If he had felt no imperatives for a scheduled arrival time in New York, he could have ordered that slower speed. There are many such *ifs* in the voyage of the *Titanic* in the final hour to the fatal encounter. These alternatives have been discussed in the extensive *Titanic* literature.

In a similar theme, *ifs* haunted the memory of First Officer William Murdoch. This was the man who looked for a moment of dismay from the starboard bridge wing, then rushed to give the helm and engine orders that failed to prevent the collision. These stories also have been the grist of earlier volumes, with some judgments passed. Perhaps he should have steered the ship head-on into the iceberg, perhaps he should have ordered reversal of the port wing shaft while maintaining full thrust on the starboard (and presumably centerline) shaft, or perhaps the case was hopeless no matter what he did.

These points of discussion and uncertainty are within the scope of our analyses. To some extent they involve matters of seamanship that already were discussed in Chapter 12. To a much greater extent, they involve the underlying characteristics of the ship. Some earlier authors have bruited criticisms of those characteristics. We believe that the criticisms require discussion, and in some cases, refutation.

The Event. The *Titanic* was steaming at 22.5 knots on a course approximately WSW according to the testimony of Helmsman, Quartermaster Robert Hitchens, through a millpond-calm sea at a point in the Atlantic that can be identified only by the coordinates 50014'W, 41046'N reported by Fourth Officer Boxhall. With midnight approaching and with no moon to help the sailor's eye, visibility was not safe for such a speed by a ship so large, and in an area known to be dangerous with ice.

The location of the ice field that had been reported by ships throughout 14 April 1912 were plotted on the chart for the area at hand. From the plots, the officer on watch before 1000, Second Officer Lightoller, expected to see ice no later than 2300.

First Officer Murdoch took the watch from Lightoller. As the fatal moment neared, he had stationed himself on the starboard bridge wing, leaving Sixth Officer James Moody on duty in the wheelhouse. At 2340 local time, lookouts Fleet and Lee, located forward and high in the Crow's Nest, sighted an iceberg dead ahead. They struck the bell three times, the signal for "object ahead," then informed Murdoch of the presence of an iceberg via telephone. After a quick look, Murdoch ordered "starboard helm" (which would produce a turn to port), and then rang "full astern" on the engine order telegraph.

Though the engineers sprang to obey the astern order, their frantic efforts had little effect in the half minute that remained before the impact; the ship apparently was still moving forward at a speed of approximately 22 knots at the fatal moment. The helm order was similarly fruitless, for the ship responded very slowly, turning only about 22.5 degrees to port before the impact—not quite enough to avoid the underwater mass of the iceberg.

The Voice of the Critics. Not quite enough, but is that phrase tantamount to "unsafe at any speed?" A later-day critic[1] charged that the *Titanic* had inadequate maneuverability. Indeed, it was if the criterion for maneuverability is a high probability of avoiding collisions with icebergs while steaming at near maximum speed when visibility is inadequate for spotting such objects. On the other hand, if the appropriate criterion is the ability to avoid collision while being operated by prudent seamen at speeds suitable for the circumstances, then the defense can argue otherwise.

How about the *Lusitania* and *Mauretania* in a similar situation? Consider the *Lusitania* in her encounter with the torpedo. That was similar to some degree, since ships have been known to evade torpedoes that have been detected in time. On the other hand, consider that the torpedo in this case was not detected at the moment of its firing, and that there were inevitable time lapses between a lookout's first spotting of a torpedo wake and the consequent realization of its danger by the Bridge personnel. With those thoughts in hand, we must conclude that the maneuverability of the victim played no part in the loss of the ship and many lives. But suppose that on a dark night either the *Lusitania* or *Mauretania* lookouts had sighted an iceberg a half minute ahead. Could either ship have avoided the hazard? If the answer is positive, then does that conclusion support an assertion that the *Titanic* and her sisters were deficient in maneuvering ability?

The Mersey report had very little to say about maneuverability. However, the final report described a test of the *Olympic* in a simulated encounter with a fixed object. That ship was steamed in the open sea at 74 revolutions, a speed of 21.75 knots. In an emergency turn, the ship turned 22.5 degrees and traveled 466 yards over a time-span estimated to equal that available to the *Titanic* on the fatal night.

The SOLAS 1914 regulations have only the enigmatic "vessels shall have sufficient power for going astern," a stipulation in Article XXII. Could those regulations have been asking for backing power sufficient to stop a ship of *Titanic* size and speed in the distance it was allowed? Almost certainly not, because every SOLAS delegate must have known that no ship so large, steaming at a speed of 20 knots, could stop in so short a distance. Further, they also realized that turning rather than backing the engines was the preferred way of avoiding open-sea collisions. They obviously found no reason to codify maneuvering standards for passenger ships, a step that would proba bly have implied criticism of the *Titanic*.

The "unsafe" critic, writing in 1972, epitomizes his judgment in one short sentence: "Such a rudder was inappropriate for so large a vessel." The *Titanic* was fitted with a plate-type rudder that was similar to that used in ships before her. It would be superceded in ships built in the 1920s and later by rudders of better hydrodynamic cross section. The implied connection between size of ship and type of rudder can be shrugged off as not pertinent, though the belief that the rudder wasn't big enough for its type also can be inferred from the critic's statement.

Regardless of the particular critic's intent, others have advanced the notion that First Officer Murdoch should not have attempted a hopeless maneuver to avoid the iceberg. Instead, he should have steamed straight on into the iceberg. Their rationale is that a collision of that nature could not possibly have caused damage sufficient to cause her loss. Their model for such a maneuver was the experience of the liner *Arizona* in 1879.

The *Arizona* was a ship of 5,000 gross tons and hit an iceberg head-on at her full speed of 15.5 knots. She survived, despite crushing 20 feet of her bow, because subsequent flooding was limited by her intact collision bulkhead. She was able to make her way to St. Johns, Newfoundland by going astern. The suggestion that the *Titanic* could have faired as well has been both supported and condemned by the *Titanic* mavens ever since the circumstances of her demise became widely known. At the British Inquiry, Naval Architect Edward Wilding stated that in her collision with an iceberg, the *Titanic* would have had 80–100 feet of her bow crushed by the collision. He reasoned that the damage would not have progressed beyond Cargo Hold No. 1. There would have been heavy casualties in the crew quarters and possibly in the forward steerage accommodations. Two or possibly three compartments would have flooded, but the ship would have remained afloat.

Joseph Conrad,[2] the intemperate critic whose comments have provided grist for Chapter 19, railed against the revisionist notion of hitting the iceberg head on. He wrote that if such a maneuver were to be favored, then those trained to command at sea should be taught to avoid collisions by steering ships *into* obstacles. He also was critical of the *Titanic*'s large size. Size was probably the most important factor in the ship's maneuverability, but Conrad did not explicitly bring out that point. What judgment can be made by those who come forward with a near-century of hindsight, as well as knowledge of analysis techniques that were unknown to naval architects at the time, or to the authorities who deliberated at the SOLAS conference in 1914? Before refuting these critics two caveats need to be introduced. The first is the remark that a definitive analysis of the *Titanic*'s maneuvering ability cannot be made because the necessary data does not exist or cannot be found. The second concerns the subjective climate that would cloud even the most precise determination of maneuvering characteristics. Arguments would surely erupt over the question of whether a precisely determined turning circle diameter or other maneuvering characteristic was "good" or "poor."

Lacking the means of providing the inputs to such arguments, we shall rely on comparisons that can be made between the *Olympic* ships and other liners in the same service. Before doing so, let us arm ourselves by discussing the factors that affect the maneuverability of ships operating at sea speed.

An Introduction to Ship Maneuverability. Many readers are likely to have a familiarity with marine steering because of their experiences with small boats, especially small motorboats. Just as with an automobile, the steering wheel is turned counterclockwise to turn the boat to the left, and the boat responds immediately much as the automobile does. The same can be said for a sailboat, though if she is fitted with a tiller, "starboard helm" is applied to turn to port. There are many similarities between small boats and larger ones, but there are differences also. Because those differences are significant, it may be best to start with the fundamentals of steering and maneuverability that illuminate the differences.

As a rudder moves through the water, it develops a resisting force, or drag, which might be considered a necessary evil, necessary because the rudder is needed for steering, and evil because the drag must be overcome by the engines in propelling the ship. The magnitude of the drag can be estimated from standard hydrodynamic or aerodynamic data on the cross sections of foils. If the rudder is turned to have a non-zero angle of attack with respect to the incoming water direction, a side force, or lift, also is developed. Its magnitude can be estimated from similar hydrodynamic or aerodynamic data. The lift, nominally acting at right angles to the ship centerline, is, of course, the force that causes the ship to turn.

The rudder force causes the ship to turn because it produces a torque around a turning center somewhere near the mid-length of the hull. It also causes the ship to move sideways at first in order to develop a hydrodynamic force to oppose the side-ways component of the force (the friction between tire and road provide this force for the turning automobile). Figure 28-1 shows turning circles, but also illustrate the initial side-ways movement of the ship. Note that this movement is in a direction opposite to the direction of the turn. For this reason, a ship alongside a pier cannot simply drive away in the manner of an automobile driving away from the curb against which it is parked. The rear end of the car obediently follows the front, but the stern of the ship initially goes the *other* way, to scrape and bump along the pier. The ship must first get her stern away from the pier, or must have a thruster at the bow to push that end out without using the rudder that will push the stern inward. This movement of the hull (especially at its stern) opposite the direction of turn can be demonstrated by the act of driving even a small boat away from a pier as if she were an automobile. The scrape and bump at the stern follow.

As the hull turns under rudder action, it also presents a non-zero angle of attack to the surrounding water, and thereby develops a side force and a torque just as the rudder does. The hull thereby augments the torque developed by the rudder. Above all, the principles handed down to us by Sir Isaac Newton come into play: the forces and moments develop accelerations which in turn develop opposing inertia forces and moments in accord with his second law.[3]

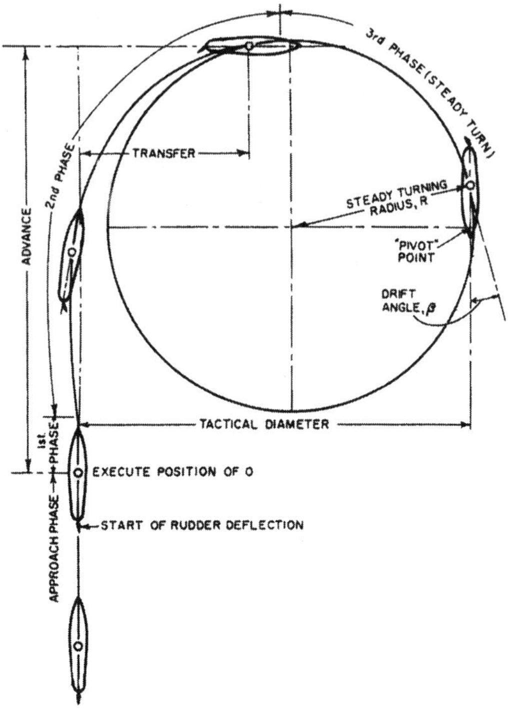

FIGURE 28-1 This is a typical path followed by a ship with a fixed rudder deflection. (*From Steering and Maneuverability,* Principles of Naval Architecture, *third edition, Society of Naval Architects and Marine Engineers*).

The balancing of forces and moments produces two equations for turning ships[4]

$$\text{Moment balance: } (I_Z - N_{\dot{r}})\dot{r} - N_v v - N_r r = N_\delta \delta_R$$
$$\text{Side force balance: } (\Delta - Y_{\dot{v}})\dot{v} - Y_v v - (Y_r - \Delta)r = Y_\delta \delta_R$$

The unknowns are the *yaw* rate-of-change (symbol r), yaw being the angle by which the ship deviates from the original straight-ahead direction, and v, rate-of-change of the sideways velocity, or *sway* rate. The superimposed dot indicates first derivative with respect to time (rate of change of r and of v). The terms in parentheses in the first term of each equation characterize the inertia of the ship (e.g., Δ symbolizes the displacement—or weight—of the ship expressed in mass units). The remaining symbols are coefficients that relate v and r to the moments and forces (e.g., N_d multiplies the rudder angle δ_R to produce the rudder torque; Y_d multiplied by δ_R is the sideways force developed by the rudder).

We shall make no attempt at a solution of these equations; their purpose here is merely to demonstrate two significant points. The first is the presence of inertia properties, i.e., the parenthetized coefficients of the respective first terms. The magnitudes of these terms are near-negligible for the small boat, and in consequence the owner of a small motorboat may be unaware of the delay in the beginning of a course change after the steering wheel is turned. A ship the size of the *Titanic* may have a mass on the order of 10 times^5 that of her miniscule cousin. If a maneuver to avoid an iceberg, say, is suddenly demanded, then initiation of a perceptible turn is surely going to seem agonizingly slow.

The second significant point is the presence of the N and Y factors whose numerical values are essential prerequisites to solutions of the equations. These values are much more difficult to estimate from the principal characteristics (e.g., length, beam, draft) than, for instance, displacement or propulsive power required for a specified straight-ahead speed. Values of any accuracy require special maneuvering tests of the ship or of its model. Such tests were unknown in the early 20th century, and hence were not performed during the time of the *Titanic*. The upshot is that we have no accurate way of using later-day design and analysis techniques to pass judgment on the ability of that ship to respond to a collision-avoidance emergency. Of course, models could yet be built from data presumably in the archives of the builder, then tested in a model basin equipped with the appropriate facilities. However, such possibilities, if undertaken at all, had not come to fruition at the time of this writing.

Solutions of the two equations would show rates of change of r and v. If a solution for a large ship were carried out for a sufficient length of time, perhaps as long as a minute, with the rudder held at a constant angle, these rates would become zero. A zero rate signifies that the ship has settled into a steady turn of fixed radius, i.e., into its steady turning circle. Note that Figure 28-1 shows not only the turning circle, but the transition from a straight course to a circle, and illustrate the meanings of *tactical diameter*, *transfer*, and *advance*. The transition happens during that first minute or so, and probably is of prime significance for the *Titanic,* given that only about 37 seconds elapsed between sighting and impact.

The magnitude of each of the parameters defined Figure 28-1 can be significant in the appropriate situation, but the diameter of the turning circle is usually regarded as the principal measure of "maneuverability" for a ship operating at sea speed, and with unrestricted sea room. The implication is that comparing this property between ships may be misleading when the event of concern occurred *before* the steady-state turning circle was established. However, it can be assumed that a ship with a smaller turning circle radius will pass through the transition to steady turning in a shorter time.

If the *Titanic* had been given time enough to get well into her steady turning circle, she surely would have missed the iceberg. Now, there's no pretense that Figure 28-1 is to scale for the *Titanic* or any other ship, but note the possibility that this remark suggests: that the initial movement to starboard—*toward* the iceberg—was just about cancelled by the beginning of the turn to port. Perhaps, we can speculate, but have no definite way of knowing.

When a ship is in a steady turn, the derivatives of r and v in the preceding equations have become zero. The equations then serve as the basis for constructing a formula[5] for the turning radius R, namely

$$R = \frac{-L}{\delta_R}\left[\frac{Y_V N_r - N_V\left(Y_r - \Delta\right)}{Y_V N_\delta - N_V Y_\delta}\right]$$

The symbol L represents the length of the ship; the other symbols were defined for the earlier equations. From this new formula we, therefore, can see, regardless of our ability to supply the numerical values for the other symbols, that short ship lengths tend to produce smaller radii. With identical rudder angles, the *Mauretania* with a waterline length of 760 feet is thus likely to be "more maneuverable" than the *Titanic* at 850 feet. Unless the Y and N quantities are greatly different—which we have no means of confirming—that conclusion will have to be accepted. Such an admission seems to play into the hands of the critics who say that the *Titanic* was "too big," and it is then but a small additional step to say that yes, the ship was too great a leap forward, and hence a challenge to the fates. We shall leave such speculation in other hands, but will attack the "too big" canard in the following paragraphs.

Comparisons. The turning radius of the *Mauretania* has been reported to have been 3.75 ship lengths,[6] or 2,850 feet, as measured on her trials at about 26 knots. The *Titanic* was presumably subjected to maneuvering tests on her trials, conducted on 2 April 1912, though quantitative results have never been published, and are apparently unobtainable. Thus, a direct comparison with the Cunard ships cannot be made except noting that the equation, cited two paragraphs previous, strongly suggests that they could have turned in a shorter radius than White Star's *Olympic*-class ships.

The best comparative approach to the *Titanic* turning ability is to cite the performance of her essentially identical sister, the *Olympic*. That ship, of course, had a much longer career, with many more opportunities to maneuver into and out of trouble. The first evidence presented comes from a test made soon after the loss of the *Titanic*.[7] (This was mentioned earlier in this chapter.) The ship was put through the same emergency maneuver that had faced her sister when the iceberg loomed ahead. Distance and time required to turn two compass points (22.5 degrees) following a hard-a-starboard helm change were measured, the two points being an estimate provided by surviving Fourth Officer Boxhall of the course change that the *Titanic* had made before impact. The *Olympic* was found to travel about 466 yards in 37 seconds. Unfortunately, the technology of the time did not allow accurate plotting of the track followed in those seconds, leaving us no way to extrapolate the results into magnitude of turning circle, nor to calculate any other parameter. About the only conclusion of value is that the iceberg must have been about 500 yards ahead when it was first sighted. This result is of little help in the present discussion.

Better evidence comes simply from the long and successful career of the *Olympic*. That liner shuttled across the Atlantic for 24 years (with some disruption of service during World War I) until obsolescence and the Great Depression sent her to the shipbreakers. Her maneuverability, whether at sea or in narrow waters, was not a significant factor in her success. Knowing critics will no doubt point to her involvement in three well-publicized collisions.[8] One incident was the intentional ramming of the German submarine *U-103* during World War I. That could be taken as *positive* evidence, though such evidence is impossible to reduce to numerical or geometrical fact. The other three incidents contribute little either way to the question of maneuverability. By implication it can be concluded that the Harland & Wolff Shipyard did not give Captain Edward Smith a faulty ship either.

Even if the critics do not dispute the charge (too fast in the presence of hazard) against the captain of the *Titanic*, they might still claim that the ship was too large for safe handling in an emergency. Indeed, they also might use as evidence for that charge the near-collision with the *New York* and *Oceanic* while undocking at Southampton (Chapter 11). The obvious fact that was illuminated both in port and at sea is that larger phenomena accompany larger vessels. For example, in other venues we find that the 747 requires a longer runway than a DC-3, and so on. No, the *Titanic* was only too large to have a turning circle as tight as that of a smaller ship, but learning to handle progressively larger ships has always been part of industrial-age seafaring. To be sure, once the digital computer had reached sufficient maturity, ship simulators were built to give the humans who were to manage the actual ship a hazard-free way of becoming acquainted with the beasts they were to drive. In 1912 such things were items of the future, and so it was that seamen had to take the next step upward as they steered the ship herself. With prudence and recognition of potential problems, they were successful. In the case at hand, Captain Smith didn't even have to take that step with the *Titanic*, since he had already commanded the *Olympic* for almost a year.

Is there no limit on size, a limit beyond which mere humans are overwhelmed? Is there a "theoretical limit" on ship size, as one critic has suggested? Somewhere, beyond the size at which any ship has been built, there may be such limits. Limits are already imposed by harbor channel widths and depths and by the capacities of shipbuilding facilities. Ships continued to grow larger following the *Titanic* experience. The "supertankers," whose size created much concern around 1970, have had accidents that have caused significant spillage of oil, are an obvious example. If the discussion should be limited to transatlantic liners, the following modest list of ocean liners in Table 28–1 will demonstrate that the *Titanic* did not tower over the ships of the same type that were soon to follow.

The Steering Outfit of the Titanic. The critics, who criticize the size of passenger vessels may point out that the later and larger ships were fitted with "more modern" steering equipment. Thereby, they could make the argument that the real fault in the *Titanic* case was that the size of the ship had increased faster than the capabilities of the machinery with which she was fitted. A specific point, made in the damning article "Unsafe at Any Speed" is that the *Titanic* was fitted with a plate-type rudder instead of the airfoil type that superceded it after a maturing aerodynamics had shown the way.

The plate-type rudder consists of a flat plate of steel, reinforced against side-ways forces by a sufficient number of arms.

The rudder of airfoil shape, later to become the almost universal type, is superior mainly in that it offers less drag. If nothing else, having its reinforcing members internal is beneficial in this way. Of course, the ratio of lift coefficient to drag coefficient is higher, but unlike the wings of aircraft, ship rudders of any kind are mounted close behind an immensely larger body, the hull of the ship. The upshot is that they are not independent foils, but are only trim tabs on the hull, and in consequence, their ability to develop the side force that produces the initial turning torque depends only trivially on their cross-sectional shape.

Regardless of the type, was the *Titanic* rudder too small? Ques-

TABLE 28–1 *Data for Notable Passenger Liners of the 20th Century*

Ship	Entered Service	Length on Waterline (Feet)	Waterline Beam (Feet)	Displacement (Tons)	Speed (Knots)
Titanic	1912	850	92	52,140	23
Aquatania	1914	880	97	49,400	23
Queen Mary	1936	1004	118	77,400	33
Normandie	1935	962	119	66,400	32
United States	1952	940	102	45,400	38.25
France	1962	984	111	51,800	35.56

TABLE 28–2 *Rudder Area as a Function of Length and Draft (From Det Norske Veritas Rules)*

Ship	Length	Rudder Area (Square Feet)	Rudder Area (Length × Draft)	Formula Area (Square Feet)
Mauretania	760	380	0.0149	340
Titanic	850	393	0.0134	380
Normandie	962	432	0.0123	378
United States	940	551	0.0188	380

tioning the size is more pertinent than quibbling about the shape because the adequacy of a rudder is judged primarily by the ratio of its size to the underwater profile area of the ship. There is even a later-day formula for minimum rudder area, based on that profile area.[9] Table 28–1 is based on the area ratio notion, showing how the *Titanic* compares to other liners, and the minimum area required by the formula.

If one compares the rudder areas to the formula areas, it then appears that all examples are satisfactory, but that the *Titanic* has the least margin over the formula. One may note, however, that alone among the four ships cited, only the *Titanic*, being triple-screw and thus having a centerline shaft, has her rudder mounted directly behind a propeller. Since force developed by a foil with a nonzero angle of attack is proportional to the square of water velocity, the location of a rudder with respect to a propeller can be of considerable importance because propeller wake adds to wake produced by the moving hull. Can the benefit of propeller wake have been important enough to have made a difference in the severity of the collision? Between a collision and a life-saving near miss? If so, the attempt of Officer William Murdoch to reverse the engines would have deprived the rudder of wake from the centerline propeller, and was hence a mistake with devastating consequences.

No conceivable astern thrust, even if instantly applied, could have stopped the *Titanic* short of the iceberg. Stopping the turbine-driven centerline shaft would have robbed the rudder of the benefit just noted,[10] and it stopped (or slowed to windmilling speed) when the reciprocating engines were stopped or reversed. Further, stopping and reversing the reciprocator-driven wing shafts would have tended to block the flow of water over the after end of the forward-moving hull.[11] In short, instant engine response to Murdoch's orders—good for whatever stopping effect that could have been generated—would have left the rudder with a greatly diminished ability to develop a turning torque. As it was, apparently insignificant reduction in speed was obtained before impact. Although the ship turned through an estimated two compass points in 30 seconds, that was far from enough. A little more—maybe enough more—probably would have been obtained if the reverse-engines order had not been given.

Turning versus Stopping. Before coming to a conclusion, let us make a detour to examine the doctrine that turning-to-avoid is preferable to stopping-to-avoid. This doctrine developed from experience as powered ships grew in size and number in the 19th century, and has been confirmed in later years by at-sea tests of the turning circle diameters and stopping distances of ships initially operating at their sea speeds. The area of applicability is the open sea, of course, since a large ship in narrow waters is likely to have little scope for turning away from collision.

The turning-to-avoid doctrine was well known enough in the time of the *Titanic* to have appeared in seamen's textbooks. An appropriate quotation from the 1912 edition of Modern Seamanship[12] has already appeared in Chapter 12, but is worthy of repeating here:

> A vessel turning away from another vessel to avoid collision should always continue at full speed, as the effort involved in this course is an attempt to cross the other vessel's bow. To turn away and slow is the surest way possible of bringing about a collision.

Figure 28-2[13] shows the typical comparative merits of stopping and turning. The ships used for this purpose are tankers with displacements roughly five times that of the *Titanic*, and hence are not comparable to that ship in many respects. Nonetheless, this illustration does show a comparison that is known from a wealth of anecdotal evidence to be qualitatively typical of all ships.

Murdoch's Attempted "Hard-a-port Around It" Maneuver. First Officer William Murdoch, watch officer when the iceberg appeared, explained to Captain Smith that he had wanted to "hard-a-port around it," i.e., do what is informally called a "fishtail" maneuver, useful in passing an object close aboard. His starboard helm order turned the bow to the left, away from the berg, but it simultaneously pushed the stern rightward, toward the berg. However, once the bow was pointing safely past the deadly hazard, a quick reversal of helm would push the stern to the left, perhaps clearing the iceberg. Shortness of time and distance foiled him. Quartermaster Robert Hitchens, who was the helmsman that such a command would have been given to, could not recall such a command being given. However, Quartermaster Alfred Olliver, who had just returned to the Bridge, not only witnessed the collision with the iceberg, but also testified at the American Inquiry that he heard an order of "Hard a-port." He also told Senator Theodore E. Burton of Ohio that the Sixth Officer James Moody saw to it that the order was carried out. Whether this order was carried out will never be known as only two of the four persons

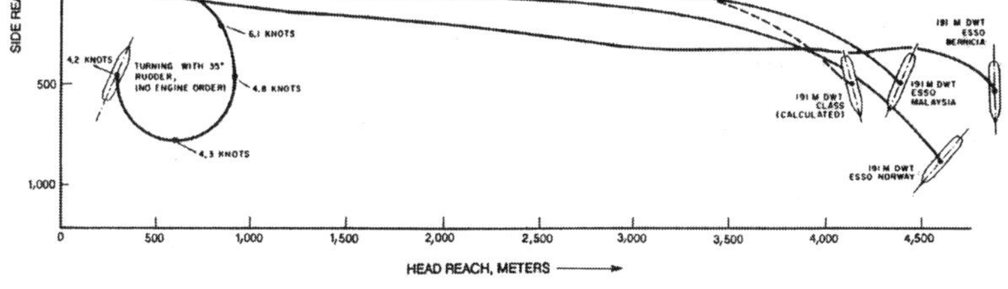

FIGURE 28-2 A comparison of stopping and turning distances. (*From Steering and Maneuverability,* Principles of Naval Architecture, *third edition, Society of Naval Architects and Marine Engineers*)

involved survived. However, the bow wreck is pointed to the northwest, but whether this was due to the port maneuver or the prevailing ocean currents carrying the vessel south during the period without motive power cannot be verified with the evidence at hand.

Conclusions. There is no evidence, either from events or from the imaginings of critics, to confirm that the *Titanic* was deficient in steering qualities. On the other hand, there is no hard evidence to say that she was outstanding in maneuvering ability, or even to convince us that it was incontestably the equal of similar ships. Both the critics and the defenders of the designers should in consequence be displeased with the cases they are able to make. However, given the comparative evidence presented here—and the strong case for the cause of the famous accident being imprudent operation—the verdict easily goes in favor of the designers.

We also might note that before the availability of side thrusters, passenger ships were not designed to be as maneuverable as warships. In harbor they always were attended by tugs; at sea they normally needed only to make small course changes to maintain heading along a navigator's track. The liner *United States* was an exception, being provided with outstanding maneuverability consistent with being designed as a high-speed troopship in wartime.

There is the question of whether the *Lusitania* or *Mauretania* would have made it safely past the iceberg. If they had been operating at the *Titanic*'s 22.5 knots (several knots less than their capability) the simple fact that they were significantly smaller in length and in displacement leads to a positive answer. To that answer a critic could say, "The *Titanic* was too big!" The answer is still no, not in the least. Large ships have to be handled with due allowance for size.

Notes for Chapter 28

1. Colin Carmichael, 1972, "Was the *Titanic* Unsafe at Any Speed?" in *Steamboat Bill,* the Journal of the Steamship Historical Society of America.
2. Joseph Conrad's two articles on the *Titanic* can be found in Joseph Conrad, 1921, *Notes on Life and Letters,* Doubleday, Page & Company, New York.
3. Newton's second law states that a force on an accelerating body equals the body's mass multiplied by its acceleration. The rotational equivalent is torque on an accelerating body equals the body's mass polar moment of inertia multiplied by its angular acceleration.
4. The two equations are derived in *Principles of Naval Architecture,* Volume 3, 1989, a publication of the Society of Naval Architects and Marine Engineers, Jersey City, NJ.

5. Derivation of this equation is also found in *Principles of Naval Architecture*, Volume 3. See Note 4.
6. The *Mauretania* turning radius is quoted from the *Mauretania* souvenir number of *Engineering*, reprinted in 1987 by Patrick Stephens, with Mark Warren as editor.
7. The *Olympic* steering test is described in the Mersey report.
8. One of the collisions was a low-speed harbor collision in which the Royal Navy cruiser HMS *Hawke* rammed the *Olympic* resulting in only minor damage. There was the ramming of *U-103* in 1918. A third collision occurred in May 1934 when the fast-moving *Olympic* ran over the *Nantucket Lightship* in a dense fog. Apparently there was no time to maneuver in that one. The fourth, but non-maneuver-related, incident occurred in 1918 when the *Olympic* was struck by a dud torpedo that the navigating personnel were unaware hit the ship.
9. The rudder formula is found in the annual shipbuilding rules of Det Norske Veritas, the Norwegian ship classification society.
10. A *stop* order required closure of the reciprocating engine throttles. An *astern* order required that exhaust steam from the reciprocating engines be diverted from the turbine to the condensers by manipulation of a three-way valve. The turbine would nonetheless continue to "windmill" in the ahead direction as long as the ship was moving ahead, but the rudder would no longer benefit from the increment in wake normally provided by an ahead-turning centerline propeller.
11. An astern order required a power-assisted changeover of reciprocating engine valve timing. To be fully effective in producing astern thrust, the reciprocating engine exhaust steam had to be diverted from the irreversible turbine directly to the condensers.
12. Austin Knight, 1912, *Modern Seamanship,* fifth edition, D. Van Nostrand, New York.
13. The turning/stopping illustration is from *Principles of Naval Architecture.* See Note 4.

CHAPTER 29

Structural Integrity

The *Lusitania* and *Britannic* sank in sequences of steps as a consequence of weapons of war. Neither of those sequences included major hull failures, though in the case of the *Lusitania* there was a severe secondary explosion. However, both ships capsized before foundering. Neither ship was criticized for its structural qualities, nor were their surviving sisters, *Mauretania* and *Olympic*. Both the *Lusitania* and *Britannic should* have survived had they not been wracked by severe explosions and downflooding through open portholes that should have been closed in a war zone. In the case of *Britannic*, she was built as a merchant vessel that was ill-suited to withstand the violence of war. The *Lusitania* and *Mauretania*, however, were not only designed as passenger ships but as auxiliary cruisers who could be armed with 6-inch guns in wartime. Although never armed with such guns, the *Lusitania* was modified early during World War I to accept this armament.

The two Cunard ships were built with the longitudinal coal bunker bulkheads that have received comment in earlier chapters. These might be regarded as part of their design for naval service, since suitably placed coal bunkers were regarded as part of the protection of contemporary warships. However, the technical literature makes no mention of this design feature being required of Cunard by the Royal Navy. Indeed, many examples can be found of longitudinally oriented bunkers in ships that were not intended for naval service.

The *Titanic* sideswiped an iceberg while traveling about 22 knots, sustaining a series of riveted seam failures that had a total area opened to the sea of less than a ordinary closet door. The fatal flooding took several hours to accomplish with the area open to flooding increasing as the ship settled deeper into the sea. The end came with a catastrophic hull fracture on the surface, brought about by the 39,000 tons of water in her bow structure. Critics by the multitude scathed those who had supposedly boasted of the ship's unsinkability, and by those who criticized the lack of sufficient lifeboats. A few critics went further, maintaining that the ship was just too big, perhaps of dimensions beyond some natural limit on ship size. Several of these critics have been quoted earlier: Joseph Conrad[1] with his implication that the *Titanic* was fundamentally weaker than a biscuit tin, and Walter Lord[2] with his statement, "If this supreme achievement (*Titanic*) was so terribly fragile. . . ." A much later (1990s) critic, Paul Heyer[3] wrote that the ship exceeded "a theoretical size limit." None[4] of these individuals, and no other critics who write in a similar vein, are ship designers or engineers. Yet, the idea of "too weak because of too big" seems to be an important part of the *Titanic* mystique.

At the time Conrad and Lord published their findings, the accepted doctrine was that the ship did not break in two while sinking. This had been the conclusion of the two 1912 inquiries in the United States and the United Kingdom. There were still those witnesses at both hearings who maintained that the ship did break apart at the surface. The belief that the ship sank intact would be shattered when Dr. Robert Ballard discovered the sundered wreck on 1 September 1985. Studies in 1996 and 1998 of the wreck's position and debris as well as finite element analysis have yielded the conclusion that the ship failed at the surface and later broke into four parts. The sequence of the actual failure mechanism may never be known, but the hull failure began at the surface and was completed during the vessel's plunge to the seabed. However, the hull rupture was a result of extreme loads placed on the hull by the manner of its sinking, and hence is not evidence of any design flaw. A critic *might* make the argument that the hull of a smaller ship, sinking in the same fashion, would not have ruptured. Perhaps not, but—again—the *Titanic* broke in two or more sections only as the consequence of extreme loads generated by the manner of her sinking, an event that provides no grounds for a broad-brush criticism of her structural design.

A special category of critic encompasses the professionals who were aboard the ship at the time of its sinking. There was only one whose analysis the authors would like to have: Thomas Andrews of Harland & Wolff, the man reputed to be more intimately acquainted with the *Olympic*-class design than any other person. However, he did not survive. The only expert eyewitness was Second Officer Charles Lightoller. His pertinent comment of present interest is his observation that as ships had become larger during years at sea leading up to the climactic 1912, they had become noticeably more vulnerable to damage from bumps and scrapes that occurred in docking/undocking—suggestive of bumps and scrapes against icebergs. He attributed the phenomenon to an inverse relationship between size and strength: as a ship gets bigger, that ship gets weaker. The implication was either neglect by designers of an important problem, or a ship whose size made it unavoidably vulnerable. His observation seemed to support the Conrad biscuit-tin analogy.

Twenty-some years after his *Titanic* experience, Lightoller published an autobiography, *Titanic and Other Ships*.[4] The book relates a fascinating story of a man who went to sea in sail, moved into steamships, graduated into a high position aboard the century's most famous ship, escaped from its sinking through several near-fatal mischances, and went on to other adventures at sea, particularly in the century's great wars. On the other hand, his book gives no hint of an engineering education or of any learning-through-practice in the arcana of ship design. In short, like Joseph Conrad, Charles Lightoller was a person whose well-publicized association with ships gave him a platform from which to express his opinions on ship design. In both cases, the opinions broadcast were faulty, but on the other hand were not without merit, and so are to be dismissed only by reasoned argument.

Before going into a plating thickness vs. size of ship argument, let's look at the thought expressed by Heyer that the *Titanic* (and her sisters as well) exceeded a theoretical size limit. Somewhere beyond the size at which any ship has ever been built, there may be such a limit, but the limits imposed by harbor channel widths and depths as well as the capacities of shipbuilding facilities, are likely to keep us from ever finding out. Ships continued to grow larger following the *Titanic* experience. The "supertankers" that caused so much concern from the 1960s through 1970s, and that have indeed caused an occasional debate by spilling oil, provide an obvious example. Ships of that type, with sizes (as measured by loaded displacement tonnage) an order of magnitude greater than the *Olympics,* have been built with several over 300,000 and one of 500,000 tons. If the discussion should be limited to transatlantic liners, Table 29–1 will demonstrate that the *Titanic* did not tower over the ships of the same type that were soon to follow. How were the smaller displacements in some achieved in spite of greater dimensions? The answer lies in use of high-strength steels, and especially of aluminum in superstructure and deckhouses. Firm evidence for the structural integrity (including thickness of plating) of the *Olympic* ships comes from an estimation of what their primary hull stresses probably were, and from further comparison with other transatlantic liners. The "primary" stresses are those produced by

TABLE 29-1 *Characteristics of Passenger Liners*

Ship	Entered Service	Length	Beam (Waterline)	Tons (Displacement)	Speed (Knots)
Titanic	1912	850	92	52,100	23
Aquitania	1914	880	97	49,400	23
Europa	1929	888	102	55,500	27.9
Queen Mary	1936	1004	118	77,400	33
Normandie	1935	962	118	66,400	32
United States	1952	940	102	45,400	38.25
France	1962	984	111	51,800	35.6
Queen Mary 2	2005	1143	140	140,000	30

the overall loading of the hull, in contrast to the loads produced by very local loadings such as a bump against a pier or a scrape against an iceberg. Figure 29-1 should make clear the distinctions.

Design for Primary Loads.[5] The ship hull is a complex structure that must be designed with regard for several competing criteria. Watertight integrity, low resistance to forward motion, acceptable response to rolling and pitching excitations, unimpeded flow of water to the propellers, and adequate strength are all criteria of high importance. Among these, the last one listed should be ranked as the most important, since inadequate strength may permit catastrophic failures that override whatever merit may be found in other qualities.

Because a hull is a complex structure, and since its failure can be so devastating, its design for strength taxes the skill of the naval architect. On the other hand, the underlying principle is simple: the hull is primarily a beam that will be bent vertically by a combination of weight forces acting downward and buoyancy forces acting upward. The necessary analysis can be learned in sophomore mechanics courses taught to all engineering students who must deal with mechanical problems. Of course, we cannot teach the theory and its accompanying analyses here, but can outline the processes to a degree that allows illuminating comparisons between the *Titanic* and other liners.

Conventional beam theory dictates that the loading of a beam be analyzed to find shear force and bending moment at all points along the beam, then that the beam structure be designed to provide a section mod-

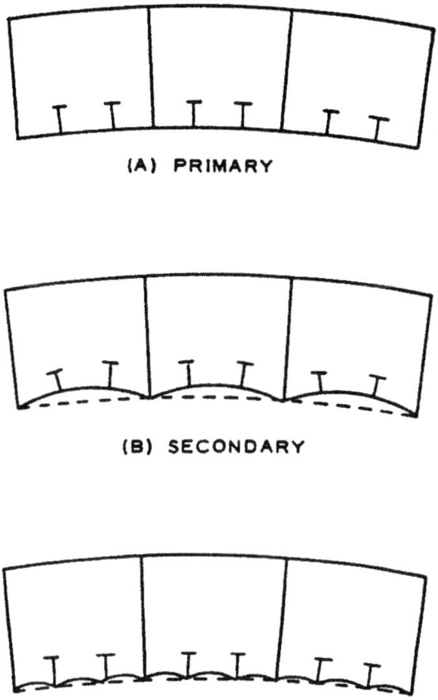

FIGURE 29-1 Primary, secondary, and tertiary hull loads. (*From* Ship Design and Construction, *Society of Naval Architects and Marine Engineers, 1980*)

ulus sufficient to keep bending stresses within allowable limits. If the beam is long compared to its length (the typical ship hull), the shear force is ignored except possibly in the design of details.

Figure 29-2 shows a very simple beam problem: a beam, uniform in cross section over its length (L), uniformly loaded by a weight of (W lbs/ft), and supported at its ends by a force w L/2 each. A reader with a mechanical–technical background should be able to calculate that the bending moment is $WL^2/8$ at the beam's mid-point, and recognize that this is the highest value experienced by the beam. Obviously, this example cannot represent a ship hull because the hull is unlikely to be the simple prismatic beam of the textbooks, is unlikely to be loaded by weight distributed uniformly, and certainly is not supported solely at its ends.

The principle illustrated is nonetheless used by means of its application piece-wise over the length of the hull through the following steps:

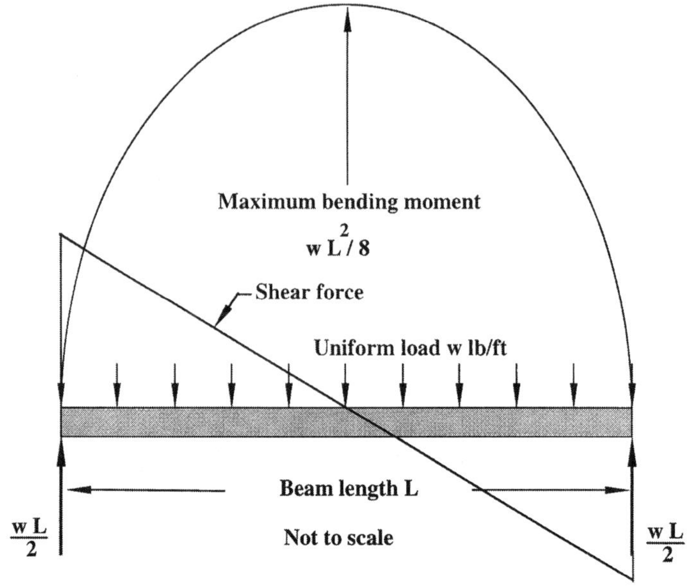

FIGURE 29-2 A simple beam with shear and bending moment curves superimposed. (*Sketch by the authors*)

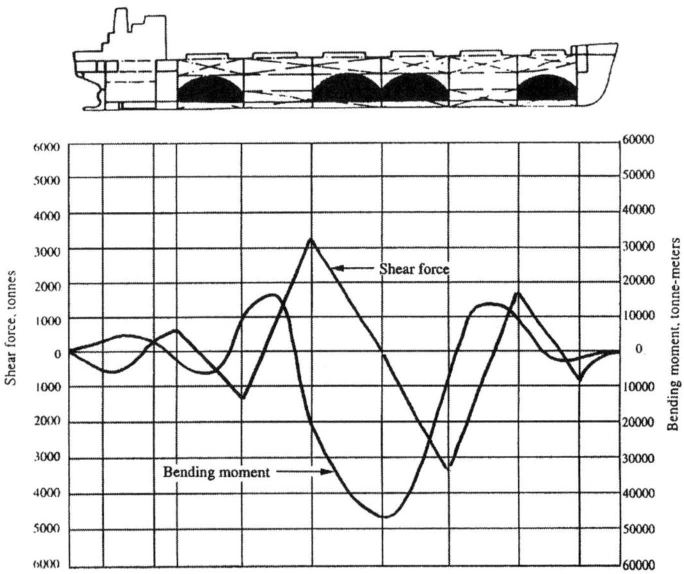

FIGURE 29-3 Typical shear and bending moment curves for an ore carrier in a particular condition of cargo loading. Holds loaded with ore are indicated by the dark rounded symbols. (*Source unknown*)

- The load on this hull-beam at any point along its length is the difference between the downward force of all weights at that point and the upward force of buoyancy at that point.
- Integration of the load along the length of the hull produces the shear force at any point.
- Integration of the shear force along the length of the hull produces the bending moment at any point.

The result is a set of curves showing load, shear, and moment over the length of the hull. Figure 29-3 demonstrates the result obtained by application of these steps to a sample ship hull.

Carrying out the other major part of the analysis—finding the beam's section modulus—is similarly a simple process at the elementary level, but one that ship hull geometry complicates. If the beam of the first illustration has a simple rectangular cross section, width b, height h, then its section modulus is $b\,h^2/6$. For a ship hull, the process begins with a structural plan of the hull's mid-ship cross section, with those members that resist longitudinal bending identified. Figure 29-4, in two parts, is an example. It shows a half section of the *Titanic* hull, accompanied by a simplified sketch that shows only the members that run fore-and-aft through at least the middle ⅗ of the hull, and gives their cross-sectional dimensions. In the calculation, the neutral axis of the hull is identified from the distribution of the cross-sectional areas of these members. Then the second moment of all areas about the neutral axis is determined; the section modulus follows by the division of the second moment by distance from the neutral axis to any point at which stress is to be found. The maximum bending stress occurs at the top or bottom. Its magnitude is given by the simple formula Stress = Bending moment/section modulus, or

$$\text{Stress} = \text{BM}/z$$

in conventional symbols.

An analysis of the *Titanic* hull that was accomplished in a 1996 paper[6] of the type just described produced a section modulus of 89,100 ft-in^2 at the strength ("B") Deck, 155,700 ft-in^2 at the bottom shell. A detailed calculation of the bending moments requires information that has not been published, but a reasonable value for the maximum bending moment can be found from the well-tried rule-of-thumb formula:

$$\text{Bending moment (ft-tons)} = 0.75\ L^2B\ d/35^2$$

where L is length, B is beam, and d is draft, all in feet.

The bending stress indicated by the formula and figures cited is 9.06

tons/square inch. Table 29–2 provides the same information for several other transatlantic liners. An immediate reaction to the data in this table may be that the bending stresses of the *Titanic* and several other ships cited were quite conservative when compared to the *Europa* and *Normandie*. However, those two ships were built with high-tensile steel for their major bending stress members, and hence could safely exploit the properties of their steel for higher stresses than the other ships. Nevertheless, it should be noted that the *Titanic* does not stand out among this group as a ship with a highly stressed hull.

Studying the data in Table 29-2 some more, we observe that the *Titanic* was by far the earliest built ship in the list. Could we, therefore, suspect that her steel was inferior, and perhaps not suited to withstand stresses as high as those of the other ships? The answer is negative, since the embrittlement at low temperatures that has attracted the interest of *Titanic* investigators was not eliminated from shipbuilding steels until after World War II. Also, a slow bending test done on a sample of *Titanic* steel in ice-cold water in the laboratories of the National Institute of Technology in Gaithersburg, Maryland in 1998 proved that the steel was *not* brittle in collision-type scenarios. In addition, it may be illuminating to single out the *Europa* for examination beyond the glimpse given by the table. The authors who assembled the table (except for the *Titanic* entry) obtained structural data from which they estimated the bending moment under design conditions to have been 2,067,000 foot-tons rather than the 1,650,000 given in the table by the rule-of-thumb formula cited here. From this, a maximum tensile stress in the Promenade Deck was 15.12 tons per square inch, which left very little margin for the additional dy-

Table 29–2 Bending Moment and Bottom Bending Stress[7]

Ship	Bending Moment (foot-tons)	Bottom Bending Stress (tons/square inch)	Year Built
Europa	1,650,000	12.10	1939
Normandie	2.440,000	11.13	1935
Rex	1,275,000	9.91	1933
Conte di Savoia	1,170,000	9.78	1931
Manhattan	712,000	7.86	1930
America	682,000	9.88	1938
Titanic	1,412,000	9.06	1912

namic stresses that could accompany severe weather on the North Atlantic. The *Europa* did have a history of cracking of her plates in the vicinity of the Promenade Deck. Although these were repaired and openings reinforced after such events began to occur, the problem continued during her North Atlantic service before World War II. Even when these weaknesses became apparent, sufficient remedial measures were not taken to insure her safety; the operators placed reliance on careful operation and seamanship."[8] In 1939, the Germans resorted to adding four doubler plates in the vicinity of the second funnel.

The *Europa* survived World War II and became the property of the United States from June 1945 to April 1946 through the process of war reparations. The United States declined its option on the ship and gave her to France as compensation for the loss of the *Normandie*. Upon study of the cracking problem, the French engineers added a doubler plate 443 feet in length on both sides of the strength deck to lower the stresses. The *Europa*, renamed the *Liberté*, continued in service until the liner *France* entered service in 1962.[9]

Riveting versus Welding. The joining of steel plates and shapes in the erection of a hull by electric arc welding is superior to the riveting alternative on several counts. For one thing, the "efficiency" of a riveted joint cannot be 100%, i.e., the ability of the joint to withstand a load cannot equal that of plates, say, that it joins. This handicap imposed by the need to join by riveting was recognized from the earliest days of iron and steel ships, and taken into account by their designers. In the present instance, all of the ships listed in our comparative table were of riveted construction, and hence riveting versus welding is not a pertinent issue.

On the other hand, there is a point of interest: experience suggests that a given trauma (torpedo explosion, for example) will cause the rupture of a significantly greater length of a plating seam if it is riveted rather than welded. This observation is of no importance in the realm of ship design because we abandoned riveting after World War II for other, more easily demonstrable, reasons. However, it is fuel for speculation. Would a welded *Titanic* have sustained the same degree of flooding or damage and possibly survived? If we knew more about the initial damage suffered by *Lusitania* and *Britannic*, would we be able to speculate about their survival as welded hulls? Pointless speculation, of course, because welding was not a viable shipbuilding technique in their day.

Secondary and Tertiary Loads. We have been speaking of primary loads and their associated primary stresses. Of course, where there is stress there must be deformation, but except in extreme circumstances, primary stresses in a ship hull (or in any engineering structure), never reach the elastic limits of the structural material. The extreme circumstances were present in the late minutes of the *Titanic* sinking; recall the events described in Chapter 16. In brief, a primary stress that exceeds the material's elastic limit is often tantamount to the destruction of the ship.

The ship hull is a complex artifact of many horizontal and vertical members, each built up from plates and stiffeners, all joined into a structural whole—by rivets for the ships of our interest. Specifying structural elements so that primary stresses remain within acceptable limits is an essential task of hull design, but there are inevitably secondary loads and tertiary loads, producing their respective stresses as well. The definitions of "secondary" and "tertiary" are best offered through Figure 29-4.

Naturally enough, a ship designer never intends that secondary and tertiary stresses exceed elastic limits, but they often do, and without fatal consequences. The collision of the *Olympic* with HMS *Hawke* is an example: the *Olympic* hull was holed by the bow of the *Hawke*, but only over a small area, and neither ship sank. *Olympic* was hit by a torpedo that failed to explode, leaving only a dent and sprung plates that were discovered after the ship was in dry dock after World War I. On the other hand, a contact with the iceberg, producing what seems to have been only tertiary failure, led to the sinking of the *Olympic*'s younger sister. Does one out of three incidents rate a major concern? The ratio must be much smaller than one-for-three because of the population of "all ships," there must be innumerable incidents of dents, and even minor punctures, that go unnoticed.

Before concluding this discussion of the Conrad/Lightoller criticism of weak sides, let's explore a principle or two that determine how tertiary (and to some extent, secondary) loads are handled. Ships became bigger, especially during the first decade of the 20th century. Within the transatlantic liners, the ratios of beam, depth, and draft did not vary much as hull size increased. If so, then hull-bending moment increased as length to the fourth power. If the all-important primary stress was to remain constant, then hull section modulus had to increase by the same ratio. Now, the hull girder is a complex assembly of decks, inner and outer bottoms, side shell plating, longitudinally continuous bulkheads, and longitudinally framed members. However, the point of interest here can be made by considering the hull to be a simple rectangular box consisting of plating, thickness t_d for top and bottom, and t_s for the sides. The section modulus is then defined to be:

$$Z = D\,(B\,t_d + D\,t_s/3)$$

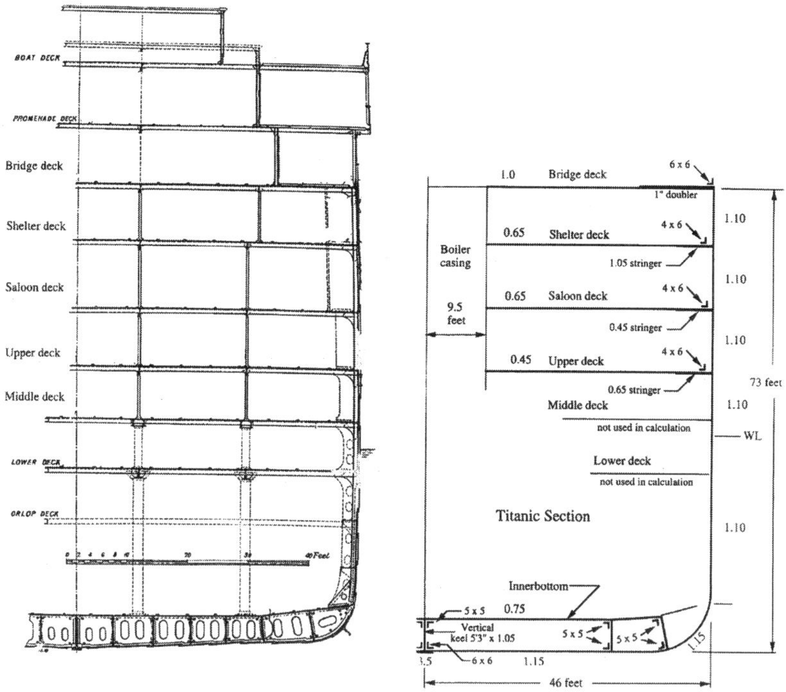

FIGURE 29-4 *Titanic* midship half-sections for hull analysis. Left half shows actual section, taken from *Engineering*, vol. 90. The right half-section is simplified to indicate only those structural members that resist fore-and-aft bending, i.e., that form the hull section modulus. The right half is adapted from William H. Garzke, Jr., et al., 1996, The *Titanic* and *Lusitania*: A Final Forensic Analysis, *Marine Technology*, vol. 33.

If we make B = 1.5 D (actual ratios were 1.44 for *Olympic/Titanic*, was 1.45 for *Lusitania/Mauretania*), then

$$Z = D\,(4.5\ t_d + t_s)/3$$

and the weight of this simple hull is

$$W = D\,(2\ t_d + 3\ t_s) \times \text{specific weight of steel}$$

If Z must increase as the fourth power of dimensions, while the unavoidable increase in hull weight must be minimized, then these simple formulas show that the best solution is to increase t_d at the expense of the least increase in t_s. Meanwhile, the kinetic energy of a ship bumping a piling at one-knot speed, or an iceberg at 22-knot speed, increases in the ratio L^3. Inevitably, the bigger ship is damaged more readily, unless t_s (and hence hull weight) is increased to an unreasonable degree.

Quite similar analyses can be made for any mobile engineering structure. Indeed, the phenomenon is the stuff of everyday experience. But are we remiss in accepting it? Yes, to a certain extent, as the need for federal standards for the survivability of automotive bumpers in low-speed collisions demonstrates. Then, should shipbuilders be required to furnish heavily plated hulls that will be immune to dings and dents? No, because the problem is quite adequately countered by watertight subdivision. The ship also is handled by highly trained operators.

In the *Titanic*, where 1,522 people perished, a piece of ice scraped or forced open riveted seams in her hull plates. This incident might not have happened if the rivets were not made of wrought iron or if the number of rivets had been two or more in the area of impact with the iceberg. The opened seams may have been less in area if smaller frame spacing had been chosen in the area of impact. Regardless of the structural considerations, this incident might not have happened if the propulsion plant had been capable of only 15 knots. It would not have happened if Captain Smith had chosen a speed suitable for a dark night in an ice field. The professionals had many things to say (SOLAS 1914, for instance; Chapters 22, 23) about post-*Titanic* safety, but requiring hull structure that could withstand the rupturing of hulls by accidents was not among them.

Joseph Conrad and Charles Lightoller were themselves professionals—but not in structural design of ships. Their criticisms were without merit.

As we also have seen in Chapter 23, the professionals who commented on the *Titanic* design at the SOLAS 1914 conference did not find fault with the structural design of the *Olympic* ships. Nonetheless, the Mersey hearings and the findings of forensic investigations of the 1990s also shed light on structural design. In concluding this chapter we will give a summary of those findings as they illuminate the structural qualities of the *Titanic*.

Structural Summary

Most interesting is the testimony of naval architect Edward Wilding at the British inquiry. He was the assistant to Thomas Andrews, and the man on whose shoulders fell the responsibility of explaining, in terms of the *Titanic*'s naval architecture, what befell the great ship. Using survivor's accounts, he was able to give a logical explanation of how the ship became flooded. His calculations showed that the initial ingress of water occurred over an area of 12 square feet. (This area would be confirmed in 1995 by a computer analysis provided by John Bedford and Christopher Hackett, also of Harland & Wolff Shipyard.) Wilding also concluded that the damage to the ship was spread over a length of 300 feet, embracing the first six watertight compartments. Unfortunately, his conclusions were wrongly interpreted by non-naval architects as a uniform 300-foot gash with a total area of 12 square feet. Also, Wilding, himself, erred in concluding that the stresses in the hull arising from the flooding were no more than the design stress used to size the structural members. This meant that these stresses would be no more than sustained during the encounter of a wave of the same length of the ship and a height of the ship's length divided by 20. Unfortunately, the analyses by Wilding, and later by Bedford and Hackett, ignored the large stress created by the weight of 39,000 tons of seawater that caused the vessel to sink. Further, the recovery of a 22-ton hull section of the *Titanic* in August 1998 revealed that no check valves were fitted to the drainage piping from wash basins and toilets that had plumbing discharges located above the margin line.

As the ship sank deeper into the water, sections of the bow were immersed above the "margin line." This circumstance, coupled with the few open portholes on "E," "D," and "C" Decks, created additional flooding areas for the water to enter the ship during the advanced stages of the sinking. While the portholes should have been closed and secured, there was no protection against the upflooding through the plumbing drains once they were submerged. All of these factors contributed to an accelerated flooding rate once the flooding waters reached "E" Deck.

A further observation comes from a finite element analysis performed on the *Olympic* hull that showed peak hull stresses moving aft as the flooding progressed. The most obvious example of this transition of stress was in the area around the forward expansion joint. An expansion joint prevents stress build up in the superstructures and allows the use of lighter materials for the superstructure scantlings. See Figure 29-5 for a typical joint. Lightoller's remembrances of that night were of extreme interest to the investigative team doing the structural analysis for the Discovery Channel documentaries, "*Titanic*, Anatomy of a Disaster" and "*Titanic*, Answers from the Abyss." They found that the forward expansion joint was in a region of high stress. That joint opened as the flooding and stress increased.[10] Radio Operator Bride remarked that between the noise of air escaping from the joint (the Marconi Room was located near it) and the steam from the three funnels, Radio Operator Jack Phillips found it difficult to concentrate. The joint continued to open as the flooding progressed, putting increased stresses on the aft wire stays of the forward funnel. Finally, when Lightoller was washed overboard, shortly after 0210, he noticed the failure of these stays. As a consequence, the forward funnel fell

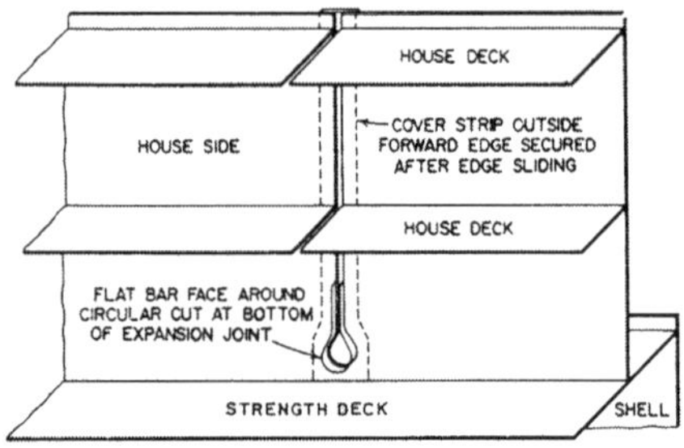

FIGURE 29-5 Typical expansion joint in the deckhouse of a ship. (*From* Ship Design and Construction, *Society of Naval Architects and Marine Engineers, 1980*)

to starboard onto the bridge wing and overboard. A 1998 examination of forward expansion joint in the *Titanic* determined that this joint was opened as much as 18 inches in the area of the forward stack's aft stays. An examination of the forward expansion joint in the *Britannic* by diving teams in 1999 revealed that her forward expansion joint also had opened slightly beyond the construction plan value of 2 inches. We believe that opening was caused by the same forces from the sinking process as occurred in the *Titanic*.

The side shell at the level of the Well Deck of the *Titanic* at the forecastle and bridge front also exhibits plate failures. These failures likely happened after the forecastle was submerged, causing the bow to take a sudden plunge, hence may have caused the "wave" that was observed on the Boat Deck by those who saw it and survived. It is also possible that the mass of water that had entered the ship by this time created such a stress on the hull that the large 10-foot bend in the port bottom shell outside Boiler Room No. 3 began to form. Further analysis is needed to confirm that conclusion. However, the validity of any such analysis will be clouded by our realization that the damage may have been exacerbated when the ship plowed into the seabed.

The *Olympic*-class passenger ships were two-compartment ships that were designed well above the standards of their day. In fact, comparisons were made by the Bedford and Hackett paper, cited earlier, that showed that these ships were superior to some passenger ships in the late 20th century. As was confirmed at the Mersey hearings in 1912, they would remain afloat with the first four compartments flooded. However, the damage to Boiler Room No. 6, allowing flooding far beyond the capacity of all the pumps that could be brought to bear, caused the ship to sink. Thomas Andrews recognized from his floodable length data the inevitable consequence to flooding of this extent. Captain Smith and Chief Engineer Bell recognized this circumstance as well, or were quickly apprised of it by Andrews. However, they agreed that the life of the ship could be prolonged by controlling the rate of flooding into Boiler Room No. 5. That condition was maintained until about 0115. At that time the intense hydrostatic pressure caused either the failure of the nontight coal bunker bulkhead at the forward end of Boiler Room No. 5 or the forcing open of the nontight coal bunker doors in that bulkhead.

If the failure was triggered by the collapse of the forward watertight bulkhead, it constitutes an example of the "secondary failure" that has been mentioned in this chapter. If so, a critic might seize upon the event as a consequence of structural weakness, since this was a bulkhead that should have withstood a hydrostatic head up to the "E" Deck level. However, we recall that this is the bulkhead that well may have been weakened by the fire in the coal bunker it bounded.

Upon hearing that Boiler Room No. 5 had flooded, Chief Engineer Bell was overhead to say, "The ship is lost!" However, it still took almost 45 minutes for the flooding to bring about the hull's failure and sinkage. It is believed the heroic, but futile, effort of the engineers of the *Titanic* achieved a measure of success in that it bought time for all lifeboats to be boarded and lowered to safety.

Notes for Chapter 29

1. Joseph Conrad, 1921, *Notes on Life and Letters*, Doubleday, Page & Company, New York.
2. Walter Lord, 1955, *A Night to Remember,* Henry Holt and Company, New York. Walter Lord published a later book in 1987 that was based on the discovery and exploration of the *Titanic* wreck by Dr. Robert Ballard in 1985 and 1986. In that book, he disputes his earlier conclusion of the 300-foot underwater damage from the iceberg.
3. Paul Heyer, 1995, *Titanic Legacy*, Praeger, Westport, CT, and London.
4. Charles Lightoller, 1912, *Titanic and Other Ships*, 1935, Ivor Nicolson and Watson, London.
5. This section gives only a sketchy treatment of the application of conventional beam theory to a ship hull. Readers who are mechanical engineers will probably recognize the treatment. Others with some technical background might consult a book such as Robert B. Zubaly, 1996, *Applied Naval Architecture,* Society of Naval Architects and Marine Engineers, Jersey City, New Jersey.
6. William H. Garzke, Jr., David K. Brown, Arthur D. Sandiford, John B. Woodward, and Peter K. Hsu, 1996, "The *Titanic* and *Lusitania*: A Final Forensic Analysis," *Marine Technology*, Volume 33.

7. Except for the *Titanic*, all data in this table come from a paper by James L. Bates and Ivan Wanless, 1946, "Aspects of Large Passenger Liner Design," published in the *Transactions*, Society of Naval Architects and Marine Engineers.
8. The quotation also comes from the Bates and Wanless paper cited in Note 7.
9. J.P. Richard, "La Réfection du Paquebot Allemand *Europa*," Bulletin de L'Association Technique Maritime et Aéronautique, 1950, pp. 405–428.
10. The finite element analysis was performed by Gibbs & Cox, Inc. for the Society of Naval Architects and Marine Engineers and the Discovery Channel. It was included in a Discovery Channel documentary entitled, "*Titanic*, An Anatomy of a Disaster."

CHAPTER 30

Titanic Stability

Had it not been for the torpedoes and mines of World War I—the banes of *Lusitania* and *Britannic*—there might be no need for a chapter on stability. However, both of those ships capsized in the act of sinking, demonstrating that at some point they became unstable as a consequence of the damage they received.

What about the *Titanic*? That ship sank with very little list until very late in the course of the tragedy, a circumstance that allowed every lifeboat to be launched successfully. The *Titanic* did not list as much as *Lusitania* and *Britannic* because there were very few open portholes to allow unsymmetrical downflooding that would exacerbate the tragedy. There was no capsizing, and the consequent sinking process allowed a safe and orderly evacuation. The ship's "damage stability" was exemplary, a commendation that need not come from calculation by the naval architect, but one that the *Titanic* earned *in extremis*. Most of the 706 survivors probably owed their lives to this fortunate circumstance. The *Titanic* sank through progressive flooding that eventually caused a loss in longitudinal stability.

Both *Britannic* and *Lusitania* capsized in the course of their sinking process, but a comparison with *Titanic* is obviously wrong. These two ships were victims of war. The *Lusitania* experienced a secondary explosion that may have caused further damage comparable to the initial torpedo damage, and that may have contributed to their capsizing and rapid sinking. Paradoxically, the capsizing of the *Britannic* also may have been caused by an improvement, the double skin that was to enhance her ability to resist raking types of damage. The principal task of this chapter is to examine the stability characteristics of both *Britannic* and

Lusitania, to see whether there was fault in design evident in their losses, fault that may have been obscured by the violence of the events.

To begin, let's abandon that quest for the moment to look at the elements of stability, both in the intact and damaged conditions.

Elements of Intact Stability. These elements—the concept of center of buoyancy, center of gravity, and metacenter—are familiar to every naval architect. This section provides a brief tutorial for readers who may not have this background.

The essentials of intact stability (stability in the absence of flooding) are shown by Figure 30–1. Part (a) shows the cross section of a ship hull as it lists (or heels[1]). Presumably some unseen influence, such as a wind blowing hard from the right-hand side of the sketch, is causing this condition. The points K, B, G, M, and Z should be noted. The letter K is symbolic for keel, the lowest point of a ship. The letter B refers to the "center of buoyancy," or the centroid of the underwater volume of the hull. This is the point through which all of the upward buoyancy force is assumed to act. In accordance with Archimedes' Principle, the magnitude of the buoyant force is equal to the displacement (i.e., the weight) of the ship. The letter G represents the "center of gravity," or the centroid of the summed weights of all the elements that constitute a ship. The "metacenter," termed M, is the point where the upward force through B crosses the centerline of the hull. Z, usually nameless, is the point where a horizontal line through G crosses the vertical line between B and M.

It is apparent in Figure 30–1 that the pair of forces, one acting upward through B and M, and the other acting downward through G, are tending to rotate the hull to the right, i.e., they are opposing the effect of the wind by creating the moment whose magnitude is displacement multiplied by righting arm. This arm is often referred to as the "GZ." If we assume that Figure 30–1 shows a condition of equilibrium, the ship is stable despite the wind force. Increasing wind strength will cause a larger list, which in turn increases the righting arm, GZ. Up to a point, the increased wind moment is equalled by the increased opposing moment. However, at some still greater wind strength and list angle, the righting arm will reach a maximum, usually at the point where the deck edge is immersed. Further increase in wind strength then will overwhelm the ability of the ship to resist capsizing.

The same effect and consequence comes from moving a weight across the hull (or placing a new weight at a location off the centerline). The point G is moved from the centerline, creating a moment that causes a list that increases until the point B is directly under G. Stability has not

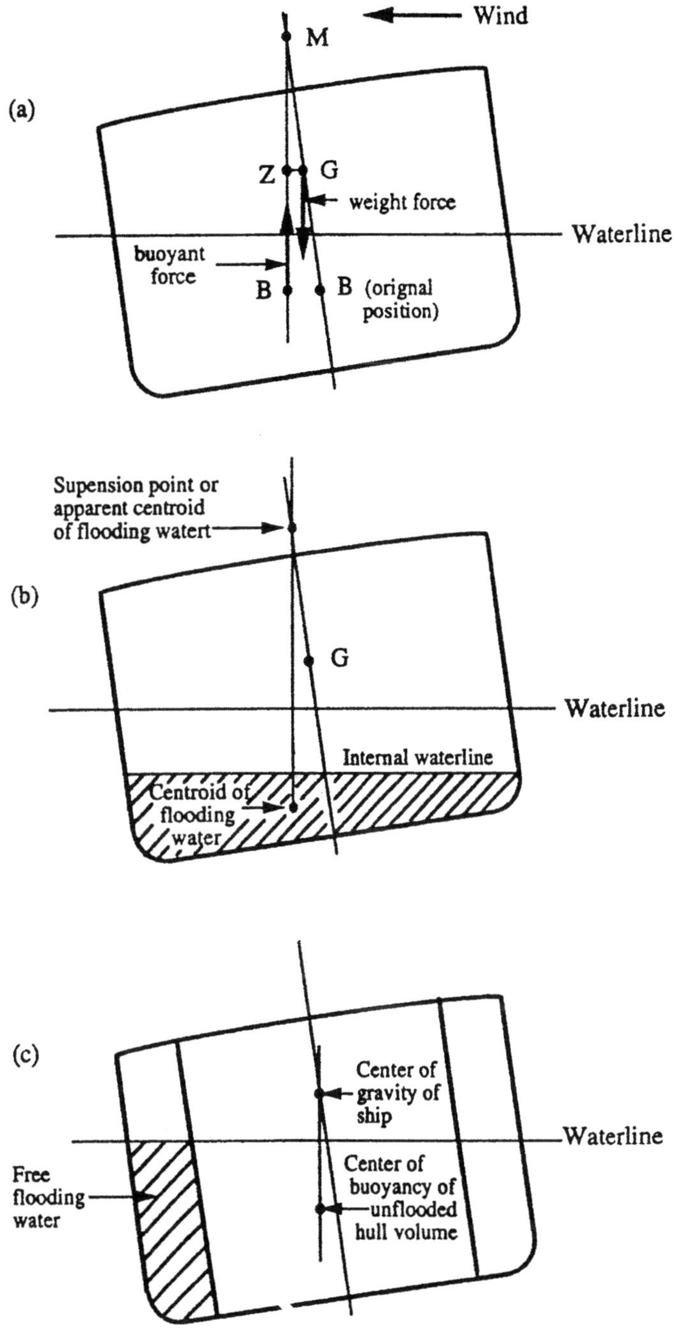

FIGURE 30-1 Basic stability illustrations. View (a) shows the forces of buoyancy and weight that constitute the basic elements of intact stability. View (b) shows how unconfined water within a hull will act as if it were suspended at a point high on the centerline, and hence act to destabilize the ship. View (c) illustrates the concept of "free flooding" that is the key factor in "damage stability." (*Sketches by the authors*)

been violated, but the list remains until the weight is moved to its original position, or some counterbalancing weight is taken aboard.

Let's assume that the ship has a list caused by the movement of G from the centerline, but is stable because B still lies directly below G, the condition described in the preceding paragraph. A vertical line passing through B will, therefore, pass through G as well, and will intersect the centerline at point M, the metacenter. It is apparent in Figure 30–1 that the righting arm, GZ, is proportional to the metacentric height, GM. The latter parameter is comparatively easy to calculate from a simple formula, and hence its value is fairly easy for the officers of a ship to track as weights are added or removed from the ship. The GM thereby serves as a valuable *indicator* of stability. For example, the statement "The ship has sufficient GM ('gee-em,' almost universally)" indicates that it will have positive righting arm throughout all expected operating conditions.

The proportionality between GM and GZ disappears as angles of list become large (above the 10 to 15 degree range, say), but GM nonetheless serves as an excellent indicator of stability throughout normal operating ranges.

A large value of GM indicates a safe margin in right arm, GZ, but paradoxically, that margin should not be too great, for there is danger in excessive stability. If a ship is too stable, it tends to have a shorter period of roll. The resulting accelerations from that quick roll could cause furniture and cargo to come loose from whatever nominally restrains them. Even much smaller accelerations can be uncomfortable to passengers, and if nothing else, give a ship a bad reputation among its potential clientele.

Now, if an internal weight that is moved is a liquid whose top surface is unconfined, a special situation, the "free surface effect," exaggerates the moment of G. In effect, the liquid acts like a pendulum, i.e., acts as if it were a solid suspended from some point above the actual surface. The point of suspension is taken as the liquid's center of gravity in the process of summation to find the point G. Some liquid-free surface is inevitable as tanks are filled or emptied. In normal circumstances, however, the consequent change in G is small, and, of course, the designer takes the change into account in any stability calculation. In abnormal circumstances, the contribution of free surface may raise G above M, making the ship unstable. Obviously, these abnormal circumstances are most likely to occur when the hull has been punctured, allowing seawater to flood spaces that normally are dry.

Elements of Damage Stability. A ship's hull is opened to the sea by collision or other disaster. Seawater floods the affected space up to the level of the sea outside. By this consequence, weight has been added, and probably at some distance off the ship's centerline. The ship's center of gravity, G, thus has been moved transversely, and the ship must list until its center of buoyancy, B, moves back into its equilibrium position directly under G (see Figure 30-2). Is the equilibrium thereby restored? The answer is No! It is not restored because the very act of listing and sinking from the weight of the water has caused *more* water to flow into the flooded space. The analyst who found the angle and sinkage then must revise the calculation to find the increased angle and sinkage caused by the additional water introduced by the initial list and sinkage. Of course, the additional list and sinkage then allows still more water to flow in, and so on, nominally ad infinitum. However, if the physical circumstances are such that the ship reaches an equilibrium position before capsizing, then this calculation process converges to the equilibrium angle and the final draft. The analyst will have completed a "damage stability" calculation, and probably will have attained acceptable accuracy with only three calculate-correct cycles.

The preceding paragraph may not describe the calculation process that a naval architect actually follows, but the essence is there: if water within the hull has free communication with the sea, the list and sinkage it causes will cause further list and sinkage. Whatever the actual steps of

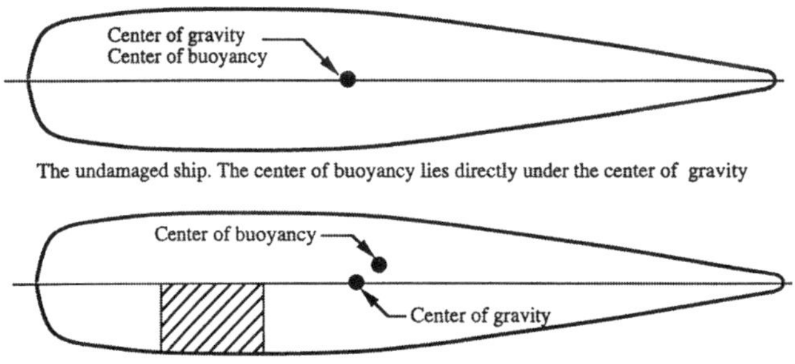

FIGURE 30-2 The damaged ship, with cross-hatching indicating a flooded compartment, open to free flooding from the sea. For the moment, the ship has neither a trim nor a list, and hence the center of buoyancy—the centroid of the underwater volume *excluding* the flooded volume—has moved to port and forward. Because the structure of the ship remains essentially intact, the center of gravity has not moved. But the hull must move toward equilibrium of forces and moments, and will do so by trimming by the stern and listing to starboard until the center of buoyancy lies directly under the center of gravity. (*Sketches by the authors*)

calculation, they must find the equilibrium position beyond which no more water enters.

Flooding water inevitably has a free surface. If the opened space has small horizontal extent, then the "free surface effect" is of minor consequence. In analyzing the effect of a flooded *Lusitania* coal bunker, a naval architect would probably ignore free surface because of the bunker's small transverse extent. But how about the consequences of free surface presented by the side-to-side flooding of a *Titanic* boiler room? The answer may seem obvious, since the boiler rooms of that ship were surely flooded, yet the ship's list was virtually zero during most of her two-hour agony. However, we do find reason to look at this question later in this chapter.

The *Titanic* damage stability question does have another aspect that has a degree of uniqueness: for a very large ship that was sunk in consequence of a collision, the *Titanic* sustained damage from a relatively small hole—the approximate 12-plus square feet area that has been found by several independent analyses. Of course, the inflow through that small fissure was quite enough to overwhelm the ship's pumping capacity, and moreover, the affected area increased as sinkage exposed nonwatertight openings. However, a stability calculation made for the early stages of the sinking would probably be quite accurate enough if it treated the weight of water in the ship as fixed (for the moment). On the other hand, modification of the calculation to include the free surface effect would be appropriate.

Stability of the Titanic *in Extremis.* The *Olympic*-class ships were fitted with transverse coal bunkers that were fitted to transverse bulkheads between the boiler rooms. This arrangement had the great operational benefit of allowing every boiler to have a bunker door close at hand. Trundling coal in wheelbarrows from bunker to furnace was not necessary. Lord William James Pirrie, the Chairman of Harland & Wolff, who helped to design these ships, did not want to repeat the longitudinal subdivision favored by their Cunard counterparts. In those ships, the coal bunker doors, of which there would be many, had to be watertight to take advantage of the watertightness of the longitudinal bunker bulkheads. However—as has been noted several times before this—residual coal in the door openings would prevent their tight closure in emergencies. Also, counter-flooding measures needed to be fitted to prevent unsymmetrical flooding if the longitudinal coal bunkers on one side flooded.

After *Titanic* was damaged, water that poured in distributed itself across the width of the hull in Boiler Rooms Nos. 5 and 6 when they flooded. The flooding of Boiler Room No. 6 did not contribute to the initial five-degree starboard list. That list took place because of differences in permeability of the coal piles in Boiler Rooms Nos. 5 and 6, and because of the initial unsymmetrical flooding in Cargo Holds Nos. 2 and 3 (caused by obstruction presented by the Firemen's Passage running fore-and-aft along the centerline). Once that list disappeared, lifeboats were easily loaded and lowered. Therefore, sinking from the consequences of an extreme list did not occur in the *Titanic*. If the ship had listed severely, and sunk more rapidly because of it, the temperature of the sea, and the distance that the *Carpathia* had to steam, might have left very few survivors for that ship to rescue.

The flooding water did indeed appear to distribute itself evenly across the hull. There were some free surface effects, though their numeric value is difficult to compute at any time during the sinking process. However, let's look at a different (hypothetical, of course) accident to the same ship. Suppose that the iceberg had holed the *Titanic* hull only over the length of Boiler Rooms Nos. 2, 3, and 4 and that the water then flooded those three rooms. From the floodable length curve in Chapter 20 with the sides of the "pyramids" extended upward to encompass those rooms, one finds that the approximate 180-foot length of those combined compartments is less than the floodable length in that area of the ship. The ship will not sink! However, there must be concern about the effect of the free surface of that amount of water entering the ship. That surface is 180 feet long by 92.5 feet wide. Suppose that at one moment the water is 10 feet deep above the inner bottom in those three boiler rooms, which means about 4000 tons of water is in the ship. As part (b) of Figure 30–1 shows, the water in running down hill acts as if it were suspended from a point much higher in the ship (a convenient way to picture the effects of free surface). The point of apparent suspension would be about 80 feet[4] above the bottom of the ship, or in the vicinity of Shelter Deck. Such a weight actually added that high would surely reduce the value of GM and hence make the ship less stable.

But how much less stable? That question cannot be answered exactly, but we will get an idea of the answer by citing the essence of a published calculation for a similar situation. In a post-*Titanic* paper reproduced in toto by *Engineering*,[5] Professor Oswald Flamm of Charlottenburg (Germany) made a much more thorough calculation of this nature for the midlength of the *Titanic*, arriving at a numerical answer. He concluded that free surface would have made the *Titanic* unstable with flooding of about the 180-foot length discussed in the foregoing paragraph. However, the ship was designed for two-compartment survival (any two adjacent compartments flooded without sinking), and he found that no two adjacent floodings could have the consequence of instability because the bulkheads were spaced so that all two-compartment combined lengths were *less* than 180 feet. The professor even put in a compliment to the Harland & Wolff designers for spacing the bulkheads in that way. Given that damage stability apparently was not given much emphasis in the pre-*Lusitania* years, this may be another case of serendipity.

Let's note a relationship between intact and damage stability. Flooding that is not sufficient to overcome the buoyancy of a ship will inevitably degrade its stability, and, therefore, places an obligation on designers to provide intact stability of such magnitude that it can accept the burden of reasonable damage. The damage to the *Titanic* was more than "reasonable"; it sank the ship. A significant consequence to stability never appeared. One may take the event to be the ultimate test of the ship's stability, and the outcome a proof of adequate *intact* stability.

Even if we withdraw the notion of serendipity in favor of crediting the designers of the *Titanic* with a high degree of foresight, they surely did not examine the stability of their ship when she reached the bizarre state in which she trimmed to 15 degrees near the end of the sinking process. At that angle of trim, the stern of the *Titanic* lifted out of the water. Apparently she had assumed a slight port list by that time. However, there was little danger of capsize. One might claim that the ship was stable even after breaking in two. In determining how much stability remained at the end, one might look at the depiction of the sinking at the minus-five minutes point, shown in Figure 30–2. The forward 40% of the hull is completely flooded, yet with a significant portion of the hull and superstructure submerged and still free of water. (The illustration comes from a computer simulation[6]). If Figure 30–2 is accepted as at least approximately correct, the flooding is confined mostly so that a simple intact stability analysis may offer some insight. Even without a calculation, it is apparent that the center of buoyancy must be abnormally high and the center of gravity (including the flooding water) abnormally low. Therefore, the center of buoyancy must be *above* the center of gravity. That state is sometimes called "pendulum stability" since it suggests a weight at G hanging from a fulcrum at a height B. Indeed, it is the stability state that a submarine must have, and here the state is that of a surface ship that is momentarily a submarine.

Stability of the Britannic *in Extremis.* HMHS *Britannic* was holed forward of her boiler rooms by a mine explosion, and may have been damaged further by a second (apparently internal) explosion forward. If the watertight door between Boiler Room No. 6 and the Firemen's Passage had been closed, the ship should have remained afloat, regardless of the damage forward. Such is the interpretation of her floodable length curve.[7] However, the vital door was probably open at the moment of explosion because of a change of watch in the engine and boiler rooms, and apparently could not be closed because of shock damage to the door and/or its operating mechanism. A whipping response of the hull (Chapter 18) easily could have caused shock damage to cast iron members known to be part of the door construction.

The *Britannic* was apparently doomed to sink by an open watertight door that could not be closed. Why did she not sink slowly and without significant list in the manner of *Titanic*? Ironically, the answer may lie in the inner skin added to both sister ships after the *Titanic* sank. Recall in Chapter 24 that an inner skin was added to the hull from the forward end of Boiler Room No. 6 through the machinery rooms aft of Boiler Room No. 1, with a 30-inch space between shells in way of the boiler rooms. The same shock that damaged the vital watertight door may have ruptured structural connections at the forward end of this protective space on the starboard side only, allowing water to enter this space. If so, the resulting list may have been sufficient to immerse open portlights on "E" Deck, causing still more listing, leading to a fatal chain of events. The most important link was the impediment to cross-flooding offered by nontight wooden joiner bulkheads. The list led, in turn, to the flooding through open portlights on "F" Deck.

Massive unsymmetrical flooding made *Britannic*'s damage stability insufficient. However, the designers are not at fault, unless we tax them with the following rules: watertight bulkheads should never be equipped with a door of any kind; a hull should never have opening portlights; all structural connections should be designed for shock loads. But we cannot expect these rules to have been applied by designers working in the peaceful years before 1914, since they would have been regarded as suitable only for a naval vessel.[8]

Did flooding take place in the 30-inch space between shells? An estimate shows that such flooding, over the length of Boiler Room No. 6, would not alone have caused a list of 20 degrees. However, if this flooding did occur, it surely contributed to the list that allowed the flooding through open portlights on the starboard side of the ship.

Stability of the Lusitania in Extremis. A torpedo explosion was quickly followed by a second explosion, causing the *Lusitania* to take a large list to starboard and then sink within 18 minutes. The longitudinal coal bunker bulkheads, praised as safety features by such authorities as Leonard Peskett and by semi-authorities such as Joseph Conrad (Chapter 19) and J. Bernard Walker (Chapter 21), apparently confined flooding to one side sufficiently to cause a list. Portlights were apparently open; submerged by the initial list, they admitted water that exacerbated an already bad situation.

Modern passenger ship design requires that longitudinal bulkheads that form part of the subdivision should be within a Beam/5 line and must follow the contour of the design waterline. Although this was achieved in the midships region of the *Lusitania* and *Mauretania*, it was not at the forward quarter point where Boiler Room No. 1 was located. With the longitudinal bulkheads being parallel to these ships' centerlines, it was necessary because of speed considerations to bring the longitudinal bulkheads, bounding the coal bunkers outside of Boiler Room No. 1, inboard to achieve some degree of protection for this space and provide sufficient coal bunker capacity. Unfortunately, the distance between the repositioned bulkhead and the shell was still less than Beam/5 and the longitudinal bulkhead became an abrupt transition near the midpoint of this boiler room, creating a structural discontinuity. This is the area where the fatal torpedo hit occurred that caused the rupture of the longitudinal bulkhead that led to the rapid flooding of Boiler Room No. 1. That flooding also led to a steam explosion in the piping and boilers that crippled the entire propulsion plant, auxiliaries, and electric plant. With the ship plunged into darkness within minutes of the torpedo hit, passengers and crew found it difficult to access the lifeboats.

The damage to the side shell was much greater in area than that of the *Titanic* and the rapid inflow of water resulted in a substantial loss of righting arm (GZ), so that the ship listed to 15 degrees within minutes after the damage had occurred. We estimate that as much as 4,000 tons of water per minute passed through the hole created by the torpedo hit and at least 22 tons of water per minute for each open or damaged porthole with a 3-foot head. A greater head would increase that flow rate unless the compartment was filled. The list of 15 degrees combined with a trim of 5 to 10 degrees made it difficult to launch the lifeboats and with the loss of steam pressure and electrical power it was impossible to steer the ship to shallower waters. Within 18 minutes this ship succumbed, capsizing to starboard during her final death throes and illustrating the hazard of longitudinal subdivision without the proper cross-flooding measures.

During the 1912 *Titanic* Mersey hearings Edward Wilding and Leonard Peskett were asked to prepare a stability analysis of RMS *Mauretania* with the same type of damage as the RMS *Titanic*. Upon completion of their studies, a joint memorandum was presented by these two naval architects at Mersey Inquiry:

> We have considered by approximate methods the flooding of the *Mauretania* in the event of an accident similar to that met with by the *Titanic*. We have assumed the watertight doors and hatches to be closed and similar deductions to those made in the calculations for the *Titanic*. From the calculations made, taking the vessel as damaged from the stem to the after end of the forward boiler room (corresponding nearly—but not quite—to the length from the stem to the after end of the No. 5 Boiler Room in the *Titanic*) the vessel would remain afloat with a considerable list, say 15 to 20 degrees, which, no doubt, could be slowly reduced by carefully flooding some after spaces on the opposite side. With the data available we do not think we can satisfactorily discuss flooding corresponding to the damage extending into No. 4 Boiler Room in the *Titanic*.—(Signed), Ed. W. L. Peskett.

Edward Wilding noted that the calculations that produced the memorandum showed that the *Mauretania* would have a considerable list. In order that the water should not rise above the top of the bulkhead, it was assumed that the bunkers flooded on the other side. This operation of counterflooding was achieved by raising the watertight coal bunker doors, but they would have to be raised high enough so that the water could get through to the port bunkers.

The loading condition was not specified in their subsequent testimony. Using the rationale that Edward Wilding did to establish the draft at the time of *Titanic*'s collision with the iceberg, the *Mauretania* would have been out of port substantially the same time as the *Titanic*. This means that her coal and provisions would be depleted in a way done to establish the draft and permeabilities that he used in the *Titanic*. That condition was much better than the loading condition of the RMS *Lusitania* on May 7, 1915 when she was struck by a torpedo.

In 1997, as part of a study for Panel SD-7, the Jamestown Marine Services ran some damaged stability cases to determine the degree of flooding as well as the effect on the GM of the ship since this has not been well documented in any previous publication on the loss of this ship. Note 13 summarizes these studies done for the *Lusitania* including the loading conditions that were assumed based on earlier studies in Note 13 and a benchmark on Leonard Peskett's studies of damage stability in these ships. The *Lusitania* had a high vertical center of gravity and the flooding of the boiler rooms actually improved her value of GM. However, the off-center flooding reduced her righting arm, GZ.

The sensitivity of this ship to off-center flooding also is demonstrated in Note 13. We assumed the coal bunkers to have a permeability of 0.70 in her arrival condition and flooded the bunkers outside of Boiler Rooms Nos. 2–4. Higher permeability of 0.7 rather than 0.5 is due to the fact that this is a midvoyage permeability. This accounts for the change in coaling port from Liverpool to New York in February 1915. With one compartment flooding, the GM is 1.00 feet; with two-compartment damage, it is 0.35 feet; and with three-compartment damage, it is -0.21. This illustrates the point made earlier that these ships were unstable with more than two-compartment damage and the wing compartments flooded. This does not take into account when the watertight deck over those bunkers is ruptured and flooding occurs there.

Free-surface effects of the early flooding were not significant since many of the larger spaces that flooded were sufficiently below the waterline as to be filled up quickly with water. The only spaces not flooded completely were those on the Lower Deck over the coal bunkers. Those spaces were minutely subdivided by nontight bulkheads forming the passenger staterooms. Despite being nontight, these divisions would have reduced the effect of free surface during the early stages of flooding. However, many of these flooded spaces were at the beam ends and significantly contributed to the off-center moment that caused the initial 15-degree list. As the ship listed further to starboard and began to trim more heavily by the bow, additional open or damaged portholes and open hatches became sources of flooding. The inability to control this flooding, combined with the absence of prompt counter-flooding measures, gradually eroded the righting arms to the point where the ship finally capsized. The elapsed time of 18 minutes from torpedo hit to the ship capsizing is astounding, but understandable once consideration is given to the ship's arrangement and subdivision. In addition, the torpedo struck only 3 meters below the operating waterline. The depth setting of the torpedo was perfect to bring about the quick demise of this ship. If the depth setting of the torpedo had been greater, there would be less likelihood of rupturing the deck over the coal bunkers in Boiler Room No. 1. This would have meant a longer sinking time. In addition, the boiler rooms were open for easy access. This meant that Boiler Rooms Nos. 1 and 2 would have been flooded rather quickly, with the list and trim delaying the flooding of Boiler Room No. 3.

In July 1915 Naval Architect Edward Wilding of Harland & Wolff, who did a review of the sinking sequence for the *Titanic* at the New York liability hearings, was asked to comment on the subdivision arrangements of the *Lusitania*. These comments are quite illuminating on the comparison of the two ships in terms of flooding in view that the *Lusitania* had sunk the previous May. Wilding commented that working the coal through watertight bunker doors would have been very difficult, considering the coal particles and dust that prevented complete closure. He also noted the absence of suitable cross-flooding measures and the fact that the resulting large list made it difficult to launch lifeboats successfully from the port side of the *Lusitania*. When asked about the longitudinal subdivision of the *Lusitania*, he commented that the inclusion of longitudinal bulkheads made it possible for unsymmetrical flooding that led to her large list, whereas with transverse coal bunkering in the *Titanic*, such an eventuality would not occur.

The stability of the *Lusitania* was obviously inadequate for the damage sustained. If one should argue the "but she wasn't a warship" defense that was suggested above for the *Britannic*'s stability, another dispassionate critic might counter with the case of the *Empress of Ireland*. That ship was fitted with longitudinal bunker bulkheads similar to those of the *Lusitania*. Her fate was to be (1914) rammed by the slow-moving Norwegian collier *Storstad* in the Gulf of St. Lawrence. This was no raking blow extending past five or six transverse watertight bulkheads, nor was it an explosive blow that would shatter cast-iron fittings and riveted seams. It was an approximately right-angled blow that could not have ruptured more than one transverse watertight bulkhead. Indeed, it was the very accident for which passenger ships should be made "unsinkable." Unfortunately, the victim rolled over and sank in about 15 minutes. Loss of life was heavy.

Rigorous damage stability calculations were not a standard design practice in the decade before World War I. Apparently criticisms of the longitudinal bunker arrangement were deflected by the notions of counterflooding, blowing down boilers, and the like, described in an earlier quotation (Chapter 23) from *Engineering*. With well-established hindsight, those notions seem fatuous. Upon some reflection, it seems that contemporary thought should have recognized them as such without the benefit of horrific real-world demonstrations.

Later-Day Regulation of Stability. Given the absence of any stability fault evident in the *Titanic* sinking, the authorities who produced the SOLAS 1914 rules did not publish anything regarding that characteristic. So far as international law was concerned, SOLAS 1929 was the next opportunity to improve the safety of passenger ships. Given the sinking experiences of many ships including, most obviously, the *Lusitania* and *Empress of Ireland* that had occurred since the earlier SOLAS, definite and specific action against watertight longitudinal bulkheads should have been expected. Reasons for the omission were not published, but a good reason is easy to find: watertight longitudinal barriers are too useful to be abandoned. For instance, liners of the 1920s were oil-fueled, and their designers used wing tanks, which were quite analogous to the wing coal bunkers of the *Lusitania*. Wing tanks were a preferred place to carry fuel. They also offered some protection against the collision with another ship or object, as well as convenience of location. They also provided better protection than bunkers, since tanks did not require those doors that might be impossible to close in an emergency.

In the realm of safety regulations, authorities of that era felt that rules should be based on limiting the consequences of an accident, with the designers left free to work as they pleased as long as they could present a convincing estimate of consequences. If the fault in longitudinal subdivision is in the list that their accidental flooding can cause, then that subdivision is acceptable only if damage stability calculations show that the list will remain within a safe range.

In addition, experiences since 1914 showed that in moments of crisis the "built-in" stability of a ship was of much less significance than a knowledge of its stability characteristics by the ship's master. After all, stability is only partially an as-built property of a ship; the loading of cargo holds and the status of fuel and ballast tanks add positively or negatively to the original value. Knowing what the value of GM is, what list the flooding of any compartment will cause, and what countermeasures[9,10] might be taken, are essential in managing the sinking of a ship so that it goes like a *Titanic* rather than like a *Lusitania*. SOLAS 1929, therefore, slanted its rule-making on specifying the stability information that must be immediately at hand to the master of a ship.

There must be *some* rules relating to design of the ship. For an example, we can look at the U.S. Coast Guard rules as they were for passenger vessels during the late stages of the liner era. Paraphrased for brevity, they were the following:[11]

- Intact stability to be sufficient to withstand the heeling effect of winds of specified strength.
- Flooding to the greatest extent allowed by subdivision rules must not constitute a stability problem.
- After damage, a heel of up to 15 degrees is permitted if the margin line is not immersed nor negative metacentric height (G above M) caused.

Rules of this substance, based on extensive experience and much debate among maritime authorities, have served the purpose adequately. However, rules continue to change as new accidents reveal weaknesses. In the *Andrea Doria* collision (introduced in Chapter 21) a list of over 20 degrees developed quickly because fuel tanks that should have refilled with ballast water after the oil was consumed, were instead empty.[12] The initial list prevented the launching of boats on the port side. In the 1990s, the capsizing and sinking of several European vehicle ferries caused a great loss of life. Such events demonstrated the danger of vehicle doors that might inadvertently admit water to a vehicle deck. Existing safety rules just didn't cover this type of passenger ship adequately.

Notes for Chapter 30

1. "Heel" and "list" are very nearly synonyms, though the former is used most typically to describe a temporary condition, such as the inclination that may occur in a high-speed turn. However, it also is used to describe the inclination of a sailboat under the pressure of wind on its sails, even though that condition may be far from temporary.
2. The *Eastland* accident, and especially its roots in the addition of lifeboats (weight added high in the ship) after the *Titanic* sinkings, is well covered by George W. Hilton,1995, *Eastland: Legacy of The Titanic*, Stanford University Press, Stanford, CA.
3. The submerged structure is itself displacing some water, and hence is exerting some buoyant force. When attempting accurate numerical work, we would include this effect.
4. Finding of this apparent point of suspension of a free surface can be done graphically in a way that the sketch suggests. Direct calculation is probably more accurate, and is what the naval architect usually employs, but explanation of how it is done is a bit more tedious than seems suitable for this book. Another way of approaching the problem is to look at the water as if it were a solid object acting as the bob of a pendulum. If, say, the ship lists 10 degrees, where on the centerline would the fulcrum of the pendulum have to be for the top surface of the bob to have remained horizontal?
5. The title of Professor Flamm's paper is "The Unsinkability of Modern Seagoing Ships," appearing in 1913, *Engineering*, Volume 95.
6. The illustration in question is based on C. Hackett and J. G. Bedford, 1996, "The Sinking of SS *Titanic* Investigated by Modern Techniques." The paper was published by the Royal Institution of Naval Architects.
7. We are actually using the *Titanic*'s floodable length curve in making the prediction here. The *Britannic*'s curve should show greater floodable length for the forward part of the hull because its watertight bulkheads in that area were watertight to one deck higher.
8. In a discussion of watertight doors in a contemporary volume of the *Transactions*, Society of Naval Architects and Marine Engineers, an anonymous designer of naval vessels remarked that he disfavored doors through bulkheads, knowing that in an emergency so many things could disrupt supposed watertightness. He added that he nonetheless always specified the allegedly watertight doors in his designs because he knew the U.S. Navy valued their convenience so highly.
9. Countermeasures? If, for example, an empty fuel tank on one side of the ship is flooding, the deliberate flooding of a similar tank on the other side of the ship will

cancel the resulting list. On the other hand, it might contribute just enough extra flooding to sink the ship. Knowing which will happen is essential.

10. Recall that in Chapter 13 we have Captain Smith asking Thomas Andrews, "Mr. Andrews! Would filling some of the aft ballast tanks help counter the bow's sinking?" Andrews' reply was negative, and for reasons that he then stated. The thrust of later-day thinking was to supply a ship's officers with means of answering such questions. (By 1990 or so, Smith would have gotten his answer from his movement of a cursor and the click of a mouse.)
11. The Coast Guard rules are to be found in the Code of Federal Regulations, Title 46, Subchapter H (the subchapter for passenger vessels). The rules are updated to stay current with U.S. federal law, and hence at the present time are likely to be different from any quotation or paraphrasing printed here.
12. How about the *Lusitania*? Her sinking was due to the torpedo hit near the end of a transatlantic passage, and the ship was presumably low on coal. Her coal bunkers could not be refilled with ballast water. If the torpedo had hit in way of a full bunker, would the bunker's low permeability have limited the water inflow enough to save the ship? Maybe, if the bunker doors were closed.
13. Garzke, William, Dulin, Robert, Hsu, Peter, Powell, Blake, and Bemis, F. Gregg, "The Saga of the RMS *Lusitania*, A Marine Forensic Analysis," New York Metropolitan Section, The Society of Naval Architects and Marine Engineers and New York Metropolitan Section, The American Society of Naval Engineers, 8 January 1998.

Appendix 1

Glossary

Nautical language is an everyday vehicle of communication, but it is larded with many field-specific words. Some of them are so common that a reader of this book should not need definitions. "Port" and "Starboard" are examples, and these we omit. However, there are words and terms used here that may benefit from definition and explication.

Bilge, bilge pump The bilge (or bilges) is the lowest level of the interior of a ship, and hence the place where any loose water will collect. A bilge pump is installed to pump such water overboard. The pump is an essential feature of any vessel not small enough to be freed of water with a sponge. On some occasions it can save a ship from sinking, but not if the instance involves a major flooding. Pumps big enough to counteract *Titanic*-size flooding are beyond practicality.

Brake horsepower (BHP or bhp) See discussion under indicated horsepower.

Bulkhead The thing called "wall" in a shoreside building. That which we call a "load-bearing wall" in a building is a "structural bulkhead" for a ship. When speaking of sinking, or the avoidance thereof, the "watertight bulkhead" (which is surely a structural bulkhead also) is of vital concern. Such a bulkhead has the strength to withstand the pressure of flooding on one side, and any penetrations such as holes for pipes and electrical cables are constructed to prevent their being leakage pathways. Passage of ventilation ducts through watertight bulkheads just isn't feasible. Doorways for human passage are a special case. The doors must be able to resist the same hydrostatic pressure as the bulkhead itself, and they must be closable by power actuators controlled from the ship's bridge. On the other hand, they must be openable locally to allow escape by anyone trapped by the closing.

Bulkhead deck The deck at the highest level of a bulkhead's watertightness.

Bunker A storage space for fuel, either coal or oil. The plural form often refers to the fuel itself ("Put into port to take on bunkers."), but also refers to an assemblage of storage spaces ("The *Titanic* had cross bunkers." (Fuel storage spaces running across the hull.) The word also serves as a verb relating to fuelling ("Put into port to bunker.")

Coal passer Boiler room crewman who passes coal from the bunker to within reach of the stokers' shovels. Also see trimmer.

Compass points We divide a circle, including the compass card, into 360 degrees. Before widespread literacy, seamen divided the compass card into 32 "points," each with a name that they could memorize: north, north-by-east, north-northeast, northeast-by-north, northeast, etc. By dividing 32 into 360 one finds that the points were separated by 11¼ degrees. This system was common well into the 20th century.

Condenser Except for some vessels that operate exclusively on fresh water, the steam propulsion plants of ships are fitted with condensers that receive the steam exhausted from the engines, and by cooling it, condense it to liquid for recycling to the boilers. Feeding the boilers with seawater would be ruinous because of the resultant rapid build-up of salt within them. As an additional benefit, by providing a closed space that can be kept under a high vacuum, a condenser allows steam to expand far beyond the limits of atmospheric pressure, and thereby to produce more work for each unit of fuel expended to make steam.

Davit Crane-like lifting mechanism for hoisting lifeboats outboard and down the side. Nearly always one for each end of the lifeboat.

Deep tank A tank that extends vertically through several deck levels. The term is often used to identify a tank that is built into the ship's structure, but is not an inner bottom tank.

Double bottom Synonymous with inner bottom.

Doubler plate (often just "doubler") A hull plate that must be pierced by a large opening is often reinforced by having its thickness doubled. The plate added to achieve this added thickness is the doubler plate.

Draw fires Raking the burning coals out of the furnace, then smothering them with wetted ashes. This was a way of stopping the production of steam by a coal-fired boiler as rapidly as possible. Done in at least some of the *Titanic* boiler rooms, apparently as a measure to reduce the likelihood of explosion when the cold sea water impinged on the hot surfaces of the boilers.

Falls (or boat falls) Lines (ropes) by which a boat (a lifeboat, usually) is lowered or raised.

Feed pump Feed or feedwater is water furnished to a boiler to be made into steam. The pump in question does the furnishing.

Fireman, stoker He who shovelled the coal into the boiler furnace, and tended the fire as needed. "Fireman and stoker" are synonymous words, though in English practice the word "fireman" seemed to predominate.

Gross tons (or tonnage) See discussion under tonnage.

Heel See list.

Indicated horsepower (IHP or ihp) A term commonly used only with reciprocating steam engines, though in thermodynamics books one sees it in connection with other types of reciprocating engines as well. It designates the power produced within the engine cylinder (measured by an "indicator," hence the name), and was the term usually used in stating the rated power of a steam engine. The power that emerges on the engine output shaft is typically about 80 to 85% of the IHP, and is known as the brake horsepower (BHP or bhp). The power rating of a steam turbine cannot be stated as IHP because the term is meaningless for a machine of that type; the output power or BHP is the usable number there. Thus it is that a reciprocating engine of 12,000 IHP might be delivering the same power to its ouput shaft as a 10,000 hp turbine. Power being transmitted by a shaft, as from engine to propeller, is shaft horsepower (SHP or shp).

If power is designated in kilowatts, then the same rules apply except that the word "horse" is appropriately omitted.

Inboard See outboard.

Inner bottom A secondary bottom, usually fitted about five feet above the true bottom in most ships of ocean-going size. The original purpose was protection against the consequences of bottom damage that might occur when a ship ran aground. After oil fuel was introduced, the space between inner and outer bottoms became tankage for carrying at least some of the fuel.

Knot A knot, as a measure of speed, is one nautical mile (6076 feet) per hour. Multiply by 1.15 to get statute miles per hour. The term "knot per hour" is incorrect since the "per hour" is built into the definition. In the time of the *Titanic* and earlier, however, "knot per hour" appears to have been acceptable.

List, heel, trim Trim is the lowering of the bow with respect to the stern ("trim by the bow"), or vice-versa ("trim by the stern"). The *Ti-*

tanic clearly assumed an increasing trim by the bow as the water flooded in. List and heel refer to side-to-side inclination, and are almost synonymous; however, heel typically implies a transient inclination such as a ship might experience in a high-speed turn.

Longitudinal See transverse.

Outboard, inboard Direction outward at right angles to the centerline. Inboard is opposite.

psi This is the abbreviation for "pounds per square inch," a common designation for pressure. It may refer to pressure above atmospheric (gage pressure), in which case it should become "psig." It may refer to pressure above zero pressure, in which case it is "absolute pressure" and its abbreviation should be "psia." Unfortunately, writers do not always make that distinction, so that a reader may not know whether absolute or gage pressure is meant. Since this book is not a technical treatise, such ambiguity is not a vital consideration; its use of "psi" should be taken to mean gage pressure. Where psia obviously must be used, that abbreviation is used, or the psi is accompanied by the word "absolute."

Range lights Two white lights on the masts of a ship, one forward, one aft. The after one is higher than the forward one. If they appear from a distance to be in the same vertical line, the ship is coming straight at you (they are shielded so you can't see them on a ship moving away). If the lower one moves to the left of the higher, then the ship is turning toward your left as you face the ship. If the *Californian* had been moving, watchers on the *Titanic* could have determined if she was coming to their rescue. Later, the hopeful survivors in the lifeboats were first assured of the *Carpathia* when they saw her range lights, and observed that they were almost one-above-the-other.

Reciprocating engine An engine that interacts with its working fluid (steam in all discussions here) by means of a piston that reciprocates within a cylinder. All marine engines had been reciprocating steam engines until just a few years before the *Titanic*, and they remained popular for approximately 30 years after. These engines made a comeback during the years of World War II when the war-built Liberty ships were powered by triple-expansion reciprocating steam engines.

Saloon As in dining saloon, it is just that—a dining room.

Shaft horsepower (SHP or shp) See discussion under indicated horsepower.

Stoker See fireman.

Strake Apparently a corruption of the word streak, and referred to any hull plank running fore and aft. When iron and steel construction supplanted wood, any end-to-end array of plates running fore-and-aft was spoken of as a strake.

Tons, tonnage Tons are big weights, or so we usually think (long ton: 2240 pounds; short ton: 2000 pounds; metric ton: 1000 kilograms). However, the size of commercial ships is nearly always spoken of in terms of register tons, and a register ton is a Volume measure. In the days of the *Titanic* such a ton was 100 cubic feet, and when one reads "She was the largest liner ever built, 46,328 tons in weight," (a common mistake). But how do you know that the tons mentioned aren't weight? You really don't because the word "register" is omitted, as it usually is, but in the absence of any modifying words, you should assume that register tonnage is meant. In the case at hand, that 46,328 tons was the gross register tonnage of the *Titanic*, determined by careful measurement of its internal volume according to a complicated set of rules.

Net register tonnage (usually just called *net tonnage*) is a volumetric measure of the parts of the ship that constitute its earning capacity. Volumes of cargo holds and passenger quarters are obvious elements of net tonnage, but in the time of the *Titanic* the rules for this also were quite complicated.

You have read that the *Titanic* was larger than her sister ship *Olympic*. No, she wasn't larger by having any dimension greater than that of her sister. The truth in the "larger" comparison comes from the enclosure of some open deck space to provide more passenger living space at the expense of passenger walking or lounging space. The enclosure meant that space had to become part of the ship's volume in the eyes of the measurer.

Now, back to weight. Real tons are indeed used to designate the weight of a ship, but when so used the number is called "displacement tons," or more commonly just "displacement." Displacement varies as fuel, cargo, baggage, passengers, etc., move on and off the ship, but gross and net register tonnages are fixed, unless a structural alteration that changes the original measurements of internal volume is undertaken.

Transverse, longitudinal Transverse implies cross-ways, as in transverse bulkhead, which would be one running from one side to (usually) the other. Longitudinal implies running length-wise, or parallel to the centerline.

Trimmer, coal passer One who worked in the bunkers to push coal toward the door from which the coal was taken to the furnace fronts. "Trimmer" and "coal passer" sometimes appear to be synonymous words.

Triple expansion steam engine An engine in which the steam passes in succession through three cylinders, first high pressure (HP), intermediate pressure (IP), and low pressure (LP). These cylinders are successively larger in diameter so that the steam may do its intended work by expansion to a low pressure. The *Titanic* engines were four-cylinder triple expansion engines, meaning the IP cylinder was followed by two (in parallel) LP cylinders.

Trunk An engine room or boiler room trunk, sometimes also called a casing. Much like an elevator shaft running from the room in question to the highest deck, or to the base of a funnel. However, it's not for an elevator, and nearly always larger in cross section than a typical elevator shaft. The trunk serves as a pathway for ventilation ducts, exhaust lines, piping, ladders for access to the room below.

Watertight A watertight boundary is capable of preventing the passage of water. The hull of a ship, especially below the waterline, is an obvious example of something that is watertight.

Watertight compartment A space within a ship hull that lies between watertight bulkheads. The term is something of a misnomer because the deck bounding such a compartment at the top is rarely watertight.

Way In one context, refers to the motion of a ship (cf "underway"), as in "The *Californian* had no way on when its watch officer noticed a brightly-lit liner in the distance."

Weathertight A weathertight boundary is capable of preventing the passage of loose water such as rain or spray, but probably would leak in the presence of "solid" water with any noticeable hydrostatic head behind it.

Web frame A transverse hull frame that has a much greater transverse dimension than a transverse "rib." It usually consists of a vertical, transverse plate that is stiffened at its edges by angle bars or channels.

Weigh Almost what it seems to mean, but "weigh the anchor" doesn't mean to measure its weight. In a general sense, it means to lift the anchor aboard. More specifically, the anchor is aweigh when it is clear of the bottom, and is stowed or housed when secured aboard the ship. When free of all solid connections to earth, the ship is under way (or underway), and not underweigh. If underway, and moving as well, it is under way with way on.

Wing tank A tank, most usually for the carriage of oil fuel, built into the side of a ship. The *Normandie* boiler room plans of Chapter 23 show wing tanks in their typical configuration.

Appendix 2

Important Persons

This appendix provides a brief synopsis of some important individuals that were referenced in this book.

Andrews, Thomas Thomas Andrews was the principal representative aboard the *Titanic* of Harland & Wolff, its builders. He was lost in the sinking. Stories told by survivors who saw him during the event seem to imply that he was too devastated to care. A biographer (Shan F. Bullock, A *Titanic* Hero: Thomas Andrews, Shipbuilder, Norman Remington & Company, Baltimore, 1913) gives his title as Chief Designer and Naval Architect. The biography, by the way, might better be called a eulogy; consider this sample: "He loved animals of every kind, and had over them such influence that they would follow him and come to his call."

Bartlett, Charles He joined the White Star Line in January 1912 as their marine superintendent at Liverpool. After the *Titanic* sank, he interviewed the surviving crew and was instrumental in carrying out the order to install more lifeboats on White Star Line vessels. He held a master's certificate and was a commander in the Royal Navy Reserve. He became the Captain of the *Britannic* and was still in command at the time of her loss. Unlike Captain Smith of the *Titanic*, Captain Turner of the *Lusitania*, or Captain Lord of the *Californian*, there was no controversy associated with his name.

Boxhall, Joseph Fourth Officer of the *Titanic*, a licensed master and a survivor. Made the calculations that determined the ship's position, conducted two surveys of the forward compartments of the *Titanic* based on instructions from Captain Smith, and fired distress rockets with the assistance of Quartermaster Rowe. Testified that he had seen the lights of another ship not far away, and that he had tried to contact it by signal lamp. Generally believed to have been the *Californian*.

Guggenheim, Benjamin Plutocratic American passenger, just the kind of person many would love to hate. But he disdained the lure of a lifeboat so that women and children could be first. Went below to change into evening clothes so he could die like a gentleman. Supposedly last seen, having re-emerged on deck in a tuxedo, puffing a cigar as he waited for the end.

Ismay, Bruce Managing director of the White Star Line. Survived the sinking, and was later buffeted by public opinion, since many of the public opined that he should have acted like a Captain, i.e., gone down with the ship after finding a prominent lady to whom he could offer his seat in a lifeboat. Was also suspected of ordering Captain Smith to maintain speed through the night. His testimonies at the American and British Inquiries provided some insight into the ship's operation.

Lightoller, Charles Second officer of the *Titanic*, a licensed master and the highest ranking officer to survive the sinking. Testified before the Smith hearings in Washington, and before the Mersey hearings in London. He was vulnerable to criticism because he had been in charge of loading and lowering the port-side lifeboats, and had not loaded them to capacity. He successfully asserted that in his judgment the boats had not been strong enough to hold their capacity loads while suspended from the davits. His conclusions on the ship sinking intact were instrumental in why both the American and British Inquiries concluded there was no hull failure.

Lord, Captain Stanley Captain of the *Californian*, the ship that was stopped for the night because of the danger of unseen ice, and allegedly was close enough to the *Titanic* to see her distress rockets. Captain Lord was too sleepy to get out of bed to look, and his one wireless operator was also in bed after a long day. The whole episode was a subject for controversy, and the captain the target of much criticism. However, he did not want to risk crossing the ice field in the dark.

Mersey, Lord He was Sir John Charles Bigham, the first baron Mersey. He appears to have been known simply as "Lord Mersey," signed his name even more simply as "Mersey." He was appointed wreck commissioner by the Board of Trade, and in that capacity conducted an inquiry into the *Titanic* disaster. His report is frequently quoted in this book. Later chaired the "SOLAS 1914" Conference on passenger ship safety. Later was wreck commissioner for the *Empress of Ireland* and the *Lusitania*.

Morgan, John Pierpoint He was an American financier, steel mogul, and railway magnate. His successes in amalgamating railroads in the United States was to be later duplicated in the shipping business. His idea was that a conglomeration of many shipping lines into one would eliminate price wars. He set upon a goal to set up a railway and shipping syndicate that would be a complete system of transportation. His aggressiveness in buying up shipping lines during the early years of the 20th century sparked much debate and concern. Morgan regarded the White Star Line as the standard bearer of his shipping organization—The International Mercantile Marine (I.M.M.). His attempt to buy out the Cunard Line failed as the Royal Navy was opposed to this takeover. With the loss of the *Titanic*, he was devastated and became a recluse, dying in Rome in March 1913.

Pirrie, Lord William Born in Quebec, Canada in 1847, he returned to Ireland when his father died. His grandfather, Captain Pirrie took a great interest in his grandson and encouraged him to seek a career in the marine industry. He became an apprentice at Harland & Wolff Shipyard at the age of 15 and rose up the managerial ranks. When Sir Edward Harland died in December 1895, Pirrie assumed leadership of the firm. He was ultimately elected the first Chairman of the Board and was soon augmented by being elected Lord Mayor of Belfast. The title of Viscount was bestowed in 1906 and that same year he purchased all the shares of G. W. Wolff when he retired. He negotiated a contract with the White Star Line to build the *Olympic* and *Titanic* and then worked closely with his Naval Architects, Thomas Andrews, Alexander Carlisle, and Edward Wilding to develop the concept into a working design. He was devastated with the loss of the *Titanic* and his nephew, Thomas Andrews, who he had been grooming to succeed him.

Rostron, Captain Arthur He was the Captain of the *Carpathia*. Ordered the coal to be poured on in an attempt to reach the *Titanic* in time, but his ship was too slow and too far away to arrive at the site of the *Titanic* before she sank. Was praised for the attempt, and for his skill in avoiding the *Titanic*'s fate as the *Carpathia* raced through the same ice field in the same darkness.

Schweiger, Kapitanleutnant Walther Commanded the German submarine *U-20* that torpedoed the *Lusitania*. A figure of controversy in that he was portrayed as a blood-thirsty Hun who joyfully killed as many innocent passengers as he could, or as a professional naval officer attacking a ship that was know to be an auxiliary cruiser of the Royal Navy. Did not survive World War I.

Smith, Captain Edward Captain of the *Titanic*. Veteran of many years and ship commands with the White Star Line. Had been Captain of the

Olympic since its maiden voyage, and was scheduled to retire after enjoying the honor of bringing the new sister through its maiden voyage. Died with the *Titanic* under circumstances not completely known. One story had it that while in the water he approached a lifeboat and asked the occupants to take a child he had in his hands. He then swam away to a certain death from hypothermia.

Smith, Senator William Alden Senator from Michigan who conducted immediate and extensive hearings on the disaster. Used the power of subpoena to detain all of the surviving crew, as well as Bruce Ismay, for interrogation.

Ida and Isidor Straus Isidor Straus, the owner of Macy's Department Store, and Ida Straus, his wife perished in the tragedy. Survivors told of her refusing to enter a lifeboat, remarking that she preferred to stay with her husband, that they had been together for many years and she did not want to be separated from him now.

Turner, Captain William Captain Turner commanded the *Lusitania* at the time of the fatal torpedo attack. A figure of controversy in that he has been portrayed as either a competent officer who was lead into danger by misbegotten instructions from the Admiralty, or blundered as he disregarded Admiralty warnings of submarines in his path.

Wilding, Edward He was the technical assistant to Thomas Andrews in charge of ship design. After the loss of Thomas Andrews with the *Titanic*, was the first authority at Harland & Wolff on design of the ship. Gave extensive testimony at the Mersey hearings and the New York liability hearings in 1915. After the conclusion of the British Inquiry, he was summoned to Lord Pirrie, whereupon he was discharged of his duties and told to leave the shipyard in one hour. His conclusions regarding the 12 square feet of damage over 300 feet of the ship's length from the iceberg-ship collision were interpreted wrongly by many writers. A computer reanalysis of his data and calculations in 1995 by Naval Architects John Bedford and Christopher Hackett of Harland & Wolff verified his conclusions.

Appendix 3

19th Century Glasgow University Curriculum on Naval Architecture

Excerpts from P. Jenkins, 1889, "On the Course of Instruction in Naval Architecture at Glasgow University," *Transactions*, Institution of Naval Architects, Volume 30.

The subjects taught may be classified under the following heads: Ship Calculations, Laying Off, Stability of Ships, Strength of Ships, Waves, Resistance of Ships, and Ship Design. The Rolling of Ships has not yet formed part of the regular curriculum, but I hope to take up this branch of the subject next session with a limited number of advanced students. During the first session the lectures embrace Laying Off and Ship Calculations. Under the latter head are included:

- The several rules for determination of plane areas, with proofs.
- Methods of combining the several rules.
- Determination of area of a plane figure formed by two radii and a curved boundary.
- Geometrical method of finding the area of a plane figure.
- Application of the rules in determining the volume of a vessel below any given water plane.
- Elementary principles of hydrostatics regarding the properties of floating bodies.
- Curves of displacement, tons per inch immersion, and area of midship section.
- Change of draught in passing from salt to fresh, or partly salt water, and vice versa.
- Properties of the centre of gravity.

- The centre of buoyancy—Determination of its position vertically in a vessel corresponding to any water plane, with proof of the accuracy of the method employed.
- Change in the height of the centre of buoyancy with change in draught, and method of drawing the locus adopted in practice.
- Longitudinal position of the centre of gravity.
- Centres of gravity of plane areas of various forms, both transversely and longitudinally.
- Transverse metacentre. The several kinds of equilibrium. Change in height of transverse metacentre in a vessel with change in draught, and method of drawing the locus adopted in practice.
- Metacentric diagram.
- Longitudinal metacentre. Uses to which the longitudinal metrecentre is commonly put.
- Change of trim produced by shifting a weight longitudinally in a vessel.
- Moment to change trim one inch.
- Effect on the trim of placing a known weight on board a vessel in any position.
- Centre of gravity of a plane area by polar coordinates, and moment about one of the radii.
- Centre of effort—Determination of its position both longitudinally and vertically.
- Estimate of position of a centre of gravity of a vessel from known data regarding a similar vessel.
- Volume and centre of gravity of cargo space.
- Centre of gravity of a vessel filled with homogeneous cargo.
- The centre of pressure.
- The tonnage laws.
- The freeboard of ships.
- Kirk's analysis.
- Methods of determining the surface of a vessel.
- The inclining experiment.
- Correction of results in respect of weights which require to go on board, or to come out, or to be shifted.
- Curve of stability, and its chief features.
- Metacentric height as a measure of initial stability.
- Construction of curve of stability for some bodies of simple form.
- Equilibrium of homogeneous prism of rectangular section.

The second session commences with the elementary parts of the Strength of Ships, and afterwards the Stability of Ships is taken up at the point to which it had advanced at the close of the previous session, and is fully discussed. The lectures embrace the following:

- Determination both mathematically and graphically of the stresses on parts of combinations subject to simple tension or compression.
- Simple shearing.
- Simple bending.
- Curves of shearing force and bending moment for beams variously loaded.
- Properties of curves of shearing force and bending moment.
- Curves of shearing force and bending moment for bodies of uniform section floating in still water and variously loaded.
- Curves of buoyancy, loads, shearing force, and bending moment for an actual vessel floating in still water with an assumed distribution of weight.
- Properties of iron and steel.
- Stress and strain.
- Effect of discontinuity of material on stress.
- Effect of punching and drilling on the tenacity of plates.
- Bearing pressure.
- Hooke's Law. Modulus of elasticity—Permanent set.
- Shearing and tensile strength of steel rivets.
- Riveted joint of maximum efficiency.
- Relation between pitch, diameter, and thickness in single and double riveted joints of maximum efficiency.
- Hydraulic compared to hand riveting.
- Diagrams of stress and strain for test pieces of iron and steel.
- Slip of riveted joints.
- Estimate of breaking strength of butt joints in ship construction.
- Moment of inertia of a plane surface and of a solid body.
- Proof of formulae for height of transverse and longitudinal metacentres above centre of buoyancy.
- Expression for stability of prism of rectangular section up to inclination at which an edge is immersed or emerged.
- Attwood's formula.
- Preparation of body plan from drawings for calculation of Statical Stability.
- Calculation of righting arm by the ordinary method at the several inclinations.
- Correction for the layer.
- Check calculation of stability at 90-degree inclination.

- Independent method of calculating stability at any inclination.
- Inclusion of enclosed erections upon deck in the calculation of stability.
- Most critical conditions as regards stability in which a merchant vessel can be placed.
- Effect of coal consumption on the stability.
- Having given the curve of stability corresponding with one displacement, to deduce that of another.
- Effect upon initial stability of adding a small weight.
- Correction of data supplied by inclining experiment, when necessary, on account of free water between the floors.
- Polar diagram of stability. Moseley's formula for dynamical stability.
- Deduction of curve of dynamical stability from curve of statical stability.
- Independent calculation of dynamical stability.
- Cross curves of stability, with several methods of calculting the same.
- Effect upon the stability of free water in the hold, with or without cargo in the compartment.
- Effect upon the stability of cellular bottom or McIntyre tank, both intact and broached.
- Effect upon the stability of bilging a compartment of the hold, both when it is empty and when it contains cargo.
- Stability of a petroleum steamer while loading or discharging, and when laden.
- Mechanical methods of determining stability of a vessel.
- Principles on which the roller and spindle and pivot planimeters are constructed.

The lectures of the third session are taken up with the strength of ships, a few problems on the Stability of Ships, the Geometry of Ships, Troichoidal and other Waves, the Resistance of Ships, and Ship Design. They embrace the following:

- Stability of a vessel partially waterborne.
- Stability of a vessel with masts and sails.
- Principle enunciated by Dr. Elgar.
- The shifting of cargoes.
- The action of the rudder.
- Mr. Crossland's proposition.
- Transverse bending moment due to water pressure on the bottoms and sides.
- Transverse bending moment due to weight of hull and cargo.
- Thrust upon deck beams due to water pressure.

- Effect of pressure of loose cargo, such as grain, upon a vessel's side in producing bending moment upon the bottom framing.
- Graphic representation of variation in transverse stress in some simple cases.
- The ellipse of stress.
- Panting and other local causes of straining in ships.
- Investigation of the stress upon a boat's davit under a given load.

Thomas Andrews was a Naval Architect, nephew of Lord Pirrie, and the managing director of the design department at Harland & Wolff. When the *Olympic*-class passenger ships' design began, Andrews worked with his deputy, Edward Wilding, who was responsible for the design calculations. Alexander M. Carlisle was the managing director of the shipyard, but when he left Harland & Wolff, Thomas Andrews succeeded him.

Appendix 4

Biographical Sketch of Thomas Andrews

Thomas Andrews was born in 1873 in Comber, Northern Ireland, a small village on the outskirts of Belfast. His mother was the sister of Lord William Pirrie, who became the first Chairman of the Harland & Wolff Shipyard in 1896. Pirrie surrounded himself with men of technical skill and devotion to service. After Andrews joined Harland & Wolff as premium apprentice in 1889, he quickly impressed the whole workforce with his technical skills. Often he would work late hours to master the theory of naval architecture and the techniques of a draftsman. After completing his apprenticeship in 1894, he was immediately appointed outside manager. His skills as a brilliant and innovative ship designer were also recognized. Lord Pirrie kept close watch of his nephew and it has been said that Pirrie was grooming Andrews to succeed him. In 1908 Thomas Andrews married Helen Reilly Barbour, a relation of the deputy Prime Minister of Ireland. They had one daughter. By 1907, Andrews had risen to the position of managing director of the design department under the direction of the Right Honorable Alexander M. Carlisle, who was the managing director of the Harland & Wolff Shipyard. He had already supervised the construction of the White Star Liners, *Baltic* and *Oceanic*.

When Lord Pirrie returned from Liverpool in 1907 after a meeting with Bruce Ismay of the White Star Line, Pirrie called together Thomas Andrews, Alexander Carlisle, and Edward Wilding to commence the design of the *Olympic*-class passenger ships. Ultimately Andrews would have the full responsibility for the design of the *Olympic* and *Titanic*. Carlisle would leave to join the Welin Davit Company before the two ships were completed. Edward Wilding was an assistant to Thomas Andrews and had the responsibility for the general management of the various drawing offices, but also had the task of preparing the calculations concerning the stability, subdivision, damage control, and safety.

After heading the Harland & Wolff delegation on the trials of the *Titanic* on 2 April 1912, he then sailed aboard her as the builder's representative on the maiden voyage. As the Naval Architect of the ship, he was summoned by Captain Smith to assess the degree of damage and determine how long the vessel would remain afloat. He gave his advice in the Reciprocating Engine Room to Captain Smith, Chief Engineer Joseph Bell, and Chief Electrician Peter Sloan. Later, he would provide Captain Smith of the grim news that his vessel would sink within 60 to 90 minutes. As the ship began her final death throes, he was last seen alone in the first-class Smoking Room, his eyes transfixed at a painting of Plymouth Harbor.

Appendix 5

Chronology of Lifeboat Launchings

Boats have odd numbers starboard, even port. Boats 1 and 2, the farthest forward, are the "cutters," boats kept ready for quick launching in an emergency such as "man overboard." Their capacity is 40 persons each. Boats 2 through 16 and 3 through 15 are the regular lifeboats, capacity 65 each. Boats A and B are collapsibles stowed atop the officers quarters on Boat Deck, C and D are collapsibles stowed on Boat Deck. These last four boats have a capacity of 47 persons each, and are intended to be launched by the davits that handle Boats 1 and 2.

Boats are listed here in order of the times at which they were launched, and with any comments that may be of interest. Times must be regarded as approximate only, since they depend on the memories of survivors, persons who are not likely to have been making notes during the event. Likewise, the number of persons in each boat are approximate; the numbers gleaned from testimony before Lord Mersey add to 854, while only 712 total were rescued by the *Carpathia*. Did some passengers decide they would prefer to be on the big ship after all, and dive from their boats to swim back, not realizing what hypothermia would do to them? Much more likely is that ancient muddler of accountings and memories—confusion.

Chronology of Lifeboat Launchings

Time	Lifeboat	Description of Principal Events
0045	7	Boat under the command of Lookout George Hogg. Only 42 persons aboard, though Officer Murdoch instructed Hogg to take his boat to the gangway doors on "E" Deck to load more. These doors were never opened. The men sent to open them were never seen again, apparently trapped by flooding before they could carry out their assignment.
0055	5	Officers Pitman, Lowe, and Murdoch, with unwelcome assistance from Bruce Ismay, supervised. Ismay, thinking process too slow, shouted, "Lower away" several times until Lowe angrily shut him up. Ismay meekly slunk away, to reappear when Collapsible C was being loaded. Only 41 persons were aboard, but jerking motions occurred as the boat was being lowered recalled by passenger Karl Bahr, causing him to think he had made a mistake in boarding the boat.
0055	6	Second Officer Lightoller supervised the loading of this lifeboat. Major Arthur Peuchen suggested removal of sailing gear so that more people could be accommodated. He also was taken aboard to reinforce a shorthanded crew commanded by Quartermaster Hitchens. Molly Brown, soon to gain fame for her supposed commanding presence in this boat, was reluctant to go, but was rudely thrust aboard. Only 24 persons were aboard. Lightoller instructed Hitchens to take his boat to the gangway doors on "E" Deck to load more. These doors were never opened. (See Lifeboat 7.)
0055	3	Officers Murdoch, Lowe, and Pitman supervised. Able Seaman George Moore ordered into boat to assist loading of women. With only a few women available, men also allowed to board. Boat launched with 50 persons. Lord Mersey's accounting shows 15 crewmen among them.
0110	8	Boat launched with only 39 persons aboard, but strict enforcement of women/children rule allowed no male passengers. Among the denied was Isador Strauss. His wife, Ida, refused to board the boat, preferring to stay on the *Titanic* with her husband.

Time	Lifeboat	Description of Principal Events
0110	1	This was one of the two 40-person cutters. Launched with only 12, of which 7 were crew. Passengers included Sir and Lady Duff Gordon. He was later accused of persuading crew not to attempt rescue of anyone floundering in the water after the ship sank, but Lord Mersey declared that the accusations were unfounded. Boat was among those instructed to pick up additional evacuees from the gangways on "E" Deck, but like others, failed because the doors on "E" Deck could not be opened.
0120	10	Launched with 55 aboard, 5 crewmen, 50 women.
0120	9	Purser Herbert McElroy supervised the loading of this boat. Launched with 56 aboard, including 6 male passengers. Some difficulty in loading because the ship's list to starboard at this time widened the gap between deck edge and boat.
0125	11	Launched with 70 aboard, 9 being crew, 1 being male passenger. Boat almost swamped by an outpouring of water from the ship's side, apparently the discharge of bilge pumps still working to delay the inevitable sinking. Trim by the bow raised the discharge point well above its normal height relative to the sea surface. Overloading caused some discomfort among those aboard, but they all lived to board the *Carpathia* a few hours later. Had all boats been similarly overloaded, approximately 550 more would have joined them on that ship. Or maybe the *Californian* would have had to take some of them.
0125	12	Launched with 40 women/children and only 2 crew. Ismay might have been a hero if he had been lucky enough to hover around Boat 12 and be pressed into service to bolster the inadequate crew.
0130	14	This boat was commanded by Fifth Officer Geoffrey Lowe. Total load was 63 persons, none being male passengers. A rush of people as lowering began prompted Lowe to fire his revolver to discourage anyone attempting to board. He fired again for the same purpose as the boat passed "A" and "B" Decks. Falls had to be cut to free boat after it was afloat.

Time	Lifeboat	Description of Principal Events
0135	13	This boat with 64 aboard, all women/children except for 5 crewmen. Like Boat 11, was threatened by bilge pump discharge. Avoided the danger, but drifted aft into path of Boat 15 as it was being lowered. Frantic cutting of 13's falls allowed it to get clear in time.
0135	16	Officer James Moody supervised. Boat lowered with 56 persons aboard, all women/children save for 6 crewmen. Among women was stewardess Violet Jessup who suffered a deja vu experience four years later since she was aboard when the third sister, *Britannic*, went down in the Aegean.
0135	15	Launched with 70 aboard, including 13 crewmen and 4 male passengers. Narrow escape from being lowered atop Boat 13, noted above.
0140	C	All numbered boats except Boat 2, 4 (see below) launched by this time. Officers Lightoller and Murdoch went forward to free their lashings and rig them to the davits that had lowered Boats 1 and 2. Large crowd followed the officers. Crew locked arms to hold back crowd. Boat launched with total of 71 (capacity 47), including 5 crew and 2 male passengers. One of latter was Bruce Ismay, who had been helping load women and children, and had jumped into an empty spot as the boat started down. Criticized on grounds that a managing director should join his Captain in going down with the ship. Lord Mersey disagreed on the grounds that while enough captains had died nobly by voluntarily staying with their sinking ships, enough to establish a tradition, no managing director had ever done so. If all boats had been overloaded to the same extent as Collapsible C, close to 1800 persons (all of the passengers and more than half the crew) could have been saved.
0145	2	Launched with only 26 aboard, 21 women/children, 1 male passenger, 4 crew. Delayed by difficulties similar to those of Boat 4 (following).

Time	Lifeboat	Description of Principal Events
0155	4	Officer Lightoller supervised, though with some assistance from Captain Smith. Launched with 40 persons, 36 women/children, 4 crewmen. John Jacob Astor and pregnant wife Madeline, and the families of Walter Douglas, Arthur Ryerson, Jack Thayer, and George Widener were among those present to board. John Jacob was denied. Boat ready for boarding at 0025, but uncertainty at which deck passengers were to board caused delay. Final decision was "A" Deck, but this was enclosed promenade, requiring windows to be cranked open, and a wooden sounding spar to be cut away.
0205	D	The other collapsible, with 40 women/children, 2 crewmen, 2 male passengers. The latter were young enough that they could have been counted as children, but were not accompanied by their mothers.
—	A, B	These two collapsibles were to have been launched by the same davits that had earlier launched Boats 1 and 2, then Boats C and D, but time and a 20-degree trim did not allow more than the freeing of their deck lashings. However, they floated away (though overturned) as the ship sank, and became ad hoc life rafts. Second Officer Lightoller was swept overboard from the *Titanic* and survived by standing on the bottom (with several other men) of one of the collapsibles, floating capsized, until the *Carpathia* came.

Appendix 6

Would More Lifeboats Have Helped?

A statement by John Bedford of Harland & Wolff in a letter to Author William Garzke in August 1997.

> It is very interesting to examine the alternative scenario if the ship had been equipped with the double-banked lifeboats for which Harland & Wolff and Welin Davit had designed the *Olympic*-class. This arrangement would have increased the lifeboat capacity to 2,168 persons, some 60 shy of the number aboard during the maiden voyage of the *Titanic*. If it is assumed that Captain Smith, when told by Thomas Andrews at 0010, that his ship had two hours to float and, knowing that there was lifeboat seats for all the passengers, had ordered an immediate evacuation with no limitation of women and children first, could it be assessed how quickly those lifeboats could have been lowered? Starting from the order to swing out the boats at 0020, the starboard side succeeded in lowering 9 lifeboats in 55 minutes. If this rate also is applied to both sides when there is 100 minutes, assuming the same preparation time, then the best that could be expected would be 32 lifeboats lowered, that is all but the collapsibles with their 188 person capacity. Thus, in ideal circumstances, it would seem barely possible to have everyone safely in a lifeboat in the time available. There are factors mitigating against this ideal as follows:
>
> - Could we expect Captain Smith to have acted so promptly given a probable fear of panic and his hope that a ship may have been contacted close enough to arrive before the *Titanic* sank, in particular the ship that was seen some five miles away?

- The difficulty of mustering the passengers, with no general alarm, depending as it did on stewards rousing passengers, explaining the seriousness of the situation, and then escorting them to the lifeboats.
- The fact that the loading rate into the lifeboats would have had to been over twice that was achieved on the slower port side.
- The necessity to retrieve the boat falls and reattach these to the lower tier of lifeboats.
- The additional hazards of accidents if the lowering of the lifeboats became too hurried, bearing in mind the newness of the crew to the ship and the absence of a lifeboat drill.
- The mistaken belief of some of the officers that the lifeboats were incapable of handling a full complement of persons.

Based on the research and experiences of author William Garzke, it appears unlikely that the full number of persons aboard could have been embarked into lifeboats before the *Titanic* sank, although it should be achievable on a modern liner with 200% lifeboat capacity and a central communications system. However, passenger attendance at lifeboat drills is not 100% in the modern day. In 1962, during the maiden voyage of the liner *France*, only 40% of the passengers showed up for the lifeboat drill at 4 P.M. on Sunday afternoon.

Index